Developing and Sustaining STEM Programs Across the K-12 Education Landscape

Jessica L. Spott
Texas Tech University, USA

Lane J. Sobehrad
Lubbock Independent School District, USA

Rebecca L. Hite
Texas Tech University, USA

A volume in the Advances in Early Childhood and K-12 Education (AECKE) Book Series

IGI Global
PUBLISHER of TIMELY KNOWLEDGE

Published in the United States of America by
IGI Global
Information Science Reference (an imprint of IGI Global)
701 E. Chocolate Avenue
Hershey PA, USA 17033
Tel: 717-533-8845
Fax: 717-533-8661
E-mail: cust@igi-global.com
Web site: http://www.igi-global.com

Library of Congress Cataloging-in-Publication Data

Names: Spott, Jessica, 1986- editor. | Sobehrad, Lane J., editor. | Hite, Rebecca, 1979- editor.
Title: Developing and sustaining STEM programs across the K-12 education landscape / edited by Jessica Spott, Lane Sobehrad, Rebecca Hite.
Description: Hershey, PA : Information Science Reference, 2023. | Includes bibliographical references and index. | Summary: "This edited volume is a collection of case studies and curriculum, sourced from the community of practice, that shares blueprints for establishing and best practices in sustaining K-12 focused STEM education outreach, fostered through partnerships and programs between K-12 school districts and STEM-education outreach organizations (e.g., universities, institutes, industries, and nonprofits). Contributors to the edited volume describe the types of STEM opportunities for K-12 students in their community, outside of the traditional classroom environment, via community-based partnerships between a STEM outreach organization and K-12 school district. The editors have a particular interest in partnerships and/or programs that provide novel STEM experiences for K-12 students from historically underrepresented groups and/or entities that are minority-serving and/or a curriculum that uniquely engages students in the cognitive and/or non-cognitive affordances of STEM learning. This edited volume is divided into four parts when community partnerships provide additional opportunities for K- 12 STEM learning: during the school day, during out-of-school time, at home, and in the community"-- Provided by publisher.
Identifiers: LCCN 2023029581 (print) | LCCN 2023029582 (ebook) | ISBN 9781668477717 (hardcover) | ISBN 9781668477755 (paperback) | ISBN 9781668477724 (ebook)
Subjects: LCSH: Science--Study and teaching--United States. | Community and school--United States. | College-school cooperation--United States. | Business and education--United States.
Classification: LCC Q183.3.A1 D486 2023 (print) | LCC Q183.3.A1 (ebook) | DDC 507.1/073--dc23/eng/20230817
LC record available at https://lccn.loc.gov/2023029581
LC ebook record available at https://lccn.loc.gov/2023029582

This book is published in the IGI Global book series Advances in Early Childhood and K-12 Education (AECKE) (ISSN: 2329-5929; eISSN: 2329-5937)

British Cataloguing in Publication Data
A Cataloguing in Publication record for this book is available from the British Library.

Advances in Early Childhood and K-12 Education (AECKE) Book Series

Jared Keengwe
University of North Dakota, USA

ISSN:2329-5929
EISSN:2329-5937

Mission

Early childhood and K-12 education is always evolving as new methods and tools are developed through which to shape the minds of today's youth. Globally, educational approaches vary allowing for new discussions on the best methods to not only educate, but also measure and analyze the learning process as well as an individual's intellectual development. New research in these fields is necessary to improve the current state of education and ensure that future generations are presented with quality learning opportunities.

The **Advances in Early Childhood and K-12 Education (AECKE)** series aims to present the latest research on trends, pedagogies, tools, and methodologies regarding all facets of early childhood and K-12 education.

Coverage

- Poverty and Education
- K-12 Education
- Performance Assessment
- Diverse Learners
- Curriculum development
- Individualized Education
- Special Education
- Early Childhood Education
- Head Start and Pre-K Programs
- Pedagogy

IGI Global is currently accepting manuscripts for publication within this series. To submit a proposal for a volume in this series, please contact our Acquisition Editors at Acquisitions@igi-global.com or visit: http://www.igi-global.com/publish/.

The Advances in Early Childhood and K-12 Education (AECKE) Book Series (ISSN 2329-5929) is published by IGI Global, 701 E. Chocolate Avenue, Hershey, PA 17033-1240, USA, www.igi-global.com. This series is composed of titles available for purchase individually; each title is edited to be contextually exclusive from any other title within the series. For pricing and ordering information please visit http://www.igi-global.com/book-series/advances-early-childhood-education/76699. Postmaster: Send all address changes to above address.

Titles in this Series

For a list of additional titles in this series, please visit: www.igi-global.com/book-series

Global Perspectives on Multilingualism and Multiculturalism in Early Childhood Education
Leonardo Veliz (Excelsia College, Austalia)
Information Science Reference • © 2023 • 300pp • H/C (ISBN: 9781668439326) • US $215.00

Handbook of Research on Training Teachers for Bilingual Education in Primary Schools
José Luis Estrada Chichón (University of Cádiz, Spain) and Francisco Zayas Martínez (University of Cádiz, pain)
Information Science Reference • © 2023 • 454pp • H/C (ISBN: 9781668461792) • US $270.00

Social Justice and Culturally-Affirming Education in K-12 Settings
Jonathan Chitiyo (University of Pittsburgh, Bradford, USA) and Zachary Pietrantoni (Florida International University, USA)
Information Science Reference • © 2023 • 394pp • H/C (ISBN: 9781668463864) • US $215.00

Theoretical and Practical Teaching Strategies for K-12 Science Education in the Digital Age
Jason Trumble (University of Central Arkansas, USA) Sumreen Asim (Indiana University Southeast, USA) Joshua Ellis (Florida International University, USA) and David Slykhuis (Valdosta State University, USA)
Information Science Reference • © 2023 • 298pp • H/C (ISBN: 9781668455852) • US $215.00

Practical Strategies to Reduce Childhood Trauma and Mitigate Exposure to the School-to-Prison Pipeline
Belinda M. Alexander-Ashley (Independent Researcher, USA)
Information Science Reference • © 2023 • 350pp • H/C (ISBN: 9781668457139) • US $215.00

Preparing Pre-Service Teachers to Integrate Technology in K-12 Classrooms Standards and Best Practices
C. Lorraine Webb (Texas A&M University, San Antonio, USA) and Amanda L. Lindner (Texas A&M University, San Antonio, USA)
Information Science Reference • © 2022 • 340pp • H/C (ISBN: 9781668454787) • US $215.00

Cutting-Edge Language and Literacy Tools for Students on the Autism Spectrum
Katharine P. Beals (Drexel University, USA)
Information Science Reference • © 2022 • 298pp • H/C (ISBN: 9781799894421) • US $215.00

Handbook of Research on Family Literacy Practices and Home-School Connections
Kathy R. Fox (University of North Carolina Wilmington, USA) and Laura E. Szech (University of North Carolina Wilmington, USA)
Information Science Reference • © 2022 • 353pp • H/C (ISBN: 9781668445693) • US $270.00

IGI Global
PUBLISHER of TIMELY KNOWLEDGE

701 East Chocolate Avenue, Hershey, PA 17033, USA
Tel: 717-533-8845 x100 • Fax: 717-533-8661
E-Mail: cust@igi-global.com • www.igi-global.com

Table of Contents

Detailed Table of Contents

Build a Better Book (BBB) is a novel STEM education initiative that engages youth in an authentic, empathy-driven engineering design experience in which they create inclusive books, games, and toys for children who are blind or have low-vision. BBB designers learn about users' needs through empathy-building and tactile activities, and iteratively test and improve designs using feedback from others, including users with visual impairments. Programs take place in formal and informal learning environments and benefit from community partnerships. This chapter shares the BBB network's trajectory from a few sites in Colorado to a national network of informal and formal education partners adapting and implementing BBB programming. Research findings offer insights about the strengths of the BBB model and how it has been able to effectively scale and be sustained across diverse settings, audiences, and program formats. The chapter also shares how the initial successes of short-term programs led to the current exploration of more intensive and career-oriented opportunities for youth.

This chapter focuses on STEM out-of-school time and describes the Mobile Making program as a university-community partnership-driven out-of-school (OST) program wherein racially and ethnically diverse university students bring materials in their own vehicles and facilitate youth's engagement and learning in maker-based STEM activities at school and community sites. The authors discuss the program model and the lessons learned about how to (a) design maker-based STEM activities for youth

STEM career information, hands-on content, and recruiting professionals to interact with students are challenging, especially within under-resourced areas. This chapter, within the STEM during the school day section, reports on a research-practitioner partnership (RPP) between a university and a rural school district. This collaboration resulted in a co-created curriculum to improve student (aged 7-8 years) knowledge of and interest in STEM careers. Among the 113 participants, 67 students participated in four modules with STEM content for up to five hours over 16 weeks. Observations, artifacts, and data from a STEM career metric were collected, including those who did not participate in the intervention (n=46). Analyses suggest that participants were more able to name STEM careers and make connections to content areas. Lessons learned are to monitor and pivot when students are struggling with technological challenges or lesson fatigue. Recommendations include devising ways to leverage an RPP and STEM professionals in the community to support STEM learning within elementary schools.

The Food, Energy, and Water Learning Module (FEWLM) workbooks are an effort by URM Ph.D. STEM faculty to decrease the leaky STEM pipeline and 'pay it forward' via the creation of high-quality out of school learning experiences for 'at-risk' URM 3rd – 8th grade students less competitive in STEM. Curriculum is based on Next Generation Science Standards, with a system thinking approach that provides hands-on, self-driven and team-building activities with a goal to increase scientific exposure and encourage more 'at-risk' URM youth to enter and remain in the STEM pipeline. Analysis of survey data of FEWLM workbooks use by over 50 elementary and middle school youths implemented at area public after-school programs, and a two-week summer STEM enrichment camp, suggest increases in scientific exposure and awareness about environmental sustainability and human health issues suggest FEWLM made a positive impression by increasing 1) their interest in STEM, 2) their perception of STEM, and 3) confidence in their ability to participate in STEM.

This chapter presents data from Black Girls Becoming Summer Research Institute (BGB), a two-week summer residential program for rising seventh and eighth grade Black girls at Vanderbilt University. The program was intentionally designed with Black girlhood celebration and joy as its centripetal force—all instructors, chaperones, volunteers, and even the photographer were Black women. The girls spent two weeks engaging in STEAM workshops that also supported their socio-emotional development while developing academic literacies. Using intersectionality and Black girlhood as lenses for understanding the various learning experiences, the authors present narratives that reclaim educational spaces in Black girls' own images. In so doing, this chapter serves as one model for higher education administrators seeking to expand their STEM footprint within their community as well as researchers interested in the research impacts on student learning outcomes from STEM-related programming.

There is a disparity between the number of men and women in STEM fields. The disparity is further exacerbated when considering the percentage of women from underrepresented backgrounds in the STEM workforce. However, early and persistent exposure to STEM through out-of-school programming can potentially increase the likelihood of a young girl, and particularly girls of color, pursuing a STEM degree and ending up in a STEM career. This chapter describes an out-of-school STEM program designed for middle school and high school girls. The design of the program is described as well as the critical role of community and corporate partnerships. Participants complete both pre- and post-program participation surveys which are used to gather data on demographics, STEM experiences, as well as their interests and motivation. The data presented provides evidence of the impact of the program. This chapter may also be a useful resource for others in the field who are designing similar programs which are specifically curated for girls.

This chapter presents a curriculum collaboration between Texas Tech University (STEM CORE and the TTU College of Education) and Lubbock Independent School District (Lubbock ISD) for a week-long STEM challenge, focused on bringing STEM into their communities and reducing barriers to accessing STEM opportunities. The goal of the summer 2021 iteration of Lubbock ISD STEM Challenge was to engage students in a summer STEM experience designed to increase their interest in STEM-related education and career pathways while positively affecting their learning about STEM topics. Readers will discover ten graphic organizers; detailed teacher notes that provide step-by-step facilitation of the event; learning objectives aligned to national standards; a materials list and preparation checklist for activities;

Preface

K-12 school districts across the United States are seeking means to extend science, technology, engineering, and mathematics (STEM) learning for their students beyond the traditional school day curriculum, which is largely driven by state standards. Enriching, out-of-school, and other non-traditional STEM learning activities not only enhance students' knowledge of STEM subjects, but also help students develop vital non-cognitive skills that are important for post-secondary success. In addition, these activities offer an increase in positive affect toward a pursuit of potential STEM secondary and post-secondary pathways. The National Research Council in 2009 challenged the K-12 community to embrace learner-defined outcomes of STEM learning that occur outside of the classroom as well as consider ways to better engage K-12 STEM learners from minoritized and/or marginalized populations. K-12 institutions should not work towards these goals alone, which has prompted greater involvement from educational stakeholders in higher education, education research, non-profit organizations, and state and local governmental entities. The hope is to co-design, co-create, and co-enact novel STEM experiences to meaningfully students in ways that can be evaluated for student-focused outcomes. Establishing and sustaining community-engaged partnerships in and for STEM education is the first and most critical step in this process. Disseminating this type of research and the curriculum developed from its partnerships will help school districts and community STEM entities replicate best practices, in their own communities, with more engaging and effective K-12 STEM learning.

This volume, unlike many other volumes that provide insight into effective instructional strategies, is not tied to specific content standards. Contributors to this volume instead focused on student-centered outcomes implemented in non-traditional settings. This makes replication of the interventions more replicable and adaptable for practitioners' local contexts. A common challenge for many organizations and institutions is establishing a collaborative, mutually beneficial relationship with their K-12 partners. In this volume, each chapter describes how the contributors built successful, collaborative STEM projects between a K-12 school district and community partners. Additionally, each chapter provides resources that may assist readers in developing their own community engaged, innovative STEM partnerships. These resources include formative and summative assessments, instructional materials, program evaluation criteria, and how to track measurable outcomes in non-traditional education settings. By combining these elements, the editors aim to demonstrate that building successful collaborations and programs between K-12 and STEM institutions is both practical and achievable, providing readers a lens to envision and replicate programs that will help them enhance their K-12 STEM ecosystem.

Developing and Sustaining STEM Programs across the K-12 Education Landscape is a collection of case studies and curriculum resources sourced from communities of practice that share blueprints for establishing and best practices in sustaining K-12 focused STEM education outreach, fostered through partnerships and programs between K-12 school districts and STEM-education outreach organizations (e.g., universities, research institutes, and nonprofits). Contributors to the edited volume describe various types of STEM opportunities for K-12 students in the classroom, outside of the traditional classroom curriculum, and in their community facilitated by community-based partnerships between a STEM outreach organization and K-12 school and/or district. By not being tied to specific state standards, authors describe how the curriculum they co-designed uniquely engages students in the cognitive, non-cognitive, and/or affective affordances of STEM learning. Each chapter describes the origination, development, and evolving purpose of the collaboration between the K-12 school district and the STEM outreach organization, the STEM activities that participating K-12 students engage in, and the impacts of the partnerships on STEM learners. Each contributor also notes how their community-engaged partnership provides novel STEM experiences for K-12 students from historically underrepresented groups and/or entities that are minority-serving as appropriate to their local context.

The chapters below were placed into two sections according to whether the primary location of STEM learning was in a school-based or community-based setting. School buildings are often a focal point of their local community. Families are familiar with the setting, have a relationship with school staff and, as a result, are more comfortable engaging in a non-school activity when the school hosts it. In Part 1 of this volume, chapter contributors show how working with schools and school districts in a school-based setting, both during and out of the school day, provides an opportunity to engage with students and families in an accessible environment. In Part 2, chapter contributors show how engaging students and families in a non-school environment provides direct access and awareness of STEM opportunities in their local contexts.

Chapters include the Build a Better Book (BBB) project, engaging youth in an authentic, empathy-driven engineering design experience in which they create inclusive books, games and toys for children who are blind or have low-vision. The authors not only provide practical insight into how and why BBB fosters an educational environment that produces positive student outcomes, but a clear example of how institutional collaborations can lead to replicable results in a wide variety of settings. Most notably, this chapter shares the BBB network's trajectory from a few sites in Colorado to a national network of informal and formal education partners adapting and implementing BBB programming. Research findings offer insights about the strengths of the BBB model and how it has been able to effectively scale and be sustained across diverse settings, audiences and program formats.

The Mobile Making Program provides engagement through after school activities that leveraged a university-community partnership that provided access and opportunity to K-12 students from predominantly disadvantaged backgrounds. By providing access to previously inaccessible making tools and materials reinforced through intentionally developed near-peer mentors, the Mobile Making Program increased student interest in STEM, improved their STEM self-efficacy, and made STEM more relevant in their everyday lives. Practitioners, in particular, should note the authors' comment on the tension between modest funding, sustainability, and scalability, wherein creative solutions to staffing and standardized program elements allowed for program expansion on a limited budget.

Alabama STEM Education (ASE) provides students with out-of-school opportunities to solve real-world problems in their community including learning about urban planning and civil engineering by planning a redesign of a city block and discussing climate change by testing and evaluating air quality in their local context. Researchers explored how students conceptualized engineering before and after these authentic program activities as well as its impact on students' STEM career identities. As other studies in this volume indicate, too, both students and parents had misconceptions about STEM-related careers and educational pathways, which were improved through participation in a STEM-focused intervention.

Given that the majority of the STEM experiences in high school are centered around science labs, the opportunity to engage students outside of the traditional science lab experience broadens their understanding of future STEM careers and pursuits. Oregon MESA offers a novel approach (human-centered design) for participating students from historically underrepresented in STEM groups. Through their partnership with an evaluator, Education Northwest, their chapter shares the impacts the collaborators have made on students' sense of belonging, confidence, and STEM identity. Encouraging results emerged from their iterative framework, as MESA students were more likely than their peers to graduate from high school, suggesting that out-of-school programming that provides engaging, relevant learning experiences has a positive relationship with academic perseverance and student motivation. Though there was little empirical evidence of MESA's impact on student academic performance, it is important to note that student success should not be limited to content-based assessments, especially in the context of a partnership between schools and other out-of-school providers.

In Virginia, at the Institute for Teaching Through Technology and Innovative Practices at Longwood University, a research-practitioner partnership (RPP) between a University and a rural school district relates their co-created curriculum to improve youths' (aged 7-8 years) knowledges of and interests in STEM careers. This chapter offers recommendations in leveraging RPPs between STEM entities and rural school districts to champion STEM learning within elementary schools.

The Food, Energy, and Water Learning Module (FEWLM) chapter describes an intervention for 3rd – 8th grade students who are underrepresented in STEM fields. Using a systems thinking approach with hands-on, self-driven and team-building activities students empirically improved their interests, perceptions, and confidence in STEM. The authors note that frequent, consistent communication protocols, efficient utilization of stakeholder expertise, and implementation of their intervention were critical components in the successful development of their RPP . Though seemingly common sense actions, effective administrative systems are critical for the long-term success and sustainability of any partnership.

The successes of a carefully and intentionally designed system is also apparent in the Black Girls Becoming (BGB) chapter, which addresses the structural disparities and barriers that have historically excluded Black girls from STEM spaces. This program worked to identify Black girls who would benefit from the intersectional intervention designed to serve and support the holistic student by disrupting, in the best possible way, the systematic oppression that participants and their parents had experienced throughout their experiences with the education system. The project team implemented their vision through the BGB Summer Research Institute, a residential summer learning experience hosted by Vanderbilt University. Their "identity-centered curriculum" successfully aimed to increase Black girls' affect and participation in STEAM, providing reconfirmation of the importance of including race, gender, and culturally responsive strategies in the development of curricula meant to increase participation in STEAM educational and career pathways.

Similarly, the Females of Color and Underrepresented in STEM (FOCUS) demonstrates that the participation of historically underrepresented student groups in STEM is impacted by their self-perception as a member of the STEM community. This chapter emphasizes the importance of mentorship and positive role models in creating a safe space for students to grow and explore in a STEM-focused environment outside of their normal schools. Using culturally responsive pedagogy and mentorship were essential in designing, refining and implementing this successful program. Though FOCUS started as a summer program, it has expanded to include ongoing student mentorship from program faculty during the school year and professional learning opportunities for teacher leaders preceding summer programming. The importance of continuing support provided by FOCUS is an important consideration to researchers and practitioners as they design their interventions. Though a week-long summer program provides a tidy, discrete environment for a case study, any positive impact on students may be temporary without an ongoing support system, which they have worked to create.

Truly collaborative partnerships between researchers and practitioners are an effective strategy for ensuring students are able to benefit from this type of ongoing support. Texas Tech University and Lubbock ISD have partnered for the last decade on their Middle School STEM Challenge, which has grown from a simple outreach program targeting students who had not excelled in the traditional STEM classroom to a summer staple that helps the school district recruit and enroll students into STEM graduation pathways as they matriculate into high school. This chapter describes a community-wide event across a single school district's middle schools to engage youth in STEM activities during the summer. Curriculum was designed and evaluated to determine improvements in students' use of 21st century skills, STEM learning, and STEM interest. Critical to the wider impact of this project is the culminating community event at the end of the challenge that demonstrates students' products in a public forum. This type of public engagement not only helps students internalize the learning that occurred during the STEM challenge, but shares students' excitement of STEM learning with their families and communities.

Effective partnerships need not be restricted to institutions of higher education, research institutions, or other educational entities. STEM practitioners in informal environments and industry may provide "real world" applications and experiences to students out-of-school in an authentic way. The Zoology Pathway Program describes a partnership for high school students to engage with a local zoo and aquarium by participants' developing projects presented to zoo administrators after career exploration and discovery. This unique program worked to help students identify the importance of STEM concepts in the real world, by teaching students about ecological systems and helping them to see the impact of human choices on animals and the environment.

Another example we see of this is in the chapter on School-Industry Partnerships, which recounts a partnership between one high school and a local STEM business. The success of the intervention was attributed to mutual trust, negotiation of needs, and investments from leaders. They provide valuable suggestions to those wanting to forge new partnerships with STEM industries that could be applied in other areas.

Covering topics from engineering design through cardboard boats in a local school district to nationwide partnership networks that collaborate on empathy-driven experiences, this edited volume highlights community engaged teaching (curriculum) and scholarship (research) produced from partnerships between K-12 school districts and STEM outreach organizations from across the United States. Locally and individually, each chapter describes unique opportunities to engage K-12 students, including those from marginalized groups, with the support of STEM outreach organizations through the co-construction and implementation of STEM activities during school, out-of-school, at home and in the community.

This volume encourages readers to share and replicate the K-12 district and community partnership models and strategies found in the chapters below; readers may utilize the experiences, measurement strategies, and activities of the contributors in their own unique context and environment. This edited volume provides technical assistance for university faculty and staff located in institutions of higher education seeking to expand their STEM footprint within their community, curriculum and instruction for school district leaders to implement opportunities for STEM programming and evaluate programmatic STEM enrichment outcomes, and for researchers interested in the scholarly and research impacts on student learning outcomes from high-quality STEM programming.

Jessica L. Spott
Texas Tech University, USA

Rebecca L. Hite
Texas Tech University, USA

Lane J. Sobehrad
Lubbock Independent School District, Lubbock, USA

Section 1
STEM In- and Out-of-School

Chapter 1
Engineering With Empathy in the Build a Better Book Project:
Meaningful Making Supports STEM Participation

Stacey Forsyth
University of Colorado, Boulder, USA

Jessica Sickler
J. Sickler Consulting, USA

Gigi Yang
University of Colorado, Boulder, USA

ABSTRACT

Build a Better Book (BBB) is a novel STEM education initiative that engages youth in an authentic, empathy-driven engineering design experience in which they create inclusive books, games, and toys for children who are blind or have low-vision. BBB designers learn about users' needs through empathy-building and tactile activities, and iteratively test and improve designs using feedback from others, including users with visual impairments. Programs take place in formal and informal learning environments and benefit from community partnerships. This chapter shares the BBB network's trajectory from a few sites in Colorado to a national network of informal and formal education partners adapting and implementing BBB programming. Research findings offer insights about the strengths of the BBB model and how it has been able to effectively scale and be sustained across diverse settings, audiences, and program formats. The chapter also shares how the initial successes of short-term programs led to the current exploration of more intensive and career-oriented opportunities for youth.

DOI: 10.4018/978-1-6684-7771-7.ch001

INTRODUCTION

Broadening participation in engineering and other STEM (science, technology, engineering and math) fields is a national priority and yet many groups – including women, students of color and those with disabilities – continue to be significantly underrepresented in STEM professions (Hynes et al., 2016; National Science Board, 2022; Wilkins-Yel et al., 2019). Many youths continue to have a limited, stereotypical perception of what engineering is, and often this fails to align with how they view their own interests and strengths (Hynes et al., 2016; Hynes & Maxey, 2018; National Academy of Engineering [NAE], 2008). Among girls, for example, some research suggests a disconnect between career motivators (e.g., making a difference, good work environment) and their perceptions of engineering, which they do not think is 'for them' (Hynes et al., 2016; NAE, 2008). Several studies have documented how gendered self-conceptions, professional identities, and the expertise areas that are valued in engineering (technical, rather than social consciousness) likely contribute to fewer women persisting in engineering and a significant wage gap in engineering careers (Cech, 2013, 2015; Wilkins-Yel et al., 2019). The lack of diversity in engineering has negative implications for the field, as greater diversity can lead to improved creativity, innovation and productivity (Smith-Doerr et al., 2017).

To address this disparity, some groups have called for a reframing of engineering to shift away from a focus on technology and toward a focus on the important work that engineers do. In 2008, for example, the NAE released *Changing the Conversation*, an effort to shift messaging from engineering as a profession that requires certain skills and yields personal benefits to one that uses creative ideas to have beneficial, societal impact. This aligns with research suggesting that focusing on engineering as a 'helping' profession may help to diversify the field by making it more attractive to young people, particularly women (Bielefeldt, 2017). In a similar vein, the People Part of Engineering (PPE) framework (Fila et al., 2014) emerged to emphasize the central importance of people in engineering: that engineering includes engineering *for* people, *with* people and *as* people. This framework shifts from an emphasis on technical and analytical training to a more holistic approach that emphasizes the "people" aspect of engineering. Engineering can be viewed as a process of creating products *for* people; engineers work *with* others, including team members and various stakeholders; and engineers bring their own experiences and perspectives *as* people to their education and their work. This humanistic framework calls for holistic engineering education experiences that enable students to work collaboratively to engineer for real clients with real needs, engaging in authentic communication, iterative design and feedback (Fila et al., 2014).

This reframing–from engineering as technology-driven to empathy-driven, with real potential to impact people's lives–may serve to attract a broader group of students who otherwise might not identify as being interested in STEM. Empathy is a complex construct with both cognitive and affective components but at its core, it is a means of understanding the experiences and perspectives of others (Leake, 2019). Empathy is typically not associated with STEM dispositions and has traditionally not served as a primary driver in STEM education programs, with many interventions continuing to focus on the technical and problem-solving aspects of engineering, often involving competition, rather than positioning the field in terms of its humanistic value (Holbert et al., 2017). Recently, however, increasing attention is being paid to the important role of empathy in engineering education and practice, with some linking empathy to a range of engineering skills, including effective communication, identifying problems and designing solutions for community partners, and addressing ethical challenges (Afroogh et al., 2021; Deobald, 2020; Hess & Fila, 2016; Wilson & Mukhopadhyaya, 2022). Integrating empathy may be particularly important for supporting the persistence of underrepresented groups in engineering. For example, one study found that evoking empathy by adding a narrative element to engineering activities supported girls' engagement, persistence and use of engineering practices (Letourneau & Bennett, 2020); another reported that integrating empathy-based lessons in a demographically diverse afterschool STEM program had a positive impact on interest and belongingness in both genders (Burns & Lesseig, 2017).

The *Build a Better Book* (BBB) project was developed as a STEM and maker education initiative that engages diverse youth in meaningful making through empathy-driven, human-centered engineering design. Youth work in small teams to design, fabricate, test and improve accessible learning materials, including books, games and apps, for community clients who are blind or low vision (BLV), or who have other physical and/or cognitive disabilities. The program capitalizes on participants' natural interests in helping others and aims to heighten interest in STEM and build positive engineering identities among youth who may not be inclined toward traditional engineering programs. Unlike engineering programs that focus on designing for a competition, BBB engages participants in a powerful engineering design experience to create innovative new products that meet the needs of real-world clients, for whom there are relatively few accessible and affordable options in the existing marketplace of toys, books, and games. As participating youth collaborate to engineer a product for a community client, they develop a broader awareness of engineering as a holistic discipline that may better align with their self-conceptions and natural inclinations, as well as an improved understanding of the importance of designing for all.

Since its inception, BBB has evolved over time, expanding to a wide range of community organizations and adapting to meet the goals of these diverse learning environments. The spread of the program model was largely organic, with organizations offering all types of education programs (informal, formal, and community-based) finding connections with the underlying premise. This expansion provided an opportunity for the present analysis, a synthesis of the program's past evidence about the spread, scaling and sustainability of the BBB model across a highly diverse set of learning environments, used to identify the critical factors that supported such scaling and sustainability. For this work, scale is considered as a process of adaptation, in which the core educational innovation is taken up by new educators and modified to align with the needs and assets of the users and community (Morel et al., 2019). This is distinct from a view of scale that focuses on adoption or replication, where the emphasis is more on fidelity of implementation to the original model. BBB's approach to sustainable scaling emphasized alignment with unique community context, with programs operating within a set of core principles that define BBB programs. The history of BBB's spread through adaptation across widely varying settings, learners and formats reveals insights about the critical levers of success for this program. Building on these findings, a new phase of the program is currently underway, one that supports longer immersion in the program and engages youth in even more attributes of real engineering design work environments.

BACKGROUND

In 2017, an interdisciplinary team of computer science and education researchers and STEM education specialists, based at the University of Colorado Boulder, launched *Build a Better Book* (formerly called the *Tactile Picture Books Project*) to address a community need: as makerspaces in public libraries rapidly proliferated across the country, many librarians who found themselves in new roles as makerspace managers sought training and support to facilitate engaging maker activities at their sites. Moreover, as public spaces with a mission to serve their communities, they wanted to connect their makerspace tools and programs with a meaningful real-world purpose. Public libraries have long played a critical role in democratizing access to information and with its integration of literacy and STEM, and its focus on making picture books more accessible for children with visual impairments, BBB was a good fit for libraries looking to expand their makerspace programming and make their collections more accessible.

The original design of BBB was grounded in a framework of design thinking and universal design for learning. Design thinking centers empathy at the heart of the design process (Alrubail, 2015). Designers learn about their end users' experiences to better understand who they are designing for and use these insights to create innovative solutions. The process is iterative, prompting designers to prototype, test and improve their designs based on users' feedback. In BBB, participants engage in a series of empathy-building activities to personally experience tactile sense-making and better understand critical design considerations. When possible, peers and community mentors who are BLV participate in workshops as co-designers, providing input and feedback throughout the design process.

Universal design for learning (UDL) values learning differences and aims to help all students achieve by reducing barriers to learning (Meyer et al., 2014; Rose & Meyer, 2002). UDL design principles focus on representing information in multiple formats and offering learners multiple means of engagement, action, and expression (CAST, 2018). UDL highlights the need for products and spaces that are accessible for a wide range of learning styles, skills and abilities. BBB reflects these values of access and participation by challenging designers to consider the usability of products across a range of abilities and engaging them as multimodal composers using different modes–including images, sound, movement and words–in order to meet clients' unique needs (Dalton & Musetti, 2018). In BBB, participants evaluate existing products (such as books or games) and weigh the affordances of different materials, designs and technologies as they learn to create more accessible products for a range of diverse needs and interests.

In supporting youth to learn about and design for end users who are BLV, BBB uses an asset-based approach, focusing on individuals' abilities rather than their limitations. For many participants, this leads to an important shift from initially viewing individuals who are BLV as dependent on others for help, to appreciating them as accomplished individuals who can also benefit from accessible materials and assistive technologies. Connecting youth participants to the BLV community, through videos, books or by inviting BLV individuals to work alongside students as co-designers, helps to reinforce an asset-based view and focuses the work on the end user's unique needs.

This chapter shares the development, expansion and sustainability of the *Build a Better Book* project over the six years since its inception. The evolution of the program can be viewed across three phases: 1) local, short-term workshops, across informal, school and community settings; 2) national spread, driven by educator interest and supported by train-the-trainer professional development workshops and online resources; and 3) an emerging Teen Internship model, in which youth work in teams to engineer completed products for community clients.

History

BBB initially ran as a series of short (~2-hour) workshops at multiple public libraries in Colorado (including Anythink, Boulder, Broomfield, and Louisville public libraries), in which youth used a variety of maker tools and technologies to create tactile, interactive art and stories. BBB participants evaluated the accessibility of existing books and games, assessing the affordances of different materials and techniques and brainstorming ways to make stories come alive for children with visual impairments. During these after school or weekend workshops, participants designed and fabricated early-stage prototypes, such as tactile fables and alphabet books, made with textured fabrics and craft materials; cardboard automata that added movement and interaction to stories; tactile elements for picture books, created with 3D-printing pens; and audio soundscapes for stories, created with coding tools Scratch and Makey Makey (Forsyth et al., 2020a; Figure 1). Although engagement in these workshops was high and participants developed new technology skills in the process (Thiry, 2017), the programs' short duration did not allow for product testing and iterative design, key elements of the engineering design process. After running multiple sessions of these workshops, it was also apparent that participants frequently tackled the same types of projects, starting from scratch each time. As a result, the quality and usability of workshop products was limited, and prototypes either went home with families as craft projects or sat on a shelf without any further use by the intended BLV audience.

Figure 1. Examples of prototypes created in short BBB workshops. Clockwise, from top left: interactive story element created as a cardboard automaton moving sculpture; tactile fable enhanced with sound effects, using Makey Makey and Scratch; tactile picture created with craft materials with different textures; story elements created with a 3D-printing pen; pop-up character created with a craft cutter; and 3D printed story pages.

Due to these limitations, the BBB team made two changes to the BBB model to jumpstart student projects and allow for more meaningful progress in iterative design. The creation of an online design gallery of past participants' designs became a resource to inspire new participants and enable them to learn from, build upon, and remix previous participants' designs. In addition, the program format shifted from a series of shorter workshops to a more cohesive summer camp format that supported more extended engagement between learners, educators and the design process. The longer duration of the summer camp format allowed participants to develop and hone their technical skills and produce a more useful product. Interest in the program grew quickly from educators across a variety of settings, from formal classrooms to out-of-school-time programs to museums.

Over its first several years, BBB experienced substantial organic spread across the country, driven by strong interest from teachers, librarians and makerspace educators, and supported by professional development led by the BBB team. Through a series of professional development workshops, conference presentations and online trainings, the project evolved from a local, library-based program of short-term workshops to a national network of more than 56 educator sites in 21 states implementing the program across a range of informal and formal learning settings. BBB network partners share ideas and best practices for creating accessible makerspaces, co-designing with BLV community mentors, and creating user-centered design projects for community members who are blind or have low-vision. The BBB team provides in-person and virtual professional development to new and existing network partners, supports the network with a resource-rich website, including educator resources and an online design gallery of youth-created designs, and facilitates communication across the network to spark and maintain collaborative connections. The BBB network currently spans the United States, with participants ranging from early elementary students through college students working in diverse settings such as summer camps, after-school programs, and formal classrooms across a range of disciplines, including science, technology, engineering, literacy, art, and English Language Learner (ELL) classrooms.

Findings from these foundational phases, discussed further in this chapter, led the team to identify a new area of need for the evolution of the BBB model to be able to serve more learners. Despite the widespread uptake of the BBB workshop strategy, it remained challenging for youth to get products past the design and prototype phase. This pointed to the need for an opportunity that would enable youth to dive deeper into the product development, iteration, client-centered fabrication and delivery processes that they were unable to experience within the constraints of typical summer camp-style workshops. These stages of the process most tangibly replicate the work of real-world engineering design professionals. This led to the development of a new BBB iteration centered on work-focused Teen Internships, currently being implemented and tested at four sites across the U.S. This engineering internship model capitalizes on the strengths of teens who have a natural interest in helping others and creates an entry-point to developing stronger engineering-related skills, identity and mindsets among students not typically inclined toward traditional engineering programs.

MATERIALS AND METHODS

This chapter synthesizes the findings, implications and guiding principles that have emerged from research, evaluation and practitioner reflection across the six-year evolution of the BBB initiative. Using findings from multiple research studies, this chapter highlights the key takeaways and principles that have repeatedly emerged as critical elements for the success, sustainability and opportunities for further innovation of the BBB approach. This section summarizes the programmatic focus of each of the three phases, as well as the evaluation and research questions that were addressed and data that were collected. The next section shares key findings and the principles that continue to guide the work going forward.

BBB Workshop Curriculum

The BBB team designed the project's core curriculum to be modular and adaptable for implementation in a variety of learning environments. The team co-designed the first series of workshops in collaboration with public library partners, who provided input on program topics, formats, stories and maker technologies. Given the challenges of ensuring consistent participation in opt-in, out-of-school-time settings like public libraries, BBB workshops were designed to function as standalone programs that focused on the development of complimentary maker skills. Ideally, youth participating in the full series of workshops would develop a variety of technology skills, but those only able to attend one or a few sessions would still be able to participate fully.

Regardless of the specific topic, all BBB workshops incorporated: (1) some type of introductory tactile experience, (2) an introduction to the tools or technologies available, and (3) a prompt to make a story, game or other product more accessible for someone with limited or no vision. For example, a workshop might begin with a tactile guessing game, in which participants use craft materials to create tactile pictures that are discernible by touch. Their blindfolded partner attempts to identify the object represented, using only their sense of touch (see Supplemental Material: Tactile Pictionary educator guide). By reflecting on their experience in this exercise, participants develop a set of guiding criteria that help make tactile images more readily understandable. The workshop then shifts to introduce a specific technology that can be used to enhance a book or game, for example coding in Scratch or 3D modeling in Tinkercad, enabling participants to develop new technology skills. Participants then apply these skills to the design challenge of creating more inclusive products, for example, 3D printing tactile objects to enhance a story, or using Scratch and Makey Makey to design an interactive soundscape for a picture book.

Train-the-Trainer Professional Development

The next phase addressed observations of rapid national spread and interest in *Build a Better Book* from educators across a variety of sites, including schools, public libraries and makerspaces. The BBB team developed a train-the-trainer model to teach BBB concepts through three professional development workshops, held annually from 2018 through 2020. Educators and librarians applied to participate in the training and were encouraged to attend with a colleague or community partner to ensure local support for the program in their home location. The multi-day training led participants through empathy-driven universal design principles and hands-on activities, as well as foundations in understanding visual impairments, tactile learning, and high- and low- tech tools to apply in creating accessible designs. Workshop participants were well-equipped to implement and adapt BBB workshop components and to share what they learned with colleagues. Many went on to present the project at local and regional conferences in their area, in addition to implementing the program with youth. The BBB team continues to support the network through virtual conferences, webinars, partner showcases, and online resources. During this phase, our project's research and evaluation team used a post-workshop survey and follow-up interviews to document outcomes for participating educators, elicit feedback about the workshop, assess participants' ability to implement their own BBB programs, and collect data regarding educators' needs, challenges and preferences in delivering STEM programming.

Empathy-Driven Internships for Teens

The most recent evolution within the BBB initiative builds upon the successes and limitations that were identified as the workshop model spread and was re-mixed in a variety of venues around the country. In particular, the Teen Internship model emerged from the continued finding that workshop participants were frequently "reinventing the wheel," as they started from scratch on their designs and were only able to produce partially complete prototypes. Although numerous projects had potential, if given the chance for more iteration, testing, and refinement, the short program format limited opportunities to collaborate with end-user clients. The Teen Internship program addresses these issues while providing teens a more extended, paid work opportunity that challenges them and enables them to experience the type of work environment, skillsets and purpose-driven work that could be available to them if they pursued a career in engineering.

As with previous BBB programs, Teen Interns work in teams to collaboratively design, fabricate, test and refine more accessible, multimodal products. However, the experience is driven by an updated theoretical framework that seeks to create a human-centered professional engineering environment. Building upon the People Part of Engineering framework (Fila et al., 2014), BBB internships are designed to intentionally activate work, mindsets, and experiences where teens are engineering *for* people, *with* people, and *as* people. To achieve this, the internship design is developed with a few key principles at the core. First, programs ensure that the youths' work is tightly connected to a real-world community client, who identifies a need for a particular product or a particular challenge they are facing. The client is engaged throughout the internship, providing feedback throughout the design process. The internship is also a longer duration (typically ~50-120 hours, spread over multiple weeks or months) which allows time for engineering teams to move through multiple iterations of their prototypes and arrive at a state of completion of their product, ideally so that it can be delivered to their client; this is a key part of the process that was frequently absent in shorter BBB programs. The internship design, including interactions with a client, technical and accessibility-centered mentor support, accountability and pay, is meant to mimic a real-world work, internship, or apprenticeship environment, where teens are expected to apply technical and 21st century STEM workplace skills, including communication, collaboration, self-regulation, creativity, and critical thinking. While this model is only in its second year of iteration, testing, and further development, the research will continue to explore the degree to which these principles of the program influence students' mindsets, identities, self-efficacy, and career thinking related to engineering.

Research Questions

Across the different phases of BBB, the project's research and evaluation efforts have focused on the program's potential to increase and support youths' motivation, interest and persistence in STEM, as well as the factors supporting the program's spread and sustainability across different learning environments. Data collected across multiple BBB programs address the following research questions:

- How does participation in BBB impact youths' STEM competencies and dispositions?
- Which elements of the BBB engineering design experience most effectively engage and support underrepresented learners, especially girls, in developing positive STEM attitudes and interests?

- Across different learning environments, what are the critical program factors that contribute to the program's effectiveness and sustainability?

Program evaluation focused on youth and educator participation in the program, with an emphasis on BBB's potential to broaden participation in STEM, particularly among girls and other underrepresented groups. Program evaluation tracked student participation, motivation and persistence in the project, including who was motivated to opt into the program (and why), and the degree to which youth persisted through the full program, as well as the degree to which participating educators adapted BBB programs for implementation at their own sites.

Participants and Data Sources

This chapter shares findings and themes from three IRB-approved studies, including two focused on youth programs and one focused on the educator professional development program. To address the research questions, the BBB research team collected data through pre-/post- surveys, program observations and video reflections collected on iPads. The survey measured the following domains: technological self-efficacy, STEM aspirations, belief that technological skills will help in the future, interest in engineering and design, and belief that technology can be used for social good (Thiry, 2017, 2018). An external evaluator observed multiple BBB programs using the Teaching Dimensions Observation Protocol (TDOP) to document pedagogical strategies used in the classroom (Hora & Ferrare, 2013). The TDOP documents instructional methods/activities and the corresponding level of student engagement (from low to very high) at 2-minute intervals throughout a BBB session. These data create a picture of the pedagogical approaches and student engagement within BBB programs. Short video reflections by students captured participants' explanations of their design process, challenges and key learnings. These responses, along with applicants' responses to open-ended questions in the program application, were coded and examined to identify common themes.

Participating BBB educators contributed feedback via surveys, interviews, program reports and video reflections, after their professional development workshop and following their implementation of a BBB program at their site. To share general information about the sustainable spread of BBB across learning settings and throughout the country, we reviewed data available from members who led and reported BBB programs over the past six years, which included 92 distinct youth programs. We also share several examples of BBB in action, to highlight how BBB can be implemented and sustained across a range of settings to benefit participating youth and educators, as well as the broader community.

RESULTS

Across a wide variety of learning environments, the challenge and opportunity to design a needed product for an authentic end user (frequently a younger child) is highly motivating for youth, including youth who might not otherwise participate in STEM programming. The BBB design task motivates youth to develop new technical skills and persist in their work in order to meet this need, while also increasing their awareness of the importance of universal design. By enabling scaling up to occur through adaptation, rather than strict adherence to a model (Morel et al., 2019), educators have been empowered with the latitude to creatively take the essential elements of BBB and mold them to address the assets, needs and contextual factors of their community, organization and learners. Through this approach, more educators have been able to find connections in this approach to align STEM and engineering design skills with literacy and community service.

Leveraging a Different Youth Motivation

BBB's mission, with its emphasis on creating accessible products for BLV youth, has repeatedly proven to be a strong motivator for youth to participate and persist in the program and importantly, it is effective in motivating youth who might not otherwise elect to participate in STEM programming. There is evidence that the project has high appeal for girls and may be an effective way to engage them in engineering and technology (Thiry, 2017, 2018). In the program's first two years, girls participated in informal, opt-in BBB programs at higher rates than in comparable non-BBB engineering and maker programs offered by the University of Colorado's STEM outreach program (Forsyth et al., 2020b). Many youths cited the opportunity to help children with disabilities as the primary reason they signed up for a BBB program, but this motivation was particularly strong among girls signing up for the program. In applications submitted by middle and high school youth for four informal BBB programs, for example, 72% of girls cited "helping others" as their primary motivation to participate, compared to 57% of boys. Conversely, 25% of boys cited an interest in technology as their primary motivation, while that was a driver for only 17% of the girls registering (Figure 2). Although the overall pattern was similar for both boys and girls, the profile indicated that this empathy-centered engineering experience might resonate with girls in a more meaningful way than other opportunities. Other motivators included interests in science, technology and art, and the potential to learn new skills and work together with friends. In many instances, these motivators intersected. As on applicant explained, for example, "I want to further my existing knowledge of 3-D printing, laser cutting, coding, and other technologies in a way that will have a purpose, and meaningful impact on other people."

Figure 2. Motivations to participate in out-of-school-time BBB programs, comparing responses by applicants' gender identity (n = 29 females and 28 males applying to four BBB programs, 2017-2018)

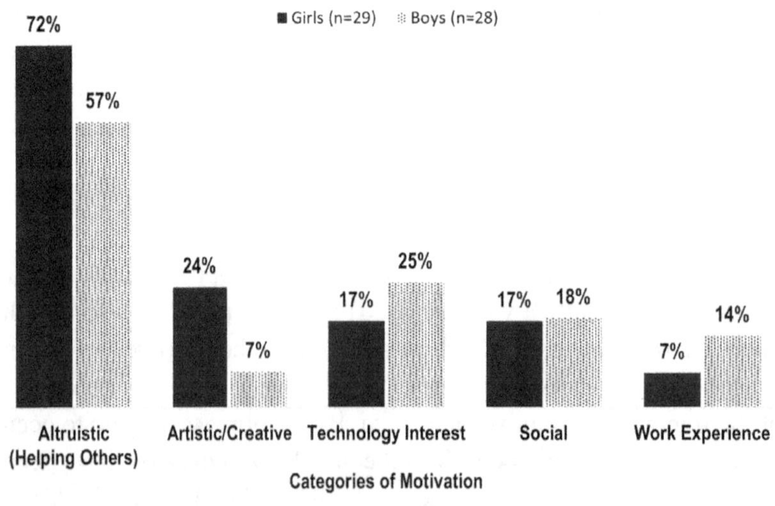

Technical Skill-Building

Across different settings, BBB fosters authentic making experiences, providing an opportunity for youths to plan, collaborate, design, and make unique products that address real-world needs (Thiry, 2018). They do not learn or apply these technological skills in isolation but learn them in conjunction with gaining a deeper understanding of how to conduct user-centered design for diverse users. They learn first-hand about accessibility and universal design, as BBB requires that they learn about the specific needs of their end users and take the perspective of others throughout their iterative design and fabrication process. Participants develop technical skills related to their project, such as 3D modeling, 3D printing or app design, and they have shown positive outcomes in their understanding of the value of technological skills for their future and their belief that they can use technology for social good (Thiry, 2017, 2018). For example, one student commented, "I'm feeling more confident about my wiring skills and more confident about my approach to making a product. After learning a lot of information about [our] clients, the design for the product and what we wanted became easier to define."

BBB research conducted in two English Language Development (ELD) classrooms found that providing students with the freedom to select from multiple modes and tools enabled students to develop skills with the tools that most interested them, which in turn supported the development of positive STEM identities (Walsh, 2017). Integrating STEM and making activities with language and literacy study allowed students to develop their skills as multimodal composers. Additionally, some ELD students connected the limited availability of tactile picture books for children who are blind with their own childhood experience of having a limited collection of picture books available in their native language, and this motivated them to design tactile books that included English, Spanish and braille. Others who had previously been reluctant contributors in class flourished as "expert composers" when given the opportunity to communicate stories with maker tools and materials (Walsh, 2017; Dalton & Musetti, 2018).

Persistence

Across different contexts, research and evaluation findings suggest that *Build a Better Book* holds promise for motivating youth to persist through challenging obstacles. Participants appreciate the authentic nature of the design task, feel like they are creating something that may improve someone's life and take a great deal of pride in their work. Community mentors play a key role in the program by sharing their personal experience of living with a visual impairment and contributing feedback on youth-created designs. Similarly, this real-world purpose and connection to a community client play a valuable role in motivating youth to persist in the design process, even when they face challenges. As one program partner expressed:

I was very proud of the progress and the level of empathy and design thinking the students gained in creating their materials. They were extremely proud of their progress and really wanted to design from the perspective of their user. Students were most intrigued in learning braille and incorporating it within their products. They used some high-tech materials in the construction of their books (3D pens, some 3D printing, circuits, Cameo cutter, Scratch, etc...); if given more time, participants wanted to improve on their designs and creations. This program was a phenomenal success.

Awareness of Disability

Consistently, across different contexts, the most impactful part of BBB is its focus on inclusive design, which has led to a profound mindset shift in many of the youth participants. Many participants reflect that the project has increased their awareness of the importance of designing books and other products that are accessible for all children, including those with visual impairments (Dalton & Musetti, 2018). The end-user audience focus, and the emphasis on instruction around universal design principles, helps students see the world in new ways. Following a BBB experience, students report an improved understanding of disability and the importance of viewing the world from the perspective of someone with a disability (Thiry, 2018). As one educator commented:

Those students, it kind of opened their eyes to something that they hadn't really necessarily thought of before, so they really looked at things from a new perspective. A lot of students were really excited to be helping out other people through their projects. It wasn't just some random project. It was something that would actually become useful to somebody else.

Similarly, after a field trip to a local center for people with disabilities, one teen intern remarked:

It was inspirational, especially in the last hour where we talked to people who have visual impairment and I got to learn many things about their preferences and how they navigate the world. It was interesting learning about the different types of vision impairment and objects they use to navigate through their environment. For example, the giant notepad or the tactile stickers or even the computer camera thing that zooms in on many things. I was inspired by how [they] overcame many difficulties as well.

In addition to youths' improved awareness of disability, many participating educators expressed a heightened awareness and understanding of the importance of accessibility in their programming and in their physical spaces (Thiry, 2020).

Importance of Community Partnerships

Regardless of setting or program format, community partnerships play a critical role in the success and impact of BBB programs. BBB partnerships include a wide variety of community organizations, including schools and school districts, public libraries, non-profit organizations, schools and centers for the blind, universities, community makerspaces and museums. Active involvement of community mentors who are visually impaired or who work with children or adults who are BLV is central to the model, enhancing participants' understanding of visual impairment and improving their designs through critical feedback. This is another example of how BBB has achieved scaling through adaptation; it emphasizes the importance of community partnerships as a core principle, but each site is able to chart its own path by determining which community partners will play which roles to advance its educational aims for the project.

Adaptation, Spread, and Sustainability

In the first two phases of the BBB initiative, collaboration among educators and across learning environments led to rapid iteration, modification and innovations of the underlying BBB workshop curriculum. Very quickly, the program moved from the initial pilot of library workshops and summer camps to a wide variety of settings, formats and audiences. This organic expansion and innovation, led by educators in our network and enabled by an emphasis on adaptation around core principles, has demonstrated the wide variety of ways that the program can be implemented in both formal and informal learning environments.

Analysis of the attributes of those programs revealed the breadth of formats, settings, and participants who have been served by these offerings. As summarized in Figure 3, BBB programs that have been implemented across the country are somewhat evenly spread among library, school and college campus settings–often in places where there was an interest in making or some type of maker facility or programming connected to the institution. While about one-third of programs have been delivered as part of the school day, the other two-thirds have been held during out-of-school times (including afterschool programs, summer camps and community events). Most participants are middle and high school students, but BBB programs have served elementary school learners, college students, adults and families. Perhaps reflecting the evolution of the BBB model, and our own reflections on the need for longer program duration for iterative design, there has been a very wide distribution of the duration of BBB programs. This includes a substantial number of one-time events or sessions (often hands-on workshops introducing the BBB concept) but more commonly, programs lasting for several sessions (e.g., a 5-day summer camp) or spanned over multiple weeks of repeated engagement. In a few rarer cases, some programs led by BBB partners lasted as long as a semester or over a full school year, typically in the form of a course elective or extended school project.

Figure 3. Summary of BBB programs implemented across the national BBB network, including the distribution of program types, settings, participants and duration, as reported by educator partners (n=92 BBB programs)

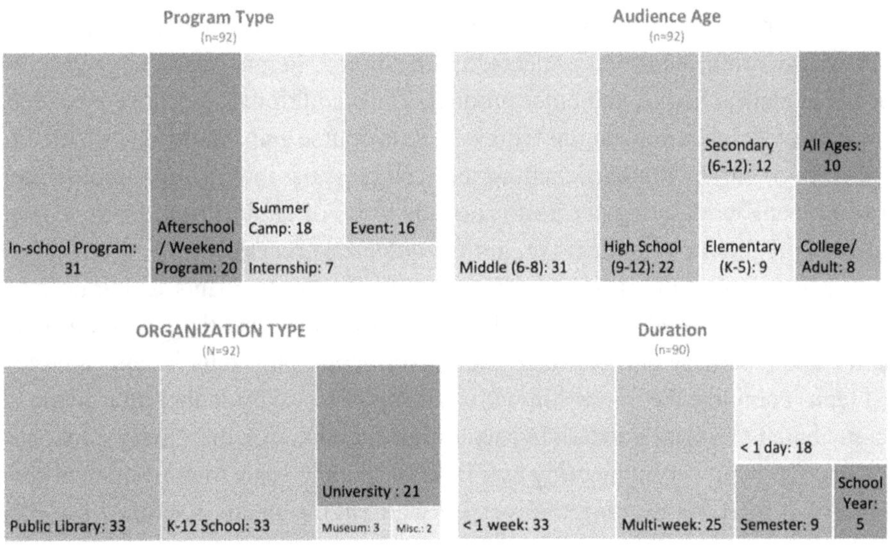

At present, numerous partner sites continue to run BBB programs. Most of these programs receive no financial support from BBB; program sustainability after initial training and implementation is maintained by each site and leader. Several sites have applied for and received other grant funding to expand their BBB programming and acquire additional maker equipment, materials and tools. Others have crowd-sourced funding from sites such as *DonorsChoose* or leveraged support and donations from community partners. Even as the COVID-19 pandemic drastically impacted informal and formal education sectors, many sites found ways to pivot, providing online or hybrid programs, sometimes supported by take-home maker kits. Anecdotal feedback suggests that the mission-aligned design of this program–knitting together makerspace technology, engineering skills, and students helping their own community–has strong appeal for partners and funders, which makes it easier to justify and sustain than programs driven only by content or academic standards.

Sharing and Celebrating the Work

As a final stage in the process, many BBB programs incorporate some type of culminating activity, event or display that serves to share participants' work with the broader community. Sample events include opportunities for youth designers to test their products with younger students at a local school, display projects at libraries or museums, share their prototypes with family members and mentors at an interactive showcase, and present projects at community festivals and exhibits. Designers can also share their work more broadly via online platforms, such as Instructables, Thingiverse and the BBB Design Gallery, enabling others to fabricate BBB products for their own use or remix and improve upon existing designs.

Affordances of an Internship

Solidifying the core principles of the BBB internship model is still underway, seeking to apply the same adaptation-focused approach to program scale and sustainability as in the original BBB program model. Our project is working to better articulate the design principles for sites to create internships that weave the PPE framework (Fila et al., 2014) throughout the experience and structure engagements that leverage the theories of persistence of interest. At this point, two partner sites have implemented their first iterations of an internship program, engaging a total of 25 teens in extended, paid, client-centered internships, in which youth identified projects, and designed, iterated and engaged with their clients to create accessible games, books, and other products. Two additional sites have provided BBB design experiences as part of a high school engineering capstone course and a multi-day spring break program. Early findings from research with participating teens suggest the internship model is particularly successful at shifting teens' underlying perceptions of the work of engineering, from general ideas about building things to being more of an iterative, human-centered process.

The extended duration of the internship program enables participants to iterate on their designs and complete more finished products (Figure 4) and creating an authentic work environment, accountability to clients, and a balance of mentoring and self-direction have thus far appeared to be critical to success. Most teens complete the internship with a strong sense of pride that their work as engineering interns has contributed to social good–doing work that they know will benefit someone with a need. This indicates that the focus on engineering as a field *for* people has strong resonance and is leading to important impacts, if the field is going to diversify who engineers are, how they see themselves, and how they see the field.

Figure 4. Inclusive products designed and fabricated by teen interns in a pilot internship program. From left to right, an interactive light-up story board designed for a toddler with CVI; 3D printed tactile tic-tac-toe game boards and playing pieces; a tactile game board for playing Wordle, designed for middle school students.

In addition to growing to perceive engineering as a field focused on iteration, initial interns have shown a significant increase in their sense of self-efficacy around the skills of engineering design over the course of their internship. To some degree, the workplace skills that are impacted may relate to more minute aspects of the design of the internship, set by the local host organization. For example, a site that took an unstructured approach to projects and student workflow saw a greater uptick in students' confidence in their creativity skills, as they were left to ideate and problem-solve independently, whereas a site that provided more structure, but also targeted interns without prior technology experience, saw greater gains in confidence using advanced technology. Qualitative data showed that most teens left the internship expressing a general sense of engineering self-efficacy, as well as gaining workplace skills, particularly interaction skills (e.g., collaboration, empathy, understanding of accessibility issues) and professional skills (e.g., communication, project management, and self-regulation).

At this stage, the number of interns is quite small and trends only emergent. As the four internship site partners develop and iterate on their models, the research team will continue to explore which attributes of internships and outcomes may have relationships and which are universal to the BBB approach.

Challenges

Across different settings, *Build a Better Book* has encountered some challenges as well. As is often the case with hands-on STEM and maker experiences, participating educators cited time constraints, lack of technical experience and lack of funding as significant challenges in implementing the program at their sites. Both youth and educators mentioned the challenge of having sufficient time to iterate on designs, which in part led to the longer-term internship model. Even in a longer summer internship program, participants still felt crunched by time. For example, one summer 2022 intern remarked,

I feel like we definitely can complete our project for our client by the time we present our work. However, the next few workdays will be stressful because we will be cutting out 400 more tiles and we have to fill in all of it with a sharpie. But I have faith in my group and I to complete this.

Implementing the program in some informal settings (e.g., public libraries) also raised new issues. For example, when participants have to opt into every session, outside factors may limit consistent participation over a period of days or weeks. Although the involvement of community mentors who are BLV has a significant and positive impact on participating youth, transportation and other constraints may limit their ability to participate on a regular basis. Additionally, although BBB products are designed to be more inclusive, many of the maker tools utilized in BBB programs, such as 3D printers and laser cutters, are not immediately accessible, although they do hold exciting potential for artists and makers who are blind.

Case Studies

Several case study examples highlight the adaptability of BBB across different settings and the program's ability to connect diverse community partners in working toward a common goal to create more inclusive learning and play materials for children.

Public Library Makerspace (New Jersey)

For public library makerspaces looking to provide a literary twist to their maker programming, BBB has proven to be a good fit. One public library in New Jersey has had success with a variety of program formats, ranging from a one-day community Make-a-Thon (in which participants crafted tactile picture books) to semester- and year-long projects completed in collaboration with local high school teachers, teen volunteers, and numerous community organizations (including local, national and international partners). The library makerspace offers a less structured experience serving a wide range of ages. Across all program formats, the library has developed its BBB programming around its strong connections to community clients, including St. Joseph's School for the Blind; Sound Start Babies, a community organization that serves infants and toddlers with hearing loss; and See3D, a non-profit organization that coordinates the design and distribution of 3D-printed models for people who are blind. These partnerships have provided the program with specific design challenges as well as constructive feedback on early-stage prototypes, ensuring that the work is meaningful and relevant for participants and the products are useful for end users.

Summer Camp for Middle School Youth (Colorado)

During a five-day summer camp, teams of middle school students worked together to make traditional board games more accessible for users who are BLV. After a series of BBB's introductory activities, campers played an assortment of board games and tested each game's potential to be played without sight. Their findings shaped their initial designs, including tactile game boards, audio dice, and braille-enhanced playing cards. Throughout the week, teams worked together to fabricate prototypes that were shared with family members at a culminating event. Following the camp, young adults from the Colorado Center for the Blind play-tested the prototypes and provided feedback. Using that feedback, a BBB Intern improved the design of several games, clarifying tactile elements and incorporating features such as high-contrast colors (Figure 5).

Figure 5. Iterative design, testing and improvement of a tactile Chutes and Ladders game board designed by middle school students. Clockwise, from top left: prototype completed in a summer camp, featuring a laser-cut game board and 3D printed pieces; updated prototype, incorporating features suggested by blind testers, such as tactile directional cues, high-contrast colors and a smoother texture for the chutes; young adults who are blind test the tactile features of a prototype board game.

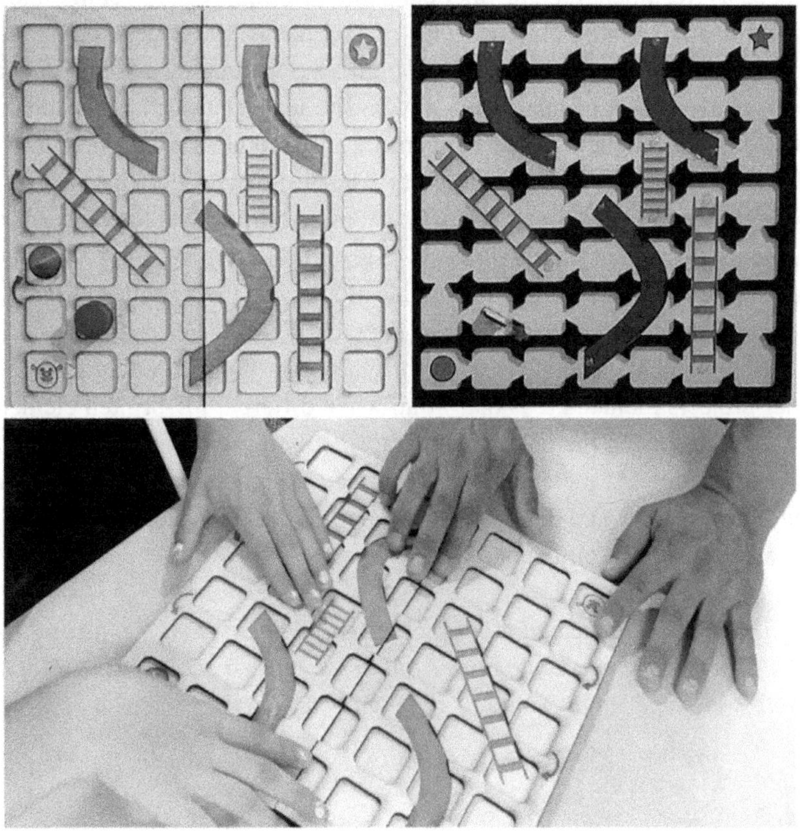

Semester-Long School Elective (Alabama)

Following a 3-day professional development workshop, two Alabama educators developed a semester-long, in-school elective course for middle school students. The program aimed to develop more accessible STEM learning materials for students in the school district with visual impairments. The teachers solicited requests from other educators and participating students selected topics that interested them. After a series of empathy-building activities, including reading literature featuring characters with disabilities, students worked in teams to develop accessible kits. Kits addressed standards-based topics, such as seasons, weather and rocketry, and included 3D-printed STEM models and audio information triggered by copper tape using Makey Makey and Scratch. Once complete, kits were housed in the school's library and made available for check-out by other teachers. The opportunity to create learning materials for other students was highly motivating for participants as was the option to make the product available to other classes.

High School STEAM Class (New Jersey)

Build a Better Book's mission has appealed to a variety of schools, including schools interested in giving back to their community. One girls' school in New Jersey integrated BBB as a core element of its sophomore level STEAM class. BBB became a year-long project for girls who applied the maker skills they developed as freshmen (e.g., 3D modeling and printing skills, circuitry and electronics, etc.) to make customized toys for young children with special needs. After developing some foundational technical skills, the teacher assigned each team of high school students a young client, most of whom were children with cortical visual impairment (CVI). The teacher tasked her students with designing a product that was "simple but elegant," to help ensure that products would be complete, high quality and able to be distributed to the identified end users. The designers used what they learned about their client's interests and abilities to design and create customized toys. Most toys incorporated electronics, such as lights and sound, as well as tactile elements (e.g., 3D printed figures). For example, a toy created for a boy who loved animals included tactile animals along with their sounds. A tactile book for a girl learning her letters introduced her to the sounds of the most common letters. The toys included features that are helpful for children with CVI, including bright colors such as red and yellow against a black background. Other projects included toys that lit up when different 3D printed elements were touched or moved, books that introduced the child's name or other common words, and toys that matched car sounds to 3D printed figures. At the end of the year, teen designers had an opportunity to share their products with their child clients, and for many it was an impactful experience. Following the conclusion of the program, the lead teacher commented:

We made a trip to [the school] at the end of May. One of my students told me it was the most meaningful experience she ever had. The whole faculty came out to see and greet our girls. All of the teachers from all of their divisions were amazed by what our girls created.

DISCUSSION

Across a wide variety of learning environments, the BBB model shows promise in heightening youths' engagement in STEM, including learners who might not otherwise elect to participate in STEM programming. The project's focus on empathy and designing for users who have a real and unmet need serves as a strong hook, motivating both youth and educators to participate. Although these motivations are strong for both boys and girls, it appears to be a particularly strong driver for girls, who are drawn both to the project's purpose as well as its strong emphasis on art and creativity.

As is often the case with lasting interventions, BBB has a 'low floor and high ceiling.' It is easy to get started with the program, no specialized equipment is required, and even short, introductory workshops (e.g., using craft materials to create a tactile alphabet book) can have important impacts (e.g., raising participants' awareness of blindness and disability). On the flipside, the project can support extended experiences (e.g., semester-long classes, weeks- or months-long internships) in which participants develop highly technical skills using 3D printers, laser cutters, electronics and computer programming skills. BBB's modular curriculum, which can readily be adapted for different settings and subject matter, supports educators in incorporating the program in their learning spaces, and its natural connections to a variety of different disciplines–including literacy, computer science and art–enables the project to spread beyond traditional STEM settings. This helps the project reach and serve a broader population of students, including youths who might not consider STEM as being 'for them.' Similarly, the breadth of opportunities and technologies encountered in BBB may also broaden student understanding of how they can integrate their diverse interests and skills into potential STEM careers that benefit society.

Across settings and regardless of tools and technologies used, the most critical factors for successful BBB programs appear to be strong community connections and sufficient time for participants to engage in an iterative design process. These factors are closely linked, as community partners are often ideally suited to provide valuable feedback in the design process. Ideation, developing and testing prototypes, creative problem-solving, and experiencing failure are all central to the engineering design process. In BBB, youth designers solicit and respond to user feedback, including BLV and sighted individuals. This encourages participants to work through failure and seek critical feedback in order to improve product design. The most successful BBB programs integrate community clientele in meaningful ways, sometimes as identified end users with specific interests and needs and other times as mentors and co-designers who provide input and feedback throughout the design process. Learning about and getting to know their community client helps increase participants' interest in and persistence with their project. Providing an opportunity for participants to share their work with their target end users is also an important element, as designers are proud of their work and excited for an opportunity to share their designs with their clients.

CONCLUSION

Authentic STEM learning experiences that enable youth to address real-world problems and design for others have potential to broaden and sustain participation in engineering and other STEM fields. The work described here suggests there is great potential for STEM learning opportunities that reimagine how we engage youth in learning about and experiencing these fields. As the People Part of Engineering framework suggests, there is an opportunity to broaden the view (and reality) of STEM professions and study by reframing the fields as human-centered endeavors. The BBB model illustrates that this does not entail incorporating art, design, or relevance in surface ways or as a "hook" that the program quickly moves beyond; rather, the human-centered attributes of the program seem to have significant power for building skills, mindsets, confidence, and persistence when they are infused throughout the learning experience. The experience is more meaningful when engineering skills and mindsets are developed in service of pursuing a purposeful, human-centered goal, benefiting a real person.

SUPPLEMENTAL MATERIALS

For more information about the *Build a Better Book* project, including program updates, a map of network partners, and our online youth Design Gallery, please visit the website: http://buildabetterbook.org.

REFERENCES

Afroogh, S., Esmalian, A., Donaldson, J. P., & Mostafavi, A. (2021). Empathic design in engineering education and practice: An approach for achieving inclusive and effective community resilience. *Sustainability, 13*, 0-0. doi:10.3390/su13074060

Alrubail, R. (2015, June 2). Teaching empathy through design thinking. *Edutopia*. https://www.edutopia.org/blog/teaching-empathy-through-design-thinking-rusul-alrubail

Bielefeldt, A. R. (2017). Is engineering about caring for and helping others? In B. Hernandez (Ed.), *Focus on educational research: Practices, challenges and perspectives* (pp. 89–118). Nova Science Publishers.

Burns, H., & Lesseig, K. (2017). Infusing empathy into engineering design: Supporting under-represented student interest and sense of belongingness. *Proceedings of the American Society for Engineering Education ASEE Annual Conference & Exposition*. Monolith. https://monolith.asee.org/public/conferences/78/papers/19857/view

CAST. (2018). *Universal design for learning guidelines version 2.2*. UDL Guidelines. http://udlguidelines.cast.org

Cech, E. (2013). Ideological wage inequalities? The technical/social dualism and the gender wage gap in engineering. *Social Forces, 91*(4), 1147–1182. doi:10.1093fot024

Cech, E. (2015). Engineers and engineeresses? Self-conceptions and the development of gendered professional identities. *Sociological Perspectives, 58*(1), 56–77. doi:10.1177/0731121414556543

Dalton, B., & Musetti, K. (2018). Tactile picture book making and multimodal composition: Students design for equity in English Language Arts. *Best Practices in Teaching Digital Literacies: Literacy Research. Practice and Evaluation, 9*, 195–213.

Deobald, J. (2020, May 1). Coding and empathy: Foundational skills for the future of work. *Forbes*. https://www.forbes.com/sites/forbesagencycouncil/2020/05/01/coding-and-empathy-foundational-skills-for-the-future-of-work/?sh=6727bffc62f5

Fila, N. D., Hess, J., Hira, A., Joslyn, C. H., Tolbert, D., & Hynes, M. M. (2014). The people part of engineering: Engineering for, with, and as people. *Proceedings of the IEEE Frontiers in Education Conference (FIE)*. IEEE. 10.1109/FIE.2014.7044106

Forsyth, S., Penzkover, K., Dalton, B., & Yang, G. (2020b). Build a Better Book: The value of adding meaning to the making. *Connected Science Learning, 2*(3), 0-0. https://www.nsta.org/connected-science-learning/connected-science-learning-july-september-2020/build-better-book

Forsyth, S., Penzkover, K., Davis, P., & Matty, I. (2020). Design thinking in public library makerspaces. *Public Libraries, 59*(4), 58–64.

Hess, J. L., & Fila, N. D. (2016). The development and growth of empathy among engineering students. *Proceedings of the American Society for Engineering Education ASEE Annual Conference & Exposition*. ASEE. 10.18260/p.26120

Hora, M. T., & Ferrare, J. J. (2013). Instructional systems of practice: A multidimensional analysis of math and science undergraduate course planning and classroom teaching. *Journal of the Learning Sciences, 22*(2), 212–257. doi:10.1080/10508406.2012.729767

Hynes, M., Joslyn, C., Hira, A., Holly, J., & Jubelt, N. (2016). Exploring diverse pre-college students' interests and understandings of engineering to promote engineering education for all. *International Journal of Engineering Education, 32*, 2318–2327.

Hynes, M., & Maxey, K. R. (2018, June 23). Investigating the fit between students' personal interests and their perceptions of engineering in a National Society of Black Engineers (NSBE) pre-college summer workshop. *ASEE Annual Conference & Exposition, Salt Lake City, Utah*. ASEE. https://peer.asee.org/30727

Leake, E. (2019). Empathy as research methodology. In P. Liamputtong (Ed.), *Handbook of research methods in health social sciences* (pp. 237–252). Springer., doi:10.1007/978-981-10-5251-4_65

Letourneau, S. M., & Bennett, D. (2020). Using narratives to evoke empathy and support girls' engagement in engineering. *Connected Science Learning, 2*(3), 0-0. https://www.nsta.org/connected-science-learning/connected-science-learning-july-september-2020/using-narratives-evoke

Meyer, A., Rose, D. H., & Gordon, D. (2014). *Universal design for learning: Theory and practice*. CAST Professional Publishing.

Morel, R. P., Coburn, C., Catterson, A. K., & Higgs, J. (2019). The multiple meanings of scale: Implications for researchers and practitioners. *Educational Researcher, 48*(6), 369–377. doi:10.3102/0013189X19860531

National Academy of Engineering. (2008). *Changing the conversation: Messages for improving public understanding of engineering*. The National Academies Press.

National Science Board. (2022). *Science and engineering indicators 2022: The state of U.S. science and engineering*. (NSB-2022-1). National Science Foundation. https://ncses.nsf.gov/pubs/nsb20221

Rose, D. H., & Meyer, A. (2002). *Teaching every student in the digital age: Universal design for learning*. Association for Supervision and Curriculum Development.

Smith-Doerr, L., Alegria, S. N., & Sacco, T. (2017). How diversity matters in the US science and engineering workforce: A critical review considering integration in teams, fields, and organizational contexts. *Engaging Science, Technology, and Society, 3*, 139–153. doi:10.17351/ests2017.142

Thiry, H. (2017). *External evaluation of the Build a Better Book program, University of Colorado Boulder, 2016-17* [Unpublished manuscript]. University of Colorado Boulder.

Thiry, H. (2018). *External evaluation of the Build a Better Book program, University of Colorado Boulder, 2017-18* [Unpublished manuscript]. University of Colorado Boulder.

Thiry, H. (2020). *External evaluation of the 2020 Build a Better Book librarian professional development workshop* [Unpublished manuscript]. University of Colorado Boulder.

Walsh, B. (2017). Texture, buttons, sound and code: Modal preference and the formation of expert identities. *7th Annual Conference on Creativity and Fabrication in Education (FabLearn17)*, Palo Alto, CA. 10.1145/3141798.3141819

Wilkins-Yel, K. G., Simpson, A., & Sparks, P. D. (2019). Persisting despite the odds: Resilience and coping among women in engineering. *Journal of Women and Minorities in Science and Engineering*, *25*(4), 353–368. doi:10.1615/JWomenMinorScienEng.2019026945

Wilson, E., & Mukhopadhyaya, P. (2022). Role of empathy in engineering education and practice in North America. *Education Sciences*, *12*(6), 1–12. doi:10.3390/educsci12060420

Chapter 2
The Mobile Making Program: Supporting Maker-Based STEM Engagement Among Youth During Out-of-School Time

Sinem Siyahhan
California State University, San Marcos, USA

Edward Price
https://orcid.org/0000-0002-0677-8476
California State University, San Marcos, USA

James Marshall
San Diego State University, USA

ABSTRACT

This chapter focuses on STEM out-of-school time and describes the Mobile Making program as a university-community partnership-driven out-of-school (OST) program wherein racially and ethnically diverse university students bring materials in their own vehicles and facilitate youth's engagement and learning in maker-based STEM activities at school and community sites. The authors discuss the program model and the lessons learned about how to (a) design maker-based STEM activities for youth (grades 4th through 8th), (b) recruit and train undergraduate students to serve as near-peer mentors, (c) work closely with school sites, (d) scale this program to multiple sites, and (d) sustain this afterschool program over the years. Further, the authors share the findings from phase two of the project and discuss implications for developing and scaling maker-based STEM afterschool programs.

DOI: 10.4018/978-1-6684-7771-7.ch002

INTRODUCTION

Making is a design-based, participant-driven endeavor based on a philosophy of 'learning by doing.' Propelled by their curiosity, Makers look for new ways to solve problems, and then share their ideas with others. Making is interdisciplinary, drawing on physics, computer science, engineering, math, and materials science, and employing both high- and low-tech tools. The open-ended, project-based nature of making has caught the attention of teachers and informal educators, who see Making as a means for engaging preK-12 students in science, technology, engineering, and mathematics (STEM). Making is a central activity in constructionist and progressive theories of learning, where learning occurs when a person engages in the process of creating and designing (a.k.a making) artifacts that are personally and socially relevant (Blikstein & Krannich, 2013). Research suggests that there are several benefits of engaging students in Making including, but not limited to, developing problem-solving, critical thinking, communication, and collaboration skills that are highly valued in today's job market (Kafai et al., 2014; Schad, & Jones, 2020).

The lack of access and diversity within Making has resulted in a desire to establish Making programs in underserved urban and rural communities (Weiner et al., 2018). Afterschool programs can play an important role, having demonstrated success in connecting girls and minorities with STEM (Afterschool Alliance, 2013; McCreedy & Dierking, 2014). The flexibility of out-of-school (OST) environments allows for stimulating and adaptive curricula that can engage underrepresented students and provide a real-world context for learning. Afterschool environments are increasingly being viewed as strategically important for bringing Making activities to low-income youth (Krishnamurthi et al., 2013; Simpson et al., 2020). It is well documented that afterschool programs that involve partnerships between higher education institutions and community organizations are mutually beneficial and effective in meeting the local needs of providing quality afterschool experiences for youth while providing opportunities for career and personal development for university students (Furco, 1996; Salam et al., 2019). However, it is unclear how university-community partnerships can be utilized to implement afterschool programs that aim to broaden the participation of underserved communities in Making and STEM.

The Mobile Making program is a university-community partnership OST program wherein racially and ethnically diverse undergraduate students bring materials in their own vehicles and facilitate upper elementary and middle school students' engagement and learning in maker-based STEM activities at school sites, libraries, and community sites. The program is designed around research-based principles for effective OST programs in underserved communities. These principles identify the importance of: (1) access to resources (Gasbarra & Johnson, 2008; U.S. Commission on Civil Rights, 2010), (2) ethnically diverse near-peer leaders (Gasbarra & Johnson, 2008, Pajares, 2006; Syed et al., 2011; Zirkel, 2002), (3) authentic activities (Archer et al., 2010; Rahm et al., 2005; Roth & Lee, 2004), (4) legitimacy within the community (Ferry et al., 2000; Gilmartin et al., 2006; Huang et al., 2000; Lemke, 2001), (5) ongoing input from participants (Honig & McDonald, 2005; Lyon et al., 2013; Sefton-Green, 2013), and (6) university-site partnerships and community engagement (Furco, 1996) to facilitate program success.

In what follows, we share the theoretical underpinnings of our program and the different iterations the program has gone through in the last seven years with funding from the National Science Foundation. Specifically, we discuss the initial development and pilot testing of the program (Phase 1, between 2014 and 2016), how we expanded the program across multiple school sites and moved the program online during the pandemic (Phase 2, between 2016 and 2022), and finally how we are currently scaling up the program across multiple institutions (Phase 3, between 2022 and 2027). From there, we describe the maker-based STEM activities we developed and refined over the years as well as our community partners. We then share research findings from the Phase 2 of the program. Our research questions focused on understanding the impact of the program both on youth participants and undergraduate students. We used multiple data sources including program attendance, pre-and-post surveys, focus groups, and session feedback back forms to understand the impact of the program on youth participants' sustained interest and self-efficacy in Making, and perceptions of relevance of Making and STEM in their everyday lives. We investigated the impact on undergraduate students' ability to teach, interest in teaching STEM informal and formal settings, and 21st century skills. We conclude this chapter with a discussion on how our research findings and lessons learned through different iterations of the program can support and sustain maker-based STEM experiences among underserved youth across different contexts.

BACKGROUND

Sociocultural and self-efficacy theories of learning inform the program design and our research efforts. Sociocultural theory takes learning as a social process within a cultural context involving the development of tools and skills (Lemke, 2001; Vygotsky, 1978). Self-efficacy is a belief in one's capability to succeed in a specific context and is developed through mastery experiences, vicarious learning, and verbal persuasion (Bandura, 2006). Self-efficacy plays an important role in motivation, interest, well-being, and personal accomplishment. Research suggests that minorities and low-income students express enthusiasm for science at a young age, yet this interest often fades by the time they reach early adolescence (Grigg et al., 2006). The way STEM content is taught in schools, stereotypes around who gets to do STEM, and lack of representation of this population in STEM fields are often cited as possible contributing factors for the dwindling interest and lack of self-efficacy of these students in STEM (Master & Meltzoff, 2020; Regan & DeWitt, 2014). Students' STEM experiences in schools are often disconnected from their interests, experiences, and learning outside of school (Basu & Barton, 2007; Bouillion & Gomez, 2001). As a result of the disconnect between the worlds of family, peers, and school, many students see little use for STEM in their lives, instead, viewing it as a subject best reserved for 'experts' (Aschbacher et al., 2009; Costa, 1995).

In contrast, children's experiences with Making can be empowering, and maker-centered learning may help them develop personal agency, self-efficacy, and a sense of community (Halverson & Sheridan, 2014; Petrich et al., 2013). Questions around equity in Making include access to resources, pedagogical approaches, the design of learning environments, and what counts as learning (Halverson & Sheridan 2014; Vossoughi, & Bevan, 2014). Even with increased participation in Making, it is unclear whether individuals and groups, especially those from traditionally underserved audiences, view themselves as Makers or part of the larger Making community (Halverson & Sheridan 2014; Kafai et al., 2014; Martin & Dixon, 2013). Weaving culturally responsive practices into after-school environments is one way to bring Making activities to low-income youth and broaden participation in STEM (Vossoughi & Bevan, 2014). While there are playbooks and implementation guides for youth maker spaces (Maker Ed, 2015), and pedagogical approaches that promote equity (Vossoughi et al., 2016), specific guidelines for engaging underserved youth are lacking. Research-based strategies, guidelines, and models would help increase the diversity of Makers and ultimately broaden the participation of underserved communities in STEM.

Program Iterations

In what follows, we describe the evolution of our program from the development and pilot phase to the scaling up of the program across California. Over the years, we have iterated the design and implementation of our program across different contexts and settings. We engaged in design-implement-redesign cycles with our maker-based STEM activities. Revisions were informed by feedback from our community partners, internal reflections on our processes, input from our external evaluator, and data collected from our undergraduate student facilitators and youth participants each semester as part of our research efforts. Our data-informed decision-making approach allowed us to analyze the information we gathered against our goal of achieving positive program outcomes for youth participants and undergraduate students.

Phase One: Developing and Pilot Testing the Mobile Making Program

The Mobile Making Program started at the California State University San Marcos (CSUSM), one of twenty-three California State University (CSU) campuses, with a small grant from the Genentech Corporation in 2013, which supported a pilot program at one middle school site. During the development and pilot testing phase, we used a small set of low-tech maker-based STEM activities. We hired and trained undergraduate students majoring in STEM to facilitate maker-based STEM activities with youth once a week for ninety minutes at the school site for nine weeks every semester. Prior research on OST programs and the Maker Movement, which was growing in education at the time, informed our program development. Initially, we lacked a set of design principles that guided our program. One of the important outcomes of our pilot work was the emergence of research-based principles that informed subsequent program design. These were incorporated into a funded proposal to the National Science Foundation (#1423612). Starting in Fall 2014, we field-tested the program at four middle schools in underserved communities in North County San Diego over a two-year period. Through this experience, we refined the design principles and their translation into practice. These principles not only inform the key elements of the expansion of our program to other contexts but also highlight what makes the Mobile Making program stand out among other OST STEM programs with similar goals. In addition to identifying key design principles for the program, we were able to demonstrate that the Mobile Making program powered by undergraduate students who take on the role of near-peer mentors can effectively engage underserved youth. The pilot program reached a total of 110 youth participants and nearly 80% were members of underrepresented minority groups. Further, pre-and-post surveys indicated a significant increase in youth participants' interest, self-efficacy, and perception of the relevance of Making and STEM.

Although we established the feasibility of our program during the development and pilot phase, we identified many areas for improvement. First, the maker-based STEM activities we implemented with youth came from other online resources such as Instructables, MakerEd, etc. As such, rather than creating our own set of activities, we found, practiced, and implemented maker-based STEM activities that were created by others. Although this was a good starting point to ensure that we engaged youth participants in maker-based STEM activities who did not otherwise have an opportunity to learn about these activities, it was unclear how these activities were socially meaningful and personally relevant to the participants, and what differentiated these activities from afterschool programs that focused on only STEM or Making.

Second, our impact measures focused only on youth participants. This was partially due to the small number of undergraduate students we hired and trained as facilitators of the Mobile Making Program (n=16). Our efforts were also mostly focused on the logistical aspects of running the program: communicating with the middle school sites, managing middle school sites and undergraduate student schedules, as well as tracking inventory for materials. However, we quickly realized that there is a need to capture the effectiveness of undergraduate training. Other than informal check-ins with undergraduate students, we had no way of knowing whether our training effectively prepared undergraduates to work with youth participants successfully. Further, we wondered what skills undergraduates were developing, if any, as part of their participation in the program. As we prepared for the next phase of the program, it was clear that hiring undergraduate students as facilitators was not a sustainable approach to expanding the program to other sites within our region. Like other OST programs, we had to address the challenge of ways to sustain and scale the Mobile Making program without the support of grant funding in the future.

Phase Two: Program Expansion and Modifications During COVID-19

With the lessons learned during the development and pilot testing phase of the Mobile Making program, we expanded the program from four to twelve middle schools in North County San Diego between Fall 2016 and Spring 2022 with additional funds from the National Science Foundation (#1612775). Our focus during this phase was on sustaining positive impacts on youth participants, measuring impacts on undergraduate students, and exploring how to sustain and scale the program through service learning courses to lower the cost of the program, reach a larger number of youth participants, and undergraduate students, extend the number of near-peers present to support youth participants at each site, and ensure institutional support for the program at CSUSM. However, like other OST programs around the country and the world, our plans for program implementation and research drastically changed due to the COVID-19 pandemic in the Spring of 2020.

Before the pandemic, we paired paid undergraduate students (called STEM Ambassadors), who led the implementation of the Mobile Making program at various school sites for nine weeks, with STEM undergraduates who participated in the Mobile Making program as part of a 3-week service learning (SL) assignment in a STEM course. The service learning students participated in a three-hour training to meet with paid STEM Ambassadors and learn about the maker-based STEM activities that they were assigned to help facilitate for three weeks at a school site. Post-surveys we administered at the end of each three-week rotation showed over 80% of the participating STEM-course SL students indicated that they enjoyed the experience enough to say they would do it again, and roughly three-quarters of the students indicated that they would volunteer for a STEM-related outreach effort at a future employer or community science fair (n = 360). While this structure allowed us to recruit additional near-peer mentors without a financial cost, it required many hours of human resources for the recruitment of undergraduate students who participated in only three out of nine weeks of the Mobile Making program. To be able to recruit these SL students, we first recruited STEM faculty who were willing to add an assignment and offer a service learning assignment in their syllabus. To recruit students, we would then give a class presentation about the program and manage a sign-up process. Every three weeks, we engaged in recruiting and training SL students and coordinating with STEM faculty to ensure that there were enough facilitators for each school site. While the service learning experience was valuable, what the undergraduate students did as part of the Mobile Making program was disconnected from other class activities and course assignments.

We stopped running the Mobile Making program in person when all K12 schools and the university shut down in the Spring of 2020 due to the pandemic. In the Fall of 2020, when both K12 schools and the university resumed education completely online, we also moved all the sessions of the Mobile Making program online and continued to work with our school sites. However, our approach to recruiting service learning students as additional facilitators of maker-based STEM activities before the pandemic was unfeasible given the online modality of the university STEM courses and the Mobile Making Program sessions.

The constraint of online modality forced us to pivot our approach to SL recruitment and setup and enabled us to pursue a new partnership within the university with the Education faculty who taught a technology-focused maker-based STEM SL course that, by design, requires 15 hours of field experience in educational settings. Additionally, the course was a prerequisite course that all teacher candidates took to enter the credential program at the university. Unlike SL undergraduate students taking a STEM course, students in this course were already committed to working with children and interested in improving their teaching skills. As such, there was a better alignment between the course goals, students' interests and career goals, and their participation in the Mobile Making program as facilitators.

We paired STEM Ambassadors (n=19) with teacher candidates (n=240) taking the service learning education course between the Fall of 2020 and the Spring of 2021. One STEM Ambassador was paired with several teacher candidates, creating teams with complementary skills related to STEM content knowledge and teaching. We were explicit about this framing with the teams, encouraging them to respect and utilize each other's skill sets. Teacher candidates learned and practiced the design challenges as part of their coursework. This eliminated the need to set up additional training days and hours for them to learn about maker-based STEM activities. Together, the STEM Ambassador-teacher candidate team met online via Zoom with groups of elementary and middle school students to explore various online technology tools including storyboarding, 3D modeling, game design, and coding once a week for 90 minutes. Youth participants engaged in brainstorming and prototyping a solution to a real-world problem framed as a 'design challenge.' The nine-week program continued throughout the Spring semester, while we also launched a virtual kit-based STEM program to engage youth participants in low-tech maker-based STEM activities. A variety of STEM activity kits were assembled in advance and then delivered to schools for student distribution. The STEM Ambassador-teacher candidate team held weekly calls with participating students and built their activities together virtually via Zoom. During the Fall of 2020, 21 elementary schools and two middle schools from three districts participated in the virtual Mobile Making program. During Spring 2021, 18 elementary schools from one district and two libraries participated in the virtual Mobile Making program. Eight middle schools from two districts participated in the virtual kit-based Mobile Making program during Spring 2021. These programs reached 670 student participants (327 during Fall 2020 and 343 during Spring 2021).

We conducted focus groups with SL education students (teacher candidates) and the STEM Ambassadors. The SL education students described the benefits of participation in a virtual version of making program that included having a 'true classroom experience' through practice lesson planning, 'being vulnerable' and not always 'having all the answers,' and applying interpersonal skills in the online environment to encourage participation and 'reach' the more shy and introverted participants. At the same time, STEM Ambassadors found working with SL education students as being a helpful and rewarding collaboration—describing the students as 'peers' who were present 'with purpose, as future teachers.' With the SL education students, STEM Ambassadors were often able to take on a subject matter expert role during the sessions, while the SL education students handled elements of classroom organization and management—all burgeoning expertise they brought to the partnership.

During this phase of the Mobile Making program, we successfully explored (a) including activities that use high tech making tools (e.g. 3D modeling/printing, coding, and robotics) along with activities that utilize circuitry and hands-on materials; (b) serving upper elementary students (4th & 5th grades) and middle school students; (c) delivering the Mobile Making program in three modalities: face-to-face, online with activities that exclusively used technology tools, and online with hands-on materials; (d) bringing together a diverse group of undergraduate students from STEM and education; (e) working with school and library partners; and (f) incorporating the facilitation of the Mobile Making program activities by undergraduate students into service-learning courses.

Phase Three: Scaling and Sustaining the Program within the University

After experimenting with offering the program through different modalities, successfully recruiting and training undergraduate student facilitators through service learning courses, expanding the age range of the youth participants we serve, and increasing the number of school sites from four to twelve, we wanted to scale and sustain the program across California State University (CSU) campuses to ensure institutional support beyond our campus. Supported by National Science Foundation funding (#2215653), we are currently in the broad implementation phase of the Mobile Making program where we are collaborating with science and education faculty at three additional CSU campuses as well as the system-level Chancellor's Office. During this phase, we are most interested in the following questions: What is the impact of Mobile Making on youth and undergraduate participants across different contexts? What processes, activities, practices, and supports contribute to participant outcomes? How do local contexts inform and impact the core program model? Our goal in the next five years is to assess outcomes for youth and undergraduate participants across different learning/teaching contexts and modalities as well as understand how to form, support, and maintain a Community of Practice (CoP) as we engage in the work of scaling and sustaining the Mobile Making program with our CSU campus, school, and community partners.

Given CSU's infrastructure for SL participation, we can involve many courses and reach large numbers across California. The four campuses implementing the Mobile Making program are located across California. Each has the capacity and community relationships to engage underserved youth in maker-based STEM activities. Through our CSU campus partners, we can offer the program in different modalities, and serve rural, urban, and suburban youth across different grade levels. Table 1 summarizes the number and type of sites we currently serve across four CSU campuses. Below, we describe the demographics and context of each campus.

Table 1. The Mobile Making afterschool program implementation sites across CSU campuses

	#School Settings	# Community Settings	#Total Sites
CSU San Marcos	30	2	32
CSU Long Beach	-	2	2
CSU Fresno	4	-	4
California Polytechnic State University	8	-	8
Total:	42	4	46

California State University San Marcos (CSUSM) is a regional university in North San Diego County, California. The region has a high concentration of technology-based firms and is home to a diverse population with 33% of its residents being Latino/Hispanic. The student body at CSUSM reflects this local diversity, and the University is a Hispanic-Serving Institution (HSI). CSUSM has strong relationships with the local K-12 school districts and has the Community Engagement Classification from the Carnegie Foundation. CSUSM's Community Engagement Division provides institutional support for CSL. The CSUSM Center for Research and Engagement in STEM Education (CRESE) operates the Mobile Making program. Currently, we are expanding the Mobile Making program from 12 to 30 afterschool and community sites, serving upper elementary and middle school students in multiple local school districts (see details in the community partners section). Paid STEM Ambassadors support key program functions, and SL students in education courses serve as facilitators.

California State University Long Beach (CSULB) has one of the largest enrollments in the CSU system (40,000) and is designated an HSI with 46% of enrolled students identifying as Hispanic/Latinx in Fall 2020. The teaching credential programs at CSULB draw students from both colleges, as well as neighboring universities. CSULB's Mobile Making Program sites include the Downey Library and the Boys and Girls Club of Long Beach (BGCLB). The City of Downey (pop. 112,000) is located 10 miles north of Long Beach and 13 miles southeast of Los Angeles. The community is 75% Hispanic, 14% White (non-Hispanic), and 7% Asian. In 2019, the BGCLB served over 3500 youth, ages 5-18. Ninety percent of these participants qualify for free or reduced lunches, and 40% come from single-family households. Facilitators for the CSULB Mobile Making program include future teachers engaged in service-learning assignments. There are two significant variations to the CSUSM-developed model. First, CSULB programs take place primarily in informal learning environments outside of school. This allows for an understanding of supporting making in sites removed from the school setting, yet still embedded in the local community. Secondly, in addition to afterschool programming, the CSULB program explores alternate formats for engagement, including Saturday family-making STEM sessions as well as weeklong youth 'camp' sessions, both conducted at the library site. This again provides an opportunity to understand broad implementation in ways unexplored in the initial models.

California State University Fresno (CSUF) is the tenth largest campus in the CSU system and served nearly 25,000 students in Fall 2021. The university is located in the San Joaquin Valley of Central California, one of the country's most culturally and linguistically diverse regions. CSUF primarily serves students from surrounding counties, including Fresno, Madera, Kings, and Tulare, and is an HSI and an Asian American, Native American, Pacific Islander Serving Institution (AANAPISI). Specifically, 56.1% of students are Hispanics/Latinx, and 12.7% are Asians, Pacific Islanders, and Native Americans. Additionally, 66.2% of the students are considered first-generation college students. The CSUF team is collaborating with four elementary schools within the Fresno Unified School District (FUSD) to operate its Mobile Making after-school program. FUSD is the largest school district around CSUF, serving nearly 75,000 students across 109 schools from preK-12 (see details below). The undergraduate facilitators involve undergraduate students enrolled in service-learning courses taught by biology and education faculty. Together, students collaborate across courses to facilitate STEM activities for FUSD students, with 90 undergraduate students reaching approximately 300 elementary students enrolled in afterschool sites in the first year.

California Polytechnic State University (Cal Poly) is a comprehensive polytechnic campus located in San Luis Obispo, CA. There are 22,000 students: 53% White; 18% Hispanic; 14% Asian; 8% two or more races; 0.8% African American. The Cal Poly team collaborates with Lucia Mar Unified School District. LMUSD is the largest school district in San Luis Obispo County and serves agricultural communities and small beach towns. Specifically, the Cal Poly team works with the Bright Futures Afterschool Program which offers free after-school care for families that qualify. The participants are in upper-elementary and middle school grades. Cal Poly's implementation consists of three phases: (1) the development of maker activities in a senior capstone course for Liberal Studies majors, (2) the implementation of those projects and other activities developed by CSU-San Marcos at Bright Future Afterschool Program by Cal Poly STEM majors, and (3) and, in year 3, post-baccalaureate credential students will help undergraduates implement and evaluate maker lessons 3 times per quarter. Cal Poly plans to have approximately 80 undergraduate students work with 8 school sites to reach approximately 600 total elementary students enrolled in afterschool sites across the five years.

Maker-Based STEM Activities

Making is inherently interdisciplinary, and most activities cross boundaries between subject areas. Existing making activities cover a broad range of STEM content areas and skills (Blikstein & Krannich, 2013). Over the years, we developed, implemented, and refined several different maker-based STEM activities that are compatible with the structure and time constraints of after-school settings. Designing activities for youth in an afterschool setting required balancing different tensions due to the informal and voluntary nature of the youth participation. As such, attendance can fluctuate considerably and can be unpredictable across different sessions in an after-school context versus a school setting. We created activities that could be completed within 60 to 90 minutes rather than having youth participants work on the same project across multiple sessions. Further, youth have a wide range of previous experiences, skills, and interests ranging from no experience to having deep knowledge, practice, and interest with a set of tools and activities.

During the development and pilot testing phase of the Mobile Making program, we focused on activities that involved low-cost materials like cardboard, rubber band, toothpicks, glue, tape, etc., and engaged youth participants in incorporating simple circuits in their design using 3V batteries, wires, LEDS, motors, buzzers, and switches. As such, our initial set of maker-based STEM activities focused on engineering skills such as troubleshooting as well as mechanics, electricity, and circuits as targeted science content. We reviewed existing Making activities found online through the MakerEd website, Makerspaces.com, the Exploratorium Tinkering Studio, Instructables, Howtoons, and others. Rather than simply taking the activities we found online and implementing them with youth as part of the Mobile Making program, we adapted them so that (a) they could be completed within 60 to 90 minutes time, (b) they do not require prior knowledge, understanding, or experience so that all participants, regardless of their attendance history, can complete them, and (c) they are simple but yet can be adjusted during the session to challenge participants who have advanced knowledge or understanding. Some of the youth participants' projects included but were not limited to building a hydraulic bridge, fan boat, automata, snow globe, etc. as seen in Figure 1.

Figure 1. Examples of maker-based STEM activities with low-cost materials

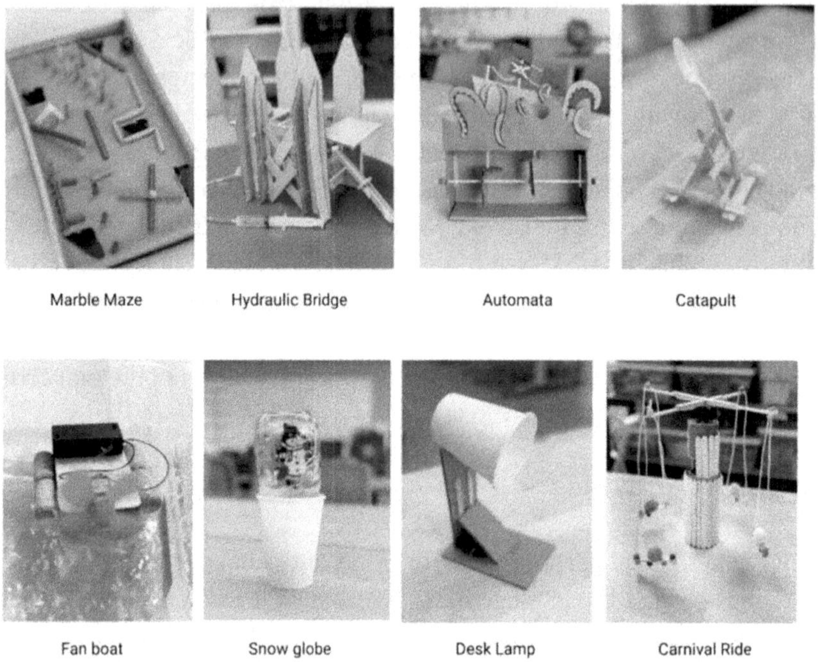

We continued to develop and implement maker-based STEM activities that use low-tech materials during the expansion phase of the Mobile Making program until the pandemic. When we moved the Mobile Making program from in-person to online during the pandemic, we developed and incorporated maker-based STEM activities that use digital tools such as Pixton, Scratch, and Tinkercad that engaged youth participants in creating digital stories, games, animation, and 3D models as shown in Figure 2. With the incorporation of digital tools and projects, our maker-based STEM activities continue to focus on engineering skills such as troubleshooting while at the same time fostering youth participants' technology literacy skills and a variety of science, math, and language arts content depending on the digital tool youth participants used to create their projects.

Figure 2. Examples of maker-based STEM activities with digital tools

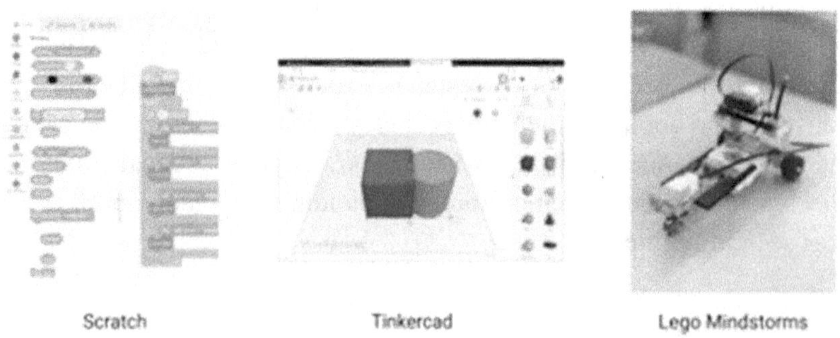

Previously, many of the maker-based STEM activities we found online and developed ourselves using low-tech tools lacked a personally or socially relevant problem context that created the need for youth participants to learn how to use the tools to design something. Learning sciences research suggests that meaningful learning and sustained engagement is more likely to take place when youth work on problems that they care about (Hidi & Renninger, 2006; Eccles & Wigfield, 2002). After the pandemic, not only have we incorporated and developed maker-based STEM activities using digital tools but also reframed our existing activities around a design challenge that introduced youth participants to a personally or socially significant real-world problem or an issue. Our goal was to engage youth participants in authentic activities that connected their making to their experiences outside of the afterschool program, increasing the relevance of Making and STEM in their lives (one of the targeted youth participant outcomes of the program). The reframing of the Making activity and STEM in terms of addressing a real-world problem positions the youth participants as agents of change in their communities and creates a need to design, create, and make something. Research suggests that learners develop self-efficacy when they have a sense of control and impact over their environment and are able to achieve their goals (Bandura, 2000). As such we expect the reframing of the maker-based STEM activities to further facilitate the youth participants' self-efficacy and interest in Making and STEM through an increased sense of the relevance of the activities in their lives and the impact their designs can have in the world.

Rather than asking youth participants to design a hydraulic bridge for the sake of designing a hydraulic bridge, we first introduce a design challenge to start them thinking about a problem and how their designs could be a prototype of a solution. For example, we reframed the hydraulic bridge activity as a design challenge where we give youth participants the following scenario: There is an area near a lake in your city where every time it rains the surrounding streets are flooded. This causes traffic jams and city workers to work overtime to manage the flood and the traffic. Brainstorm ideas about how you can build a bridge over the lake and raise the streets to address the problem. Once youth participants brainstorm ideas about the design of their bridge then we introduce them to the materials and how they can create the hydraulics. Similarly, they address real-world problems when they engage in making with digital tools. For example, in one activity youth participants are asked to consider the problem of food insecurity in the local county. Following the introduction of the problem, including relevant statistics, youth participants are asked to brainstorm ideas for a solution and then use Pixton, a comic and storyboard-building application, to create a comic to raise awareness in their community about their solution; see Figure 3.

Community Partners

The Mobile Making Program requires a successful university-community partnership. Throughout the last decade, we have built and maintained positive relationships with our school, library, and community partners. Cultivating long-lasting partnerships between the university and the community takes a lot of time and work on both sides. The initial idea for the Mobile Making program came from our meetings with teachers and school principals who were looking to bring Making and STEM programming to their sites. We wrote our first grant proposal in consultation with our four school partners to ensure that we shared the vision of the Mobile Making Program and that the program met a need in the communities the schools served.

Figure 3. A comic designed by a youth participant to address food insecurity using Pixton

At the core of our university-community partnership is a shared goal of meeting the need for low-cost quality STEM afterschool experiences for minority and low-income youth in the region. Our school district, library, and community partners trust the university with expertise in Making and STEM and creating maker-based STEM activities for youth. Further, our community partners view the relationship they have with the university as supporting their youth in pursuing a college degree and potentially a STEM career. As such, not only have they had confidence in us creating the curriculum but also welcomed the facilitation of maker-based STEM activities by undergraduate students who they see as serving as role models to their youth. For the university, the program supports undergraduates and helps fulfill the university mission and strategic plan, which emphasize community engagement.

As we expanded the program across different school, library, and community sites, we learned that clear, consistent, and frequent communication with our partners is key for the successful implementation of the Mobile Making Program. We adjust program specifics based on their needs and setting. We reach out to our partners about dates and times for our program three months before the semester. This gives them enough time to secure a room, assign a staff member, and recruit youth participants for the Mobile Making Program. As we get closer to the first session of the program at each site, we share the description of each maker-based STEM activity with them so that they are informed about our weekly activities and plan to provide additional resources if needed. At the end of the semester, we share a report that includes the number of youth participants in the program as well as a summary of our research findings (discussed in the next section). We follow up this report with a meeting where we discuss what works and what needs to be improved for the next semester.

As we scale the Mobile Making program across different socio-cultural contexts, we are acutely aware of the diversity of the communities each school, library, and community site serve, and the norms, values, and needs of our partners. Table 2 provides the demographic information for the school districts who are currently implementing the Mobile Making Program at their elementary and middle schools.

Table 2. Demographics of school districts

School Districts	Escondido	San Marcos	Vista	Oceanside	Bonsall	Fallbrook	Temecula	Fresno	Lucia Mar
% Hispanic/Latino	77.4	45.9	65.5	56.4	37.5	65.3	35.3	68.7	47.7
% Low income	78.2	43.1	75.4	70	43.8	79.3	33.6	94	57.8
% English language learners	39.7	13	17	11.9	14.2	28.8	4.2	18	12.3

We will continue to take a collaborative approach to work with our school and community partners as we scale the program. In addition to working with our partners to implement the Mobile Making program, we support them in other ways that are important to them. For example, over the years, in addition to running the Mobile Making program, we supported family nights, open houses, and professional development meetings at various sites. It is important that our community partners trust us in providing quality maker-based STEM learning experiences for youth and that they see us as a resource to serve their communities.

Program Impact Measures and Outcomes

Across different phases of the Mobile Making program, our research efforts focus on examining outcomes for youth participants and undergraduate students. Table 3 lists the research questions and measures we used to answer them. Our research activities were approved by the CSUSM Institutional Review Board (IRB). Below, we share our findings from Phase 2 of the program except for Year 5 due to the COVID-19 pandemic.

Table 3. List of research questions, outcomes measures, and data sources

Youth Participants	Measures
Does the after-school Making program support increased interest in Making, specifically, and in engineering and science, more generally? (Youth Outcome 1: Interest in Making/STEM)	A, P
Does the after-school Making program support an increase in youth's self-efficacy in Making specifically, and in engineering and science, more generally? (Youth Outcome 2: Self-efficacy in Making/STEM)	P
Does the after-school Making program support a change in youth's perceived relevance of Making specifically, and engineering and science more generally, in their lives? (Youth Outcome 3: Perceived Relevance of Making/STEM)	P, FF
Undergraduate Participants	**Measures**
Do undergraduate students develop their content knowledge and ability to teach as a result of facilitating the after-school Making program? (Undergrad Outcome 1: Improved Knowledge and Skills)	P
Does facilitating Making activities promote and extend undergraduate students' 21st century skills, specifically, perseverance, problem-solving, collaboration, and communication? (Undergrad Outcome 2: Improved 21st-century skills)	FG
Does facilitating Making activities impact undergraduates' interest in teaching STEM in informal and/or formal settings? (Undergrad Outcome 3: Interest in Teaching)	FG

A = Attendance P = pre/post instrument FG = focus groups FF = Feedback forms.

Youth Participant-Related Outcomes

As a result of continued participation in Making, we hypothesized that youth participants (grades 4th through 8th) will (1) express sustained interest in Making, and perhaps even engage in Maker-style activities beyond the after-school setting; (2) demonstrate increased self-efficacy in Making-related contexts; and (3) recognize how Making and/or STEM is relevant to their everyday lives. These three outcomes are consistent with those outlined in recent OST literature, which states that youth should develop an interest in STEM, enhance their capacities to productively engage in STEM learning activities, and acquire an understanding of the value of STEM in society (Afterschool Alliance, 2013, 2015). Similarly, they reflect the potential benefits of Making (Kafai et al., 2014; Vossoughi & Bevan, 2014).

The selection of these outcomes is also supported by research findings, which suggest that young people's self-efficacy beliefs are better predictors of motivation, future academic choices, and career decisions than preparation, knowledge, competence, or interest (Pajares, 2006). Self-efficacy also has a strong influence on persistence, effort, and other behaviors that are important for college and career readiness (Bandura, 2006). Likewise, students' perceived relevance of science instruction, and its connection to their personal needs and goals, is an important factor in motivation and future actions (Wolter et al., 2013).

MATERIALS AND METHODS

We used two data sources to measure the program's impact on all youth participant outcomes (interest, self-efficacy, and relevance of Making/STEM): (1) session feedback surveys and (2) pre-and-post surveys. Youth participants filled out a pre-and-post survey at the beginning and at the end of the Mobile Making program. We also administered a session feedback survey at the end of each nine-minute session. On both session feedback surveys and pre-and-post surveys, youth participants used a five-point Likert scale that ranged from 1=Strongly Disagree to 5=Strongly Agree to rate their interest and self-efficacy in Making and STEM, as well as perceptions of the relevance of Making and STEM and the project they completed. We used paired t-tests for pre-and-post-survey comparisons and descriptive statistics (i.e., mean scores, percentages) to analyze part of the post-survey and session feedback data.

Overall, the Mobile Making program has a positive impact on youth participants' engagement and learning. The maker-based STEM activities were rated moderately as being relevant to the community's youth participants belong to while youth participants rated highly their interest in the project they completed and learning something new as part of their participation in the Mobile Making program (see Table 4).

Table 4. Mean Scores tracked over multiple years with session feedback surveys

	Year 1 (n=494)	Year 2 (n=2213)	Year 3 (n=2950)	Year 4 (n=2399)
Project kept me interested	4.52	4.28	4.20	4.30
I learned something new	4.31	4.06	4.00	4.10

Note: n represents the total number of session feedback surveys collected across all activities during each program year.

In the post-survey, we asked youth participants to offer their perception regarding the presence of various design principles-related elements using a five-point Likert scale of 1=Not important, 3=Somewhat important, to 5=Very important. The scale assessed the importance of each contributing program feature. Table 5 represents the total pre-and-post-matched surveys received during each program year to ensure the responses of youth participants who participated in all activities were captured. The sample size for year four is low due to the interruption of the COVID-19 pandemic.

Table 5. Mean scores for the importance of different design principles

	Year 1 (*n*=82)	**Year 2 (*n*=136)**	**Year 3 (*n*=200)**	**Year 4 (*n*=27)**
Having voice in my maker project	4.20	4.27	4.03	3.93
Having project mentors that help me be successful with the maker project	4.18	4.23	4.06	4.04
Having ownership of the maker projects	4.16	4.17	4.09	4.08
Having maker projects that connect technology to how it is used in the real world	4.15	4.13	4.03	4.04
Having maker projects that connect science to how it is used in the real world	4.06	4.10	4.04	3.93
Having project mentors that I can relate to	4.01	4.04	3.92	3.96
Having opportunities for my parents to see the maker projects I've completed	4.07	3.99	3.91	4.07
Showing people in the community the maker projects I've completed	3.75	3.63	3.63	3.81

RESULTS

Youth participants' self-efficacy in Making and STEM was measured in the session feedback survey and on the pre-and-post surveys. On the session feedback survey, youth participants were asked to rate their feeling of success after completing each project. The mean scores for each year were 4.29 in Year 1 (*n*=494), 4.09 in Year 2 (*n*=2213), 4.02 in Year 3 (*n*=2950), and 4.12 in Year 4 (*n*=2399), suggesting youth participating felt highly successful after completing the projects. Table 6 shows the results for the pre-and-post surveys with significant gains from pre-to-post surveys in Year 1. In Year 4, we noted a marked difference in scores. Contrary to all previous years, each Year 4 item regressed from pre-to-post. The limitations of the data due to the COVID-19 pandemic provide some probable explanations for the reversal of trends. The shortened program year, along with a very small number of students having matched pre-to-post responses, has very likely influenced the findings.

The relevance of Making and STEM in youth participants was measured in the session feedback survey. Using a five-point Likert scale, students were asked to rate their agreement with four relevance-related statements following each session. In the items, 'project' refers to the making activity the participants just completed. Table 7 presents mean scores for each program year.

Table 6. Mean scores for pre-and-post survey items for self-efficacy in Making and STEM

	Year 1 (*n*=82)		Year 2 (*n*=136)		Year 3 (*n*=200)		Year 4 (*n*=27)	
	Pre	**Post**	**Pre**	**Post**	**Pre**	**Post**	**Pre**	**Post**
I am good at science	4.02	4.24	3.69	3.87	3.74	3.87	3.91	3.71
I am good at solving problems	4.11	4.29	3.75	3.96	3.98	4.07	4.17	3.84
People like me can have a career in science	3.86	4.10	3.77	4.03	3.80	3.88	4.03	3.84

Table 7. Mean scores for post-survey items for the relevance of Making and STEM

	Year 1 (*n*=494)	Year 2 (*n*=2213)	Year 3 (*n*=2950)	Year 4 (*n*=2399)
The project helped me apply Making to my everyday life	3.85	3.60	3.65	3.71
This project helped me learn something I will use in school	3.77	3.62	3.66	3.71
The project helped me understand how science is important in the world around me	3.90	3.76	3.81	3.85
People in my community would think this project is important	3.74	3.55	3.64	3.71

Undergraduate Participant-Related Outcomes

We hypothesized that undergraduate facilitators, specifically service-learning students would (1) demonstrate evidence of improved content knowledge and teaching skills, (2) enhanced 21st-century skills, and (3) increased interest in teaching STEM in informal and/or formal settings. Research suggests that service-learning participation is a means to promote 21st-century skills in undergraduate education for all students (Finley & McNair, 2013; Felten & Clayton, 2011). Similarly, these skills reflect the sort of personal growth and civic engagement thought to be associated with persistence toward degrees (Knight et al., 2011), as well as skills valued by potential employers (Finley & McNair, 2013).

Technical, leadership, and 21st-century skills for service learning students were measured using a pre-and-post survey. We asked SL undergraduate students to assess their current skills prior to beginning their field experience and again later upon completing their field experience hours at the sites. We used paired t-tests for pre-and-post-survey comparisons and descriptive statistics (i.e., mean scores, percentages) to analyze part of the post-survey. Table 8 highlights findings for service learning undergraduates who assisted STEM Ambassadors in the implementation of the Mobile Making program between Years 1 and 4.

We measured service learning students' projected interest in ongoing STEM outreach activities. This was accomplished with three items that posed potential avenues for future involvement. Students responded by indicating whether they 'will not participate,' 'unlikely to participate,' 'probably would participate,' or 'certainly would participate' (see Table 9).

Table 8. Mean scores for pre-and-post survey for service learning students

	Year 1 (n=77)		Year 2 (n=19)		Year 3 (n=110)		Year 4 (n=169)	
	Pre	Post	Pre	Post	Pre	Post	Pre	Post
I have/applied strategies to successfully communicate with middle school students	3.96	4.30*	3.94	4.26*	4.30	4.49	4.27	4.48
I have/applied strategies to coach middle school students to success with maker projects	3.96	4.29*	3.84	4.26	4.22	4.45*	4.19	4.43
I am/was able to relate to middle school students	3.70	4.01*	3.84	3.47	4.14	4.21	4.13	4.25
I feel comfortable working with middle school students	4.02	4.31*	4.10	4.0	4.21	4.43*	4.31	4.51
I have/applied strategies to help middle school students successfully collaborate in the afterschool program	3.98	4.23	3.73	4.21*	4.27	4.41*	4.20	4.40

*Statistically significant p<.05

Table 9. Percentages of students who indicated 'probably' and 'certainly' would participate

	Year 1 (*n*=77)	Year 2 (*n*=19)	Year 3 (*n*=110)	Year 4 (*n*=169)
Participate in STEM-related outreach effort at future employer	87%	79%	81%	91%
Participate in a community science fair	70%	53%	66%	79%
Volunteer in your child's or a relative's science class	79%	68%	76%	89%

DISCUSSION

The Mobile Making program is a university-community partnership driven OST program wherein racially and ethnically diverse undergraduate students bring materials in their own vehicles and facilitate youth's engagement and learning in maker-based STEM activities at school, library, and community sites. In this chapter, we shared the design and implementation of the Mobile Making program over a seven-year period. Specifically, we discussed the iterations the program has gone through from the initial pilot phase to the broad implementation phase to support positive youth and undergraduate student outcomes. Further, we shared the design of the maker-based STEM activities we developed and described how we maintained relations with our school, library, and community partners over the years. Finally, we described the impact of the program on youth participant and undergraduate student outcomes during Phase 2 of the program.

One important issue that weaves across different phases and aspects of the Mobile Making program is sustainability and scalability of our program, especially in the absence of grant funding support. As we discussed in this chapter, the university-community partnership provides the foundation for sustainability and scalability of the Mobile Making program as we can optimize the infrastructure and resources provided by both partners. We successfully reduced the cost of the program by involving a large number of service learning undergraduate students and folding the training of these students into their regular coursework. Incorporating high-tools into maker-based STEM activities and engaging youth participating in creating digital artifacts and projects also reduced the material cost of running the program.

However, at the same time, a tension exists between scalability and university- community partnership. Operating a larger number of school, library, and community sites is easier with more standardization of the program while making it harder for us to tailor the program implementation to meet specific needs and context. For example, we implement the Mobile Making program for nine weeks each semester across different sites and contexts. Some sites consider the length of the program short and while others consider it to be long. Another point of tension exists between sustainability and responding to youth participants' needs. As our research findings suggest, youth participants value having opportunities for their parents to see the maker projects they completed. This is only possible when they use low tech tools to create a hands-on Maker project that they can take home at the end of the session. This is not to say youth participants do not value making with digital tools. Instead, it highlights the need to find ways to ensure youth continue to share their projects outside of the session with family and friends even when digital tools are used to reduce the cost of the program.

Secondly, there is a need to balance research (contributing to knowledge base) and programmatic goals (evaluation). We continue to draw upon multiple data sources to inform our decision making for program improvement and research. We heavily rely on surveys to capture the impact of the program activities on youth participants and undergraduate students for research purposes. Although survey data provides insights into trends, it fails to capture the rich learning experiences of youth participants and undergraduate students. Further, as our research findings suggest pre-and-post survey comparisons minimally rarely result in statistically significant change before and after the program. There are many reasons for why it is challenging to capture statistically significant change over time. However, more importantly, there is a need to systematically incorporate other data collection methods that can better shed light onto the learning processes of youth participants and undergraduate students. This introduces a challenge in terms of the timeliness of data analysis of qualitative data gathered through observations and interviews, and how soon we can apply what we learn into practice as analysis of qualitative data takes longer to analyze and disseminate. Much of the program improvement comes from informal observations and information sharing among us, the undergraduate students, and our community partners. Making and STEM OST programs like ours need to engage in both formative and summative evaluation and research to better meet the needs of diverse communities and ultimately broaden their participation in Making and STEM while contributing to the literature.

Finally, college students are important role models and near-peer mentors in youth Making and STEM OST programs (Maker Ed, 2015) and can support university-based outreach and service learning (White House, 2014). SL is a high-impact practice involving real-world problem-solving, making a difference in communities (Furco, 1996), and is sustainable because students receive course credit rather than payment. The undergraduate students who participate in the Mobile Making program are usually first-generation students who come from the same communities we serve as part of the program. Therefore, youth participants in the Mobile Making program are more likely to identify themselves with the undergraduate student facilitators than college professors or site staff.

There are many OST programs that focus on Making and STEM. Providing access to these programs is important to reach underserved youth. However, as we have learned, access alone does not ensure broadening participation of underserved youth in Making and STEM if there are no role models that the youth interact with and learn from as part of the program. We found that having a role model in proximity, a connection between the Making and STEM activities to real world problems, and lower barriers for entry and successful completion of projects are key in supporting youth's engagement and learning in Making and STEM.

CONCLUSION

Making is an activity that engages learners in STEM through tinkering, creating, and designing using low-and-high tech tools. However, there is a lack of ethnic, gender, and socioeconomic diversity among Makers and in STEM. The Mobile Making program is designed on research-based principles for engaging underserved communities. Undergraduate facilitators, STEM and education majors who serve as near-peer mentors, are key to the success of the program. In the past seven years, the program went through many cycles of iterations as we expanded the program from four middle schools to many school, library, and community sites with diverse student populations, using activities compatible with the format and time constraints in afterschool settings. We have documented positive impacts on youth participants' interest and self-efficacy in Making and STEM and their perceived relevance of Making and STEM in their everyday lives. We have shown that the program has a positive impact on undergraduate students by helping them develop teaching and 21st century skills and motivating them to engage in STEM teaching and outreach in the future. Our work is currently being expanded into additional afterschool programs across California. The continuing inquiry will involve how to scale and sustain the Mobile Making program in underserved communities, and for underserved K12 students through university-community partnerships.

Supplemental Materials

None

REFERENCES

Afterschool Alliance. (2013). *Defining youth outcomes for STEM learning in afterschool.* Afterschool Alliance. http://www.afterschoolalliance.org/STEM_Outcomes_2013.pdf

Afterschool Alliance. (2015). *Afterschool programs: Inspiring students with a connected learning approach.* Afterschool Alliance. http://afterschoolalliance.org/documents/Afterschool_and_Connected_Learning.pdf

Archer, L., DeWitt, J., Osborne, J., Dillon, J., Willis, B., & Wong, B. (2010). 'Doing' science versus 'being' a scientist: Examining 10/11-year-old schoolchildren's constructions of science through the lens of identity. *Science Education, 94*(4), 617–639. doi:10.1002ce.20399

Aschbacher, P. R., Li, E., & Roth, E. J. (2009). Is science me? High school students' identities, participation and aspirations in science, engineering, and medicine. *Journal of Research in Science Teaching, 47*(5), 564–582. doi:10.1002/tea.20353

Bandura, A. (2000). Self-efficacy: The foundation of agency. In W. J. Perrig & A. Groub (Eds.), *Control of human behavior, mental processes, and consciousness: Essays in honor of the 60th birthday of August Flammer* (pp. 16–31). Taylor & Francis.

Bandura, A. (2006). Guide for constructing self-efficacy scales. In F. Pajares & T. Urdan (Eds.), *Self-efficacy beliefs of adolescents* (Vol. 5, pp. 307–337). Information Age Publishing.

Basu, S. J., & Barton, A. C. (2007). Developing a sustained interest in science among urban minority youth. *Journal of Research in Science Teaching*, *44*(3), 466–489. doi:10.1002/tea.20143

Blikstein, P., & Krannich, D. (2013). The makers' movement and fablabs in education: Experiences, technologies, and research. *Proceedings of the 12th International Conference on Interaction Design and Children*. Association for Computing Machinery. 10.1145/2485760.2485884

Bouillion, L. M., & Gomez, L. M. (2001). Connecting school and community with science learning: Real world problems and school–community partnerships as contextual scaffolds. *Journal of Research in Science Teaching*, *38*(8), 878–898. doi:10.1002/tea.1037

Costa, V. B. (1995). When science is 'another world': Relationships between worlds of family, friends, school, and science. *Science Education*, *79*(3), 313–333. doi:10.1002ce.3730790306

Eccles, J. S., & Wigfield, A. (2002). Motivational beliefs, values, and goals. *Annual Review of Psychology*, *53*(1), 109–132. doi:10.1146/annurev.psych.53.100901.135153 PMID:11752481

Felten, P., & Clayton, P. H. (2011). Service-learning. *New Directions for Teaching and Learning*, *2011*(128), 75–84. doi:10.1002/tl.470

Ferry, T. R., Fouad, N., & Smith, P. L. (2000). The role of family context in a social cognitive model for career-related choice behavior: A math and science perspective. *Journal of Vocational Behavior*, *57*(3), 348–364. doi:10.1006/jvbe.1999.1743

Finley, A. P., & McNair, T. (2013). *Assessing underserved students' engagement in high-impact practices*. Association of American Colleges and Universities.

Furco, A. (1996). Service learning: A balanced approach to experiential education. In B. Taylor & Corporation for National Service (Eds.), Expanding boundaries: Serving and learning (pp. 2-6). Corporation for National Service.

Gasbarra, P., & Johnson, J. (2008). *Out Before the Game Begins Hispanic Leaders Talk about What's Needed to Bring More Hispanic Youngsters Into Science, Technology and Math Professions* (Public Agenda Report). IBM.

Gilmartin, S. K., Li, E., & Aschbacher, P. (2006). The relationship between interest in physical science/engineering, science class experiences, and family contexts: Variations by gender and race / ethnicity among secondary students. *Journal of Women and Minorities in Science and Engineering*, *12*(2-3), 179–207. doi:10.1615/JWomenMinorScienEng.v12.i2-3.50

Grigg, W. S., Lauko, M. A., & Brockway, D. M. (2006). *The Nation's Report Card, Science 2005: Assessment of Student Performance in Grades 4, 8, and 12*. U.S. Department of Education, National Center for Education Statistics.

Halverson, E. R., & Sheridan, K. (2014). The maker movement in education. *Harvard Educational Review*, *84*(4), 495–504. doi:10.17763/haer.84.4.34j1g68140382063

Hidi, S., & Renninger, K. A. (2006). The four-phase model of interest development. *Educational Psychologist*, *41*(2), 111–127. doi:10.120715326985ep4102_4

Honig, M., & McDonald, M. (2005). From promise to participation: Afterschool programs through the lens of socio-cultural learning theory. *Afterschool Matters, Occasional Paper Series*. WCW. https://niost.wcwonline.org/pdf/afterschoolmatters/asm_2005_op5_fall/asm_2005_op5_fall-1.pdf

Huang, G., Taddese, N., & Walter, E. (2000). Entry and persistence of women and minorities in college science and engineering education (Research and Development Report NCES 2000-601). National Center for Education Statistics.

Kafai, Y., Fields, D., & Searle, K. (2014). Electronic textiles as disruptive designs: Supporting and challenging maker activities in schools. *Harvard Educational Review*, *84*(4), 532–556. doi:10.17763/haer.84.4.46m7372370214783

Knight, D. B., Mappen, E. F., & Knight, S. L. (2011). A review of the literature on increasing the representation of women undergraduates in STEM disciplines through civic engagement pedagogies. *Science Education & Civic Engagement: An International Journal*, *3*(1), 36–47.

Krishnamurthi, A., Bevan, B., Rinehart, J., & Coulon, V. R. (2013). What Afterschool STEM Does Best: How Stakeholders Describe Youth Learning Outcomes. *Afterschool Matters*, *18*, 42–49.

Lemke, J. L. (2001). Articulating communities: Sociocultural perspectives on science education. *Journal of Research in Science Teaching*, *38*(3), 296–316. doi:10.1002/1098-2736(200103)38:3<296::AID-TEA1007>3.0.CO;2-R

Lyon, G. H., Jafri, J., & Louis, K. S. (2013). Beyond the pipeline: STEM Pathways for youth development. *Afterschool Matters*, *16*, 48–57.

Honig, M. & McDonald, M. (2015). *Youth Makerspace Playbook*. Maker Ed. http://makered.org/wp-content/uploads/2015/09/Youth-Makerspace-Playbook_FINAL.pdf

Martin, L., & Dixon, C. (2013). *Youth conceptions of making and the Maker Movement*. Exploratorium. https://www.exploratorium.edu/sites/default/files/pdfs/IDC_2013_Martin_Dixon.pdf

Master, A. H., & Meltzoff, A. N. (2020). Cultural stereotypes and sense of belonging contribute to gender gaps in STEM. *International Journal of Gender, Science and Technology*, *12*, 152–198.

McCreedy, D., & Dierking, L. D. (2014). *Cascading Influences: Long term impacts of informal STEM experiences for girls*. Paper presented at the 27th Annual Visitor Studies Association Conference, Albuquerque, New Mexico.

Pajares, F. (2006). Self-efficacy during childhood and adolescence. In F. Pajares & T. Urdan (Eds.), *Self-efficacy beliefs of adolescents* (Vol. 5, pp. 339–367). Information Age Publishing.

Petrich, M., Wilkinson, K., & Bevan, B. (2013). It looks like fun, but are they learning? In M. Honey & D. Kanter (Eds.), *Design, Make, Play: Growing the Next Generation of Innovators* (pp. 50–70). Routledge.

Rahm, J., Martel-Reny, M.-P., & Moore, J. C. (2005). The role of afterschool and community science programs in the lives of urban youth. *School Science and Mathematics*, *105*(6), 283–291. doi:10.1111/j.1949-8594.2005.tb18129.x

Regan, E., & DeWitt, J. (2014). Attitudes, interest and factors influencing STEM enrolment behaviour: An overview of relevant literature. In E. K. Henriksen, J. Dillon, & J. Ryder (Eds.), *Understanding Student Participation and Choice in Science and Technology Education* (pp. 63–88). Springer.

Roth, W.-M., & Lee, S. (2004). Science education as/for participation in the community. *Science Education, 88*(2), 263–291. doi:10.1002ce.10113

Salam, M., Awang Iskandar, D. N., Ibrahim, D. H. A., & Farooq, M. S. (2019). Service learning in higher education: A systematic literature review. *Asia Pacific Education Review, 20*(4), 573–593. doi:10.100712564-019-09580-6

Schad, M., & Jones, W. M. (2020). The maker movement and education: A systematic review of the literature. *Journal of Research on Technology in Education, 52*(1), 65–78. doi:10.1080/15391523.2019.1688739

Sefton-Green, J. (2013). *Learning at not-school: a review of study, theory, and advocacy for education in non-formal settings.* MIT Press.

Simpson, A., Burris, A., & Maltese, A. (2020). Youth's engagement as scientists and engineers in an afterschool making and tinkering program. *Research in Science Education, 50*(1), 1–22. doi:10.100711165-017-9678-3

Syed, M., Azmitia, M., & Cooper, C. R. (2011). Identity and academic success among underrepresented ethnic minorities: An interdisciplinary review and integration. *The Journal of Social Issues, 67*(3), 442–468. doi:10.1111/j.1540-4560.2011.01709.x

U.S. Commission on Civil Rights. (2010, October). *Encouraging minority students to pursue science, technology, engineering and math careers.* USCCR.

Vossoughi, S., & Bevan, B. (2014). *Making and Tinkering: A review of the literature.* National Research Council. https://sites.nationalacademies.org/cs/groups/dbassesite/documents/webpage/dbasse_089888.pdf

Vossoughi, S., Hooper, P. K., & Escudé, M. (2016). Making through the lens of culture and power: Toward transformative visions for educational equity. *Harvard Educational Review, 86*(2), 206–232. doi:10.17763/0017-8055.86.2.206

Vygotsky, L. S. (1978). *Mind in society: The development of higher mental processes.* Harvard University Press.

Weiner, S., Lande, M., & Jordan, S. S. (2018). What have we 'learned' from maker education research? A learning sciences-base review of ASEE literature on the maker movement. *American Society for Engineering Education Annual Conference & Exposition.* Salt Lake City, UT.

Wolter, B. H., Lundeberg, M. A., & Bergland, M. (2013). What makes science relevant?: Student perceptions of multimedia case learning in ecology and health. *Journal of STEM Education: Innovations and Research, 14*(1), 26–35.

Zirkel, S. (2002). Is there a place for me? Role models and academic identity among white students and students of color. *Teachers College Record, 104*(2), 357–376. doi:10.1177/016146810210400206

Chapter 3
The Design of an Out-of-School Program Focused on Community-Centered Engineering Challenges

Joni M. Lakin
https://orcid.org/0000-0002-0546-0554
University of Alabama, USA

Daniela Marghitu
Auburn University, USA

Edward W. Davis
Auburn University, USA

Virginia A. Davis
Auburn University, USA

ABSTRACT

This chapter describes a partnership of a non-profit offering afterschool K-12 STEM programming and a team of academic researchers in both education and engineering departments. The partners designed out-of-school programs that allowed students to learn about engineering through the lens of important societal challenges, such as access to fresh water and climate-change related to carbon emissions. The experiences were designed to be engaging for students who have pro-social career values, which are more prominent values among girls and the predominantly Black community the non-profit served. Evaluation questions focused on how pro-social framing of engineering might enhance the perception of goal congruity and support students' formation of a STEM career identity. Qualitative and quantitative data were used to explore how students conceptualized engineering before and after the programs and how they described their career plans with respect to STEM. Both residential camps and community-based afterschool programs demonstrated important benefits to students' STEM career identity.

DOI: 10.4018/978-1-6684-7771-7.ch003

INTRODUCTION

This chapter describes a partnership between a community-based non-profit focused on preK-12 STEM education and a team of academic researchers coming from education and engineering departments. We designed a one-week, residential summer camp program and later a 10-week Saturday Academy program that taught students about engineering through the lens of how that career field addresses important societal challenges, including access to fresh water, local infrastructure concerns, climate-change related to carbon emissions, and other topics. The goals of the program were to increase awareness and interest in engineering as a potential career while improving STEM skills among participants. This chapter presents some of the program design choices and findings to guide other professionals designing informal STEM learning opportunities.

Our theoretical framing was ***goal congruity theory*** where identification with a career field is believed to occur when there is alignment between students' career goals or values and their perceptions of the values a career field meets (Allen et al., 2015; Belanger et al., 2017; Diekman et al., 2011). Literature shows that girls and students from historically marginalized communities, including Black youth, students from low-income communities, and those who would be first-generation college students, are all more likely to hold altruistic or pro-social values for their career and therefore would be seeking out aligned career fields (Allen et al., 2015; Diekman et al., 2011; Estrada et al., 2016; Jones et al., 2021; Porfeli & Lee, 2012). However, stereotypically, engineering is perceived to align with status values, such as high pay or prestige, and to involve individual or isolated work. In reality, engineering is a collaborative enterprise that solves important social challenges and requires strong communication skills and an understanding of social complexities (Chan & Fishbein, 2009; Lakin et al., 2020). Furthermore, Black youth are more likely to be interested in entrepreneurship (Jones et al., 2021), which is an important, though less recognized, way that engineering impacts society (Swamidass & Schnittka, 2017). This mismatch in how engineering is portrayed compared to its more complex reality may be a detractor for students from underrepresented communities to consider engineering as a career. Thus, our collaborative work focused on showcasing engineering as a profession that can meet students' altruistic and community-centered career goals (Lakin et al., 2022).

The goal of this specific project was to showcase the altruistic and community-centered career potential of engineering in a culturally responsive way that would align to the interests and values of our focal audience, grade 8-10 predominantly Black students in an urban area. We did this via out-of-school programs as well as campus-based week-long summer camps. Our research activities were framed as a design-based research process (Easterday et al., 2014), which entails phases of conceptualizing, designing, and testing of a designed solution. We collected a variety of quantitative and qualitative data from the conceptual phase through evaluating the designed solution in order to base our design on data from the community of interest and to evaluate the curriculum in that same community.

Students were recruited from Bessemer, which is outside of Birmingham, AL, and is very low income with low achieving schools. Despite this, Bessemer also has economic opportunity through a variety of industries entering the city as well the entrepreneurship of its citizens. To learn more about these students, in the conceptualization phase, the research team sought to learn from the successes of a partnering organization that was already benefiting students in this area. The following section describes this partnership.

The Partnership

The non-profit partner is Alabama STEM Education (ASE), which serves students in an urban context who are predominantly Black and from low-income families. They primarily provide out-of-school time programs, including afterschool care and weekend programs. We worked specifically with their *Raise the Bar* (RTB) program, a 10-week program that meets on Saturdays. Students can range in grade level, but we designed activities targeting 8th - 10th grade students. In addition to their existing programs, we also designed a one-week residential summer camp experience targeted to their students. We used our research on that summer camp experience to redesign the RTB program to better present engineering as a career pathway that offers opportunities to help their community and meet their own career goals. This chapter describes that 10-week program, which included a variety of curriculum materials and guest speakers from local STEM industries. The timeline of these projects and the partnership is shown in Table 1.

Table 1. Timeline of program and evaluation activities

Timeline		Activity
2018	Fall/Spring	Evaluation of traditional program, Raise the Bar
2019	Summer	Week-long camp, Tomorrow's Community Innovators
	Fall/Spring	Collaboration on Raise the Bar program, weekend programs on Virtual Reality
2020	Summer	Week-long virtual camp (due to COVID), Tomorrow's Community Innovators
	Fall/Spring	Collaboration on Raise the Bar program, weekend programs on Virtual Reality
2021	Summer	TCI 2019/2022 participants offered choice of STEM camps
2022	Spring	Implementation of modified Raise the Bar program

The formal partnership began in 2018 as part of a grant from the National Science Foundation. However, a co-PI on the grant was already a partner with this organization and served on their Board of Directors. The university-based team worked closely with ASE throughout the grant and continues to partner with the organization to provide STEM experiences, including events on the college campus, providing curriculum for on-going programs, and supporting the organization's grant-writing efforts.

Each partner organization contributed to the design and implementation of the project. ASE already had the structure of a weekend STEM program and deep ties to the Bessemer, AL, community they serve. Raise the Bar weekend programs were held at their sites. The partner team at ASE primarily consisted of the executive director and program manager for RTB, although different staff members participated in the RTB program over time. Many community members were also involved in the program through guest speakers, field trips, and other support for the program.

The university team included faculty and graduate students from chemical engineering, materials engineering, and computer science and software engineering. These disciplinary experts partnered with STEM education researchers in the design and implementation of the curriculum and research activities. The grant was also able to fund graduate students who collected data on-site at all events as well as provided some STEM activities for the ASE programs.

BACKGROUND

Researchers have demonstrated that the public widely holds misconceptions of engineering (Cunningham et al., 2006; Sullivan, 2006). Professional organizations have also raised the alarm that the lack of public awareness of the role of engineering and its value to society leads to fewer students pursuing the field as well as weaker investment in the field (Vest, 2008). In one response to this need, the National Academy of Engineering introduced the *Grand Challenges for Engineering* (NAE, 2008), fourteen challenges facing modern society that reinforce the message that engineers use their creative problem-solving skills to improve our world and shape the future. The challenges include access to clean water, personalized learning, carbon sequestration, and personalized medicines.

As defined earlier, *goal congruity theory* predicts that students will be more interested in a career field if they believe the *affordances* of a career field (i.e., the values a field allows individuals to reach) align to their own career values or goals (Belanger et al., 2020; Capobianco & Yu, 2014). Therefore, in designing STEM educational programs, we may be able to broaden participation in STEM by acknowledging the varied career goals and values that students hold and showing their alignment to various career options. This is especially important relating to engineering, which—despite the important impacts it can have—is primarily seen as a career field for people who like to work alone or want high-status jobs. Thus, this program was conceptualized as a *goal congruity intervention* to provide students with a more accurate and inclusive view of engineering to support more students recognizing alignments of their values to the field of engineering.

Goal congruity is also consistent with career-level decisions through the *Social Cognitive Career Theory* (SCCT; Lent et al., 1994, 2000), which also predicts that students' experiences will shape their career goals. Career identity development for youth has often been categorized along two dimensions: whether students are exploring possibilities and how informed those choices are by their self-awareness (Porfeli & Lee, 2012). We used this framework to consider students' responses about their future career in terms of whether they had specific ideas for their career and whether those ideas seemed informed by the interests and values they may have learned about in camp. Students with clear career possibilities aligned with skills or interests can be classified as "achieved" career identity, students who are exploring based on their self-awareness (but without a specific interest) are in "moratorium". Students can also be classified as "diffused", which may reflect not engaging in much planning or thinking about the subject.

The Program

Our goals for designing the program included framing engineering as a career field that was (a) **broad** in its impact on society, (b) **relevant** to students' own lives and communities, (c) **aligned** with community-related or altruistic career values. Some of the lessons incorporated our existing camp activities, including the *What's the Challenge?* brainstorming activity, water filtration, and city block design. We also introduced new activities around carbon dioxide (CO_2) and another infrastructure-related activity of designing highway interchanges. The topics and schedule of the program are shown in Table 2.

Table 2. Raise the Bar weekly topics

Week	Topic
1	What's the Challenge? (Introduction to Engineering Grand Challenges)
2	Designing a City Block (Civil Engineering; infrastructure)
3	Guest speakers related to Entrepreneurship and computer engineering (cryptomining)
4	Water filtration (Chemical and Biosystems Engineering)
5	Field Trip to Southern research labs
6	Career values exploration activities, engineering challenges
7	CO_2/Air quality (Chemical Engineering)
8	Guest speakers related to Aerospace Engineering (flight simulators)
9	Alternative Energy (Mechanical and Electrical Engineering)
10	Vision boards, family celebration

What's the Challenge?

"What's the Challenge?" introduces the NAE's Grand Challenges of Engineering to students in a way that aligns the list of social challenges to students' own observations of the world. The activity is essentially a series of group brainstorming exercises where students identify problems in the community, identify technological solutions, and then collectively list the most important problems they felt society needed to solve (see Lakin et al., 2021, for a detailed description). The activity is designed to mimic the approach that the National Academy of Engineering (2008) used to generate their list of 14 Grand Challenges of Engineering. This process involved canvassing global stakeholders and going through a process of refinement and prioritization of the challenges.

We have used this activity with a wide variety of students (from elementary students to undergraduate students). Students may be slow to warm up, but become invested in the process when we start to sort and distill into the key challenges that the students had identified. In this case, as in most instances where we used this activity, many of the challenges identified by students mapped onto the NAE's own list. The list of challenges was also affected by current local and global events: systemic racism, gun violence and school shootings, animal welfare, COVID, mental health, supply chain issues, medical technology, pollution, as well as local and world hunger. In a final step of the process, students voted on the most important challenges and rank ordered their own Grand Challenges list. Figure 1 shows some of the top-ranked challenges generated by these students, which reflected some of the current events at the time.

Designing a City Block

The goal of this unit was to engage students in thinking about the design choices made by civil engineers in designing infrastructure and engaging in city planning. In this unit (adapted from Foresite Group, 2014), students learn about urban infrastructure by working as a team to gather information from community stakeholders and then planning a city block and using their limited budget to place necessary buildings and roads. The buildings are represented by color- and size-coded pieces of construction paper. Students work on a poster-sized piece of paper to lay out the buildings, roads, etc. See Figure 2.

Figure 1. Top-ranked challenges identified by 2022 RTB students on oversized sticky notes, "racism in the system," "gun violence/school shootings," "animal overpopulation," "supply chain/needed materials," "COVID," and "mental health"

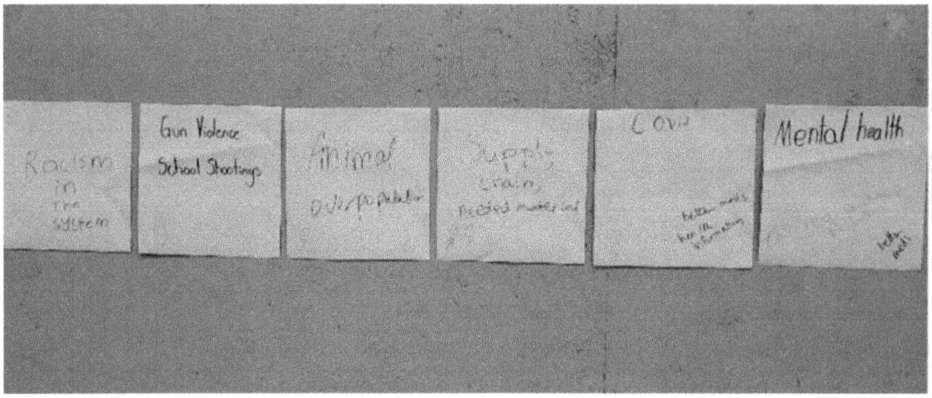

Figure 2. Students present their design and explain the choices made

Facilitators ask targeted questions about traffic, noise, and the stakeholder concerns about the design. To further emphasize the community connections of infrastructure, we connected the design challenge to the Amazon fulfillment center recently built in their community. Its impacts include traffic and pollution as well as the obvious economic opportunities.

Student groups, especially two all-girl groups, were very engaged in the activity during RTB 2022 and created their own goals and constraints based on their life experiences. One city block was designed for pedestrian traffic only and organized around a human and animal medical center. Another group included parking lots designed to maximize traffic flow. Heated discussions arose. During the discussion, students asked questions about other groups' design choices and all groups shared their budget spending and main considerations in their designs.

Water Filtration

One of our most successful and adaptable lesson plans is the water filtration activity (Karimi et al., 2022). Part of its success is that students immediately recognize the personal relevance of clean water and are aware of water contamination in both their community and more broadly based on news reports.

We used a variety of household items to explore pore size and its importance in water filtration. The core concept is that size-based filtration relies on physical removal of suspended contaminants by porous media. Pollutant particles have different sizes, and they can be eliminated from the water by passing through media with adequately small pores to prevent their passage (Karimi et al., 2022). In this activity, participants are asked to predict and then test which group of materials would clarify a container of muddy water containing debris such as crushed leaves, dirt, and food coloring. In this version, we included t-shirt fabric, surgical face masks, coffee filters, cotton fibers, cheese cloth, and pebbles that varied in size. The face masks, which are now ubiquitous, were an especially interesting variation because students thought they would be very effective, but their hydrophobic properties made it nearly impossible for water to pass through. Several students were fascinated by the process and continued to run their water sample until the end of camp, removing nearly all of the food coloring. See Figure 3.

Figure 3. A student closely studies their filtration column

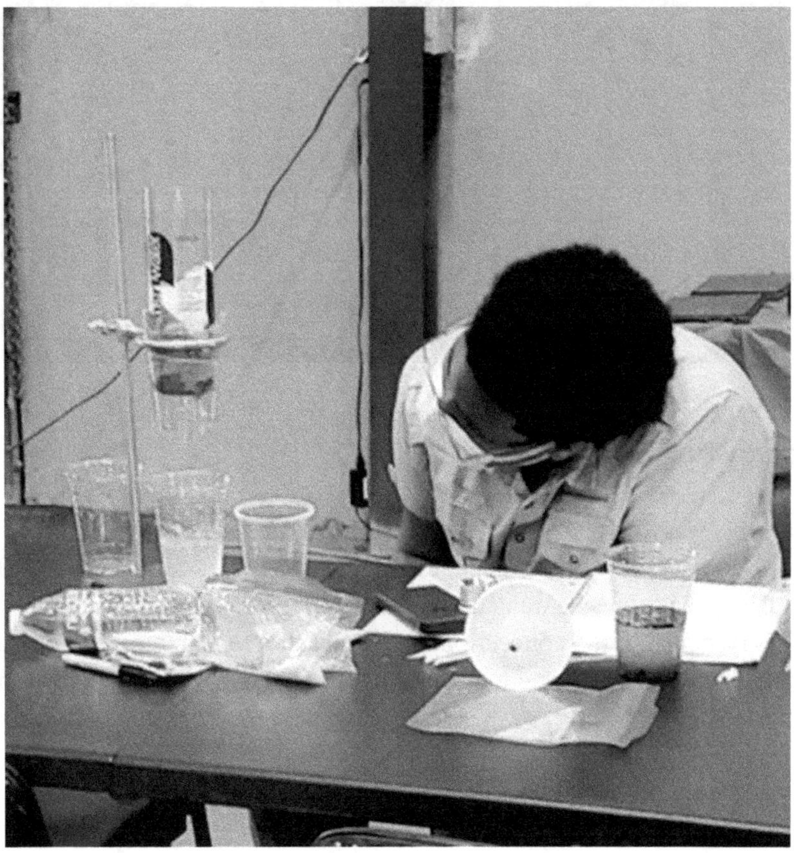

CO_2 Activity

Two activities connected to carbon emissions and its relationship to climate change (including ocean acidification). First, students used a baking soda/vinegar reaction to test CO_2 production. A CO_2 probe detected the change in the classroom air compared to the air around the reaction. See Figure 4. The second activity involved the same chemical reaction inside a closed system with water containing a pH indicator (adapted from Exploratorium Teacher Institute, n.d.). Students were able to see that atmospheric CO_2 affects ocean acidity by observing that the release of gas preceded a change in the water due to the indicator. All of these activities were related to climate change impacts and students engaged in a discussion of pollution and the impact of local industry on the environment.

Figure 4. On the left, the indicator solution shows when the acidity of the water goes up as CO_2 dissolves in the water. On the right, the commercial CO_2 detector.

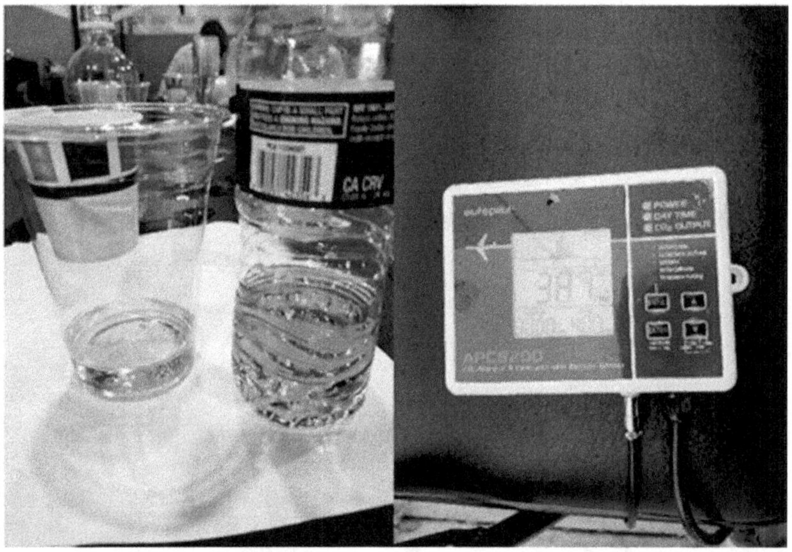

Alternative Energy Sources

The goal of this activity was to explore sources of alternative energy—a solar cell, hand-crank, and wind turbine. Our state is a leader in automotive manufacturing and is seeking to become a leader in electric vehicle manufacturing. Therefore, we wanted to highlight the promises and limitations of alternative energy for conversations about industry connections and local entrepreneurship. Students used multi-meters to measure the output of each source and to make decisions about building an emergency radio station, considering the tradeoffs of the three power sources compared to traditional batteries. See Figure 5.

Figure 5. To maximize their wind turbine output, students realize the AC vents provide a great source of wind energy

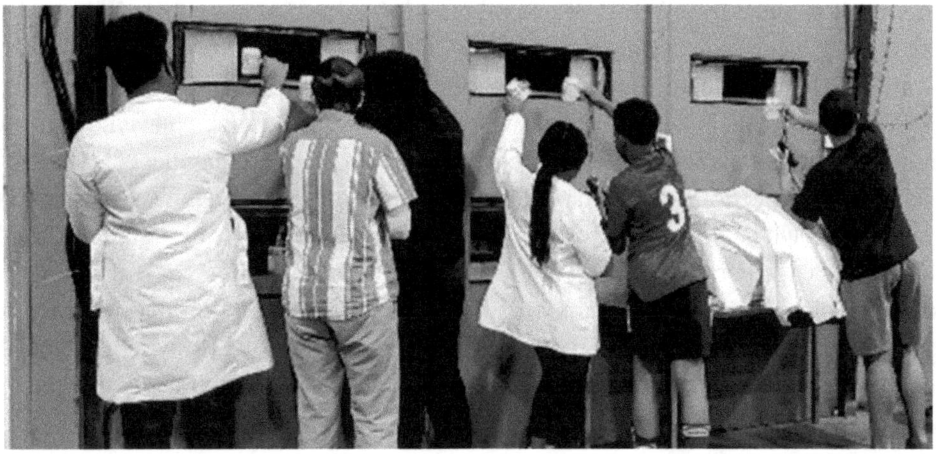

FINDINGS FROM SUMMER 2019 CAMP

Beginning in 2018, the university-based team observed and worked with the community organization to understand how the more traditional RTB was organized and how we could create a summer camp program that engaged the interests and values of students in the focal community. We used a design-based research process (Easterday et al., 2014) in our work, which entails phases of conceptualizing, designing, and testing of a designed solution (in this case, an out-of-school curriculum on engineering). We will briefly review our findings of the intensive camp experience in 2019 in order to provide a contrast with the results of the program when translated to the partner organization's on-going program in 2022. These findings suggest what program features may be necessary to sustaining program impacts.

Our collaboration with ASE included participating in educational activities, regular observation, and evaluation of the effects of the RTB program on the students. In the DBR process (Easterday et al, 2014), this addressed our *focus*, *understanding*, and *define* portions of the design process. This enabled us to become familiar with this population of students, learn about their career interests and values, and explore options for designing our summer camp program. Since we wanted to avoid serving students already connected to a STEM education ecosystem, none of the students in the RTB program participated in the summer camp. However, our observation and evaluations of RTB taught us a great deal about the topics and types of activities that were most engaging for this population of students. It also confirmed our hypothesis that helping others and serving the community was a prominent value for many students.

We the designed the week-long camp (*conceive* and *build* phases of DBR) so that three days of the five-day camp had a specific engineering theme related to one of the Grand Challenges for Engineering. We chose three challenges that we expected would be particularly interesting based on our surveys of the RTB students: providing access to clean water, making solar energy economical, and restoring urban infrastructure. Details on the specific activities and interventions are described in detail in prior work (Bhattarai et al., 2021; Karimi et al., 2022).

Similar to the methods described later, to *test* our initial designed solution, we evaluated the summer camp program with a variety of tools, including interviews asking about students' career interests and their definitions of engineering. We asked several questions aimed at getting students to define "engineering", such as "How do you define engineering?" and "What does an engineer do?"

At pre-camp, definitions tended to be non-specific, such as people who "build the roads and make the bridges and … they make power" (Brianna, 13) and "an engineer is someone who works with mathematics and things of that nature… Like a mechanical person, they are engineers" (Hailey, 14). Some students had broad and accurate definitions of engineering at the outset. However, when asked what a job in engineering would be like, we were able to discern that, before the camp, students had little sense of the daily work of engineers.

At post-camp, many students gave us a broad, encompassing definition that included features we intended to highlight, including the breadth of problems engineers solve, solving problems that affect the lives of other people, and creating technology related to our everyday lives. One student, pseudonym Jaylen, encapsulated our intended framing of engineering:

"It's more than just one specific thing...it's a large variety of things… science has math, technology and you can do multiple different things about almost everything in the world. … [Engineers] help make the world a better place."

When asked about their own career interests, several students reinforced their interests while others found new interests related to engineering or STEM more broadly. For example, Hailey (age 14) stated, "…but now that I've came [sic] to this camp, I've become more interested in engineering. So, I think I might be looking into engineering now". She later went on to express how altruistic framing was the cause of this change: "now that I hear about different parts of the world not having clean water and water being contaminated and everything, it makes me wanna go out and help people with dirty water get clean water." Thus, we found significant evidence that the camp's Grand Challenges framing achieved its goal of demonstrating how engineering is a helping profession that benefits society. We also found that community-relevance enhanced students' engineering interest.

2022 PROGRAM AND EVALUATION

After the success of the 2019 camp, we implemented a virtual camp in 2020 with a focus on hands-on lab activities that could be done from home (including a water testing kit, water filtration materials, and engineering design challenges) due to the COVID pandemic. In 2021, we partnered with existing university camps to provide a modified program. Finally, as a culminating activity for our design-based research cycle, we worked with ASE to redesign the RTB program in spring 2022 based on our experiences and lessons from the summer camp evaluations. Our evaluation of this program was intended to assess the quality of the framing intervention we had created throughout the design process.

MATERIALS AND METHODS

In the 2022 RTB session, we worked with 20 students (10 attended consistently and provided data). Students in the program were predominantly Black, with a number of students who were Latine. Most students were in 5th to 8th grade, although a few were in 9th or 10th grade. We sought to address these evaluation questions:

1. As a result of the program, did students perceive engineering as a career that can address important societal needs?
2. To what degree did students develop a greater career identity, particularly a STEM career identity, during the program?

The camp enrolled 20 students, but we had sufficient data on 10 students who consistently attended *and* were present at the final day of camp and completed the culminating creative project. Three older boys and one girl chose not to participate in the final day of the camp (present but not participating for various reasons). The other six were not in attendance by the end of camp. All of these students attended similar schools as our participants who provided complete data. Table 3 provides an overview of our 10 focal students, their demographic information, and public information about the schools they attended. For all 20 students, the range of grade levels was 5th to 10th, 45% were girls, and 15% were Latine.

This human subjects research project was conducted with the approval of the PI's Institutional Review Board. At the start of camp, students completed a survey of their interests in science and engineering. Students were encouraged to answer all questions, but many completed only the first page of questions. Low response rate, as well as a considerable number of students who joined the program after week 1, meant that our planned pre-/post-camp survey data were not particularly useful. We focused instead on observations of the program, brief reflection surveys from camp each week, and their values exploration results, and creative products created the last week of the program.

Table 3. Participants' demographic and school information

Pseudonym	Grade level	Race	Gender	School poverty level [a]	School performance [b]
Amya	5th	Black	Girl	Low poverty (<25%)	Above average
Deandra	5th	Black	Girl	Low poverty (<25%) private school	Well above state avg.
Journey	5th	Black	Girl	Homeschooled	N/A
Tremaine	6th	Black	Boy	High poverty (>75%)	Below state avg.
Deonte	6th	Black	Boy	Low poverty (<25%) school	Well above state avg.
Latrice	6th	Black	Girl	Mid-high poverty (50-75%)	Below state avg.
Aurora	8th	Latine	Girl	High poverty (>75%)	Below state avg.
Julian	9th	Latine	Boy	Homeschooled	N/A
Jamari	9th	Black	Boy	Homeschooled	N/A
Tyrell	10th	Black	Boy	Mid-high poverty (50-75%)	Above state avg.

[a]Based on Office for Civil Rights data from the U.S. Department of Education (2017)
[b]Based on Alabama State Department of Education (2022)

A rich source of data was the "vision board" activity in the final week of the program, where students were able to present their current thinking about their career interests and goals ("*My Future in STEM*"). Students were given two options for creating a project on their future self: a five-minute FlipGrid video or a poster. All but one student chose the poster. To guide their product, they were given a worksheet they could first complete with these questions:

- Introduce yourself
  - What's important to you? What do you like to do?
- What is engineering? What does an engineer do?
  - Why is it important?
  - Why would being an engineer be a good job to have?
- When you think about your future and career, what do you want to do?
  - If you were going to be in STEM, what kind of jobs would you choose?

We found that their notes for the posters and videos were rich with information for our analyses alongside the creative products. The posters and videos also provided a product for the family celebration at the end of the day.

Parent Survey

In the first week of camp (or when the student joined) we collected parent surveys asking about their perceptions of STEM careers, perceptions of their schools, and their own STEM identity (specifically if they had completed degrees related to STEM or worked in a STEM field). Information about the parents' education and training were also collected. The parent survey was adapted from a tool used by a statewide evaluation program (Weld et al., 2015). We matched parent survey data to student profiles to explore whether parents with STEM careers or education might be associated with student outcomes.

Analysis

Mixed methodology (Creamer, 2017) provided theoretical framing for this research and guided data synthesis, where the research team focused on integrating quantitative and qualitative data rather than separately analyzing those data and merging the conclusions. Consistent with this approach, we adopted a structured system for qualitative analysis that is designed for team-based analysis: the Sort and Sift, Think and Shift approach (Maietta, 2006). This approach encourages researchers to engage with the data and reflect on findings that emerge through memo-ing, creating *episode profiles* to capture the essence of individuals in the dataset, and identifying bridging concepts and patterns through topic monitoring, a more dynamic form of coding.

To create the episode profiles, we compiled the survey data (when available), our observations, parent surveys, and vision board data into a series of slides encapsulating each student. A team of three evaluators compiled the profiles, created memos, and discussed our findings and patterns to generate the results presented here. Both instructors from the 2022 RTB program were part of this data analysis team and assisted by a more expert qualitative researcher. The triangulation of findings as well as peer debriefing approach supports the trustworthiness of our classifications. Definitions of engineering were also coded following definitions established by our prior work which also showed acceptable inter-rater reliability (Lakin et al., 2020).

RESULTS

Definitions of Engineering

Consistent with our previous work on this grant, we looked at the accuracy and nature of the definitions students gave for engineering at the end of the camp. These categorizations were based on prior research which established these definitions (Lakin et al., 2020) and where we established inter-rater reliability. In the table below, we have categorized student definitions into one of these four categories. Most (5/10) students had accurate definitions of engineering. Two students gave answers that were focused on repairing and building, which may reflect the misconception (Capobianco et al., 2011; Cunningham et al., 2006) that engineers work as mechanics or electricians rather than creating new technologies and solutions for cars, electrical systems, etc. Intriguingly, two others gave answers that felt consistent with our messaging, but were grandiose and therefore may not indicate a precise understanding of engineering. These *aspirational* answers suggested they took the lessons of the camp to heart, as one of these students mentions problems with pollution, hunger, and drugs as the kinds of problems engineers could solve, which were some of the top challenges they identified on the first day of the program. However, solving all the world's issues with "just one piece of technology" may not be realistic enough to sustain this students' interest in the field.

Table 4. Quality of definitions provided

Category	Number	Representative quote (with pseudonym)
Specific, accurate definition	5	"Engineering is the development of things through math and science. Engineers build things. It is important because without engineering we wouldn't have much of the daily items we use today and life would be more difficulty." (Tyrell)
Grandiose impacts	2	"Engineering would be a good job because you could solve all the world issues in just one piece of technology." (Deandra)
Focused on fixing/ building	2	"Being an engineer is important because they build things like houses, cars, and more things like that." (Journey)
Non-specific	1	"There's different kinds of engineering. It is important because some of it can help us." (Aurora)

Career Identity

Career identity development for youth has often been categorized along two dimensions of whether students are exploring possibilities (achieved, diffuse, moratorium) and how informed those choices are by their self-awareness of their interests, values, or skills (Porfeli & Lee, 2012). Some students were classified as "achieved" career identity, where they shared clear career possibilities aligned with skills or interests. A large portion of students expressed that they were exploring (but without a specific interest) and were classified as "moratorium". Just one student was classified as "diffused" (in this case, not engaging in much planning) and one did not give us enough information to know whether his career goal was based on reflection on his skills or interests. See Table 5. Only one rater classified these results, but classifications were based on well-established definitions.

Table 5. Career identity development

Possible career identity status	Evidence of career exploration	Quotes on future and career, future in STEM (with pseudonym)
Achieved	Connecting to values/ interests	"I would want to be a owner of a no-kill dog facility"… "I would choose to specialize in technology because in this generation we mostly work with technology so why not make it into a career?" (Deandra)
		"When I grow up, I would like to be a baker or an artist. Why? Because I am very interested in both of those things." (Journey)
	Connecting to strengths	"When I grow up I wanna be a mathmatician [sic]. I'm very good at math." If I would be in STEM my job I would choose to do is make things with different chemicals." (Amya)
Achieved or foreclosed	Unclear	"I want to be a civil engineer." (Julian)
Moratorium	Connecting to values/ interests	"Make people happy and make games for people." (Tremaine)
		"I love building and using my imagination. I like learning new things and helping others." (Jamari)
	Exploring possibilities	"Engineer, designs things like rubber. Scientist, study cells, biology…figure out how cells react to chemicals. … soccer player." (Deonte)
		"I want to do something in or around the field of medicine. …. If I were going into STEM, I would choose technician jobs such as a radiology technician or a surgical technician." (Tyrell)
		"Mechanic, computer engineering, that's all and water softeners… work on technology." (Latrice)
Diffused	Unclear	"I don't always think about my future, but when I do think about joining the army." (Aurora)

When students talked about their career interests, many gave answers that reflected self-awareness in terms of their interests, values, and strengths. Although they developed these interests and self-awareness throughout their life experiences, we also saw evidence throughout the camp of students exploring these interests. The camp activities provided opportunities for students to project their own values and interests into the engineering process. For instance, Deandra consistently related her care for animals to the projects, nominating animal welfare as an important challenge, ensuring that an animal shelter was included in her group's city block, and including working with animals in her future career on her vision board.

Similarly, Tyrell, a young man interested in medical and technical careers, was one of our most engaged participants in the water filtration activity, creating a focused procedure for continuing to filter and test his water beyond the end of the activity. He was also highly attentive on the field trip, which involved laboratory skills.

Another impact of the program was connecting the students to other opportunities. Three of our participants took part in a Drone and 3D printing camp with Southern Research Company, based on the experience during the field trip to that facility. Other students continued participating in the ASE summer camps as participants or as counselors to younger students.

Parent Perceptions

Seventeen parents or family members (e.g., grandmother, aunt) completed surveys for students in the program. The adults ranged in education from less than a high school degree to one parent who was in graduate school in early childhood education. Most parents reported having an associate's degree or some college. Just two parents reported that they have a degree or advanced training in STEM. Five (including those two) reported working in a career related to STEM. The parents' education and STEM employment suggest this group of students came from families typical of the Bessemer area, which reported in the last census that just 15% of the adult population had a college degree (U.S. Census Bureau, 2022).

Most parents (64%) reported their student had "a lot" of interest in STEM while three (16%) reported their student had little or no interest in STEM. The rest were reported to have "some" interest. When asked about students' performance in various academic areas (math, science, civics, English, or music/arts), parents reported similar levels of achievement across domains. This suggests that, aligned with our intentions, we did not recruit students with unusually strong preparation in STEM fields and they ranged in their pre-program interest in STEM. In contrasting parent beliefs to student engagement in the camp, two students rated as less interested in STEM were among our most active participants. We also found no association between having parents in STEM careers or attending better schools with students' episode profiles or interest in STEM.

DISCUSSION

Some of the lessons we took away from our experiences with RTB include the following.

First, it was difficult to recruit students without pre-existing interest in STEM or connections to STEM education. In our summer camp, we purposefully omitted mentions of STEM or engineering in our advertising to try to recruit students who may not have seen themselves as "STEM types". This was intended to broaden participation in the STEM education ecosystem. However, at the pre-camp interviews, we found that all of the students had been told they were going to a STEM camp with a new focus of the program. Not all of the students were interested in STEM personally, but recognized that their parents (often their mother or a female relative) wanted them to develop STEM interests.

Students benefit from exploring their values and interests. Students reflected a range of career identity formation stages (Porfeli & Lee, 2012), from diffused, without any specific interests, to achieved, with a clear career goal. All students were still open to exploring new ideas and interests. We used a values card sorting activity as well as informal discussions with students to help them delve into their interests and values. The vision board activity asked them to integrate their learning with their self-knowledge to imagine their future career. This seems to be an effective activity for many students that allows for meaningful interviews around career identity using the artwork as an artifact to guide discussion.

It was easy to convey that engineering is an exciting field with broad impacts, but harder to convey a realistic view of the work of engineers. Consistent with prior research (e.g., Cunningham et al., 2006), many students came to the programs with misconceptions about engineering or without much conception of what engineers do. Our lessons were designed to integrate the engineering design process, particularly constraints and criteria, and to highlight career pathways related to the topic. These activities are not as flashy or eye-catching as some other STEM activities, but a continued focus on activities with authentic engineering design and explorations is more likely to show a realistic view of the work of engineers. Future innovations in the RTB program should focus on tempering some of the unrealistic perceptions while continuing to reinforce how engineers do create technological solutions that help society.

We also found that making local connections to the students' community and everyday lives was relatively easy to do, but required consistent attention to ensure those connections were made explicit in the activities. The students' vision boards suggested most students found personal connections and interest in engineering. In observing students, we also noticed that these personal connections led to greater enthusiasm for the activities.

STEM programs should provide students with a variety of opportunities to find goal alignments in order to engage more students' interest in STEM. Goal-congruity theory (e.g., Belanger et al., 2020) predicts that students will be more interested in or identify with a career field that aligns to their own career values and goals. We sought to integrate many opportunities to connect to values, including entrepreneurial and pro-social impacts. By introducing a wide range of topics and fields of engineering, we were able to appeal to students who were not previously interested in engineering fields.

Sharing that engineering can lead to a steady job and good salary is helpful to some students but not enough to engage the interest of other students. In addition to employment information, out-of-school programs can frame engineering as a means for helping the community, pursuing interests in medical technology and treatments, helping animals, and more. Engineering can also be framed as inventing cool things and solving important problems through creativity and entrepreneurship. Therefore, assessing student interests and offering opportunities to explore interests and values is important to program impacts. There is no one way to frame engineering to appeal to all students, but students can be offered a variety of ways to see themselves pursuing engineering.

This work would not have been possible or sustained without a partnership between university researchers and the community organization. Each partner brought unique expertise. The researchers have extensive experience with teaching students about the engineering career field. ASE brought deep knowledge of the students and connections to the community. By co-planning all programs, we were able to successfully engage students. With the partnership on-going, ASE is poised to continue to broaden students' horizons and expose them to meaningful engineering experiences.

CONCLUSION

Our research design was not intended to test the causal impact of altruistic framing on student identity formation, but it was highly informative of how our program influenced the students. The lessons learned from our program are intentionally adaptable, however, and therefore likely to scale to other programs. We found that students from a wide range of family backgrounds and school quality could meaningfully engage in the reimagined Raise the Bar program. The framing intervention can occur by adapting readily available curriculum to highlight the engineering design process, connect engineering careers to important societal challenges, such as the Grand Challenges for Engineering or the United Nations' Sustainable Development Goals, and emphasize the community relevance of the engineered technology and solutions. Students engaged in reflection on their career identity and showed evidence that their identity was engaged and evolved during the program.

An interesting next step in this research would be to implement a true experimental design where similar engineering content is provided to students but with framing for one group focused on the community relevance of engineering while the other focuses more on traditional ideas such as engineering as a reliable career. In many cases, we were able to take fairly standard STEM activities and add on meaningful framing, suggesting that it is a feasible intervention for existing STEM education programs. A direct test would strengthen the causal relationship of framing to goal congruity perception.

Another interesting strategy would be to compare local vs. global societal challenges. Many engineering curricula seek to connect the field to global challenges, but the prior literature suggests that locally relevant curriculum and connections to the students' own communities are important. The relative effectiveness of global/broad impact compared to local impact should be studied, although each may appeal to some students more than others.

Although not all students expressed accurate perceptions of engineering, most did offer detailed definitions that connected engineering to creating new technology and creative problem solving. Importantly, all of the definitions expressed a positive view of engineering as a career field that had positive impacts on life and society. We also found that students held age-appropriate career identities and that our program promoted their development as well. This supports the use of community-relevant and altruistic framing for providing out-of-school engineering opportunities.

ACKNOWLEDGMENT

This material is based upon work supported by the National Science Foundation under Grant No. **1826181**. We gratefully acknowledge the contributions of Juanita Graham and Lea Steele from Alabama STEM Education and Jessica Taylor and Sydney Riley from Auburn University's Samuel Ginn College of Engineering.

We also extend our appreciation to Joni Corbin and Kat Bornhoft as collaborators.

REFERENCES

Allen, J. M., Muragishi, G. A., Smith, J. L., Thoman, D. B., & Brown, E. R. (2015). To Grab and To Hold: Cultivating communal goals to overcome cultural and structural barriers in first generation college students' science interest. *Translational Issues in Psychological Science*, *1*(4), 331–341. doi:10.1037/tps0000046 PMID:26807431

Belanger, A. L., Diekman, A. B., & Steinberg, M. (2017). Leveraging communal experiences in the curriculum: Increasing interest in pursuing engineering by changing stereotypic expectations. *Journal of Applied Social Psychology*, *47*(6), 305–319. doi:10.1111/jasp.12438

Belanger, A. L., Joshi, M. P., Fuesting, M. A., Weisgram, E. S., Claypool, H. M., & Diekman, A. B. (2020). Putting Belonging in Context: Communal Affordances Signal Belonging in STEM. *Personality and Social Psychology Bulletin*, *46*(8), 1186–1204. doi:10.1177/0146167219897181 PMID:31928327

Bhattarai, B., Davis, E. W., Lakin, J. M., Davis, V. A., & Marghitu, D. (2021, July), *Poster: Use of Augmented Reality (AR) and Virtual Reality(VR) to tackle 4 amongst the "14 Grand Challenges for Engineering in the 21st Century" identified by National Academy of Engineering.* Paper presented at 2021 ASEE Virtual Annual Conference Content Access, Virtual Conference. https://peer.asee.org/37589

Capobianco, B. M., Diefes-dux, H. A., Mena, I., & Weller, J. (2011). What is an engineer? Implications of elementary school student conceptions for engineering education. *Journal of Engineering Education, 100*(2), 304–328. doi:10.1002/j.2168-9830.2011.tb00015.x

Capobianco, B. M., & Yu, J. H. (2014). Using the construct of care to frame engineering as a caring profession toward promoting young girls' participation. *Journal of Women and Minorities in Science and Engineering, 20*(1), 21–33. doi:10.1615/JWomenMinorScienEng.2014006834

Chan, A. D., & Fishbein, J. (2009). A global engineer for the global community. *The Journal of Policy Engagement, 1*(2), 4–9.

Creamer, E. G. (2017). *An introduction to fully integrated mixed methods research.* Sage Publications.

Cunningham, C., Lachapelle, C., & Lindgren-Streicher, A. (2006, June), *Elementary Teachers' Understandings Of Engineering And Technology* Paper presented at 2006 Annual Conference & Exposition, Chicago, Illinois. 10.18260/1-2--200

Diekman, A. B., Clark, E. K., Johnston, A. M., Brown, E. R., & Steinberg, M. (2011). Malleability in communal goals and beliefs influences attraction to stem careers: Evidence for a goal congruity perspective. *Journal of Personality and Social Psychology, 101*(5), 902–918. doi:10.1037/a0025199 PMID:21859224

Easterday, M. W., Lewis, D. R., & Gerber, E. M. (2014). *Design-based research process: Problems, phases, and applications.* International Society of the Learning Sciences.

Estrada, M., Burnett, M., Campbell, A. G., Campbell, P. B., Denetclaw, W. F., Gutiérrez, C. G., Hurtado, S., John, G. H., Matsui, J., McGee, R., Okpodu, C. M., Robinson, T. J., Summers, M. F., Werner-Washburne, M., & Zavala, M. (2016). Improving Underrepresented Minority Student Persistence in STEM. *CBE Life Sciences Education, 15*(3), 1–10. doi:10.1187/cbe.16-01-0038 PMID:27543633

Exploratorium Teacher Institute. (n.d.). *Ocean Acidification in a Cup.* Exploratorium. https://www.exploratorium.edu/snacks/ocean-acidification-in-cup

Foresite Group. (2014). *How to teach your students about civil engineering in 5 short steps.* Foresitre Group. http://www.foresitegroup.net/civil-engineering-class/

Jones, K. S., Newman, D. A., Su, R., & Rounds, J. (2021). Black-White differences in vocational interests: Meta-analysis and boundary conditions. *Journal of Business and Psychology, 36*(4), 589–607. doi:10.100710869-020-09693-5

Karimi, Z., Lakin, J. M., & Davis, V. A. (2022). Water Quality: Adaptable Modules for Engaging K-16 Students. *2022 IEEE Frontiers in Education Conference.* IEEE. 10.1109/FIE56618.2022.9962639

Lakin, J., Marghitu, D., Davis, V., & Davis, E. (2021). Introducing engineering as an altruistic STEM Career. *Science Teacher (Normal, Ill.), 88*(4), 29–35.

Lakin, J., Mastrogiovanni, M., Davis, E., Marghitu, D., & Davis, V. (2022). It makes me wanna go out and help people": Enhancing African American Youth's Perceptions of Engineering. *Journal of Women and Minorities in Science and Engineering, 28*(1), 41–74. doi:10.1615/JWomenMinorScienEng.2021036696

Lakin, J. M., Wittig, A. H., Davis, E. W., & Davis, V. A. (2020). Am I an engineer yet? Perceptions of engineering and identity among first year students. *European Journal of Engineering Education, 45*(2), 214–231. doi:10.1080/03043797.2020.1714549

Maietta, R. C. (2006). State of the art: Integrating software with qualitative analysis. In L. Curry, R. Shield, & T. Wetle (Eds.), *Improving aging and public health research: Qualitative and mixed methods* (pp. 117–139). American Public Health Association and the Gerontological Society of America.

Office for Civil Rights. (2017). *Civil rights data collection [Database].* U.S. Department of Education. https://ocrdata.ed.gov/

Porfeli, E. J., & Lee, B. (2012). Career development during childhood and adolescence. *New Directions for Youth Development, 2012*(134), 11–22. doi:10.1002/yd.20011 PMID:22826162

Sullivan, J. F. (2006). *Broadening Engineering's Participation — A Call for K-16 Engineering Education.* National Academies Press.

Swamidass, P., & Schnittka, C. (2017). Finding and preparing teachers to meet the needs of US student innovators in the making. *Technology and Innovation, 18*(4), 331–342. doi:10.21300/18.4.2017.331

Vest, C. M. (2008). Context and challenge for twenty-first century engineering education. *Journal of Engineering Education, 97*(3), 235–236. doi:10.1002/j.2168-9830.2008.tb00973.x

Weld, J., Heiden, E., & Kemis, M. (2015). Characteristics and effects of a statewide STEM program. *K-12 STEM Education, 1*(4), 167–180.

APPENDIX

INTRODUCTION TO CIVIL ENGINEERING: DESIGN A CITY BLOCK

Lesson Plan

Goal: Introduce students to civil engineering as a career field, emphasize the role of constraints in engineering design, and promote collaborative problem solving and presentation skills. Students will role-play different stakeholders in a community that is designing a city block.

Objectives:

1. SWBAT name at least one way that their city block design meets the needs of their team role and name at least one way the solution was a compromise to their role.
2. SWBAT create a group-determined solution with a design cost that is under budget.
3. SWBAT define a design constraint and give at least two examples of constraints from the engineering project.
4. SWBAT name at least three types of infrastructure designed by civil engineers.

Context: Students have been learning about engineers as inventors and designers and were introduced to an engineering design process. They have practiced brainstorming as part of the design. They are ready to learn about a new field of engineering that is not commonly known but impacts their everyday lives through civil infrastructure.

From their everyday experiences, students will be used to working in groups to solve problems, but may not have as much experience working with constraints and negotiating competing needs. Giving names and importance to these activities as part of a DESIGN process that is central to engineering is the goal of this lesson.

This activity can be fit into a one-hour class, but we benefited from additional time to present and discuss city block designs.

Figure 6 shows an example of a city block design.

Materials:

- Activity worksheet describing the task and each student's role
- City block print outs
- Cut outs of buildings on construction paper
- Glue sticks
- Markers for roads and other details

Safety precautions: Usual classroom courtesy and safety behaviors for scissors and glue.

Figure 6. A city block in progress with buildings of varied size and colors based on type of resource

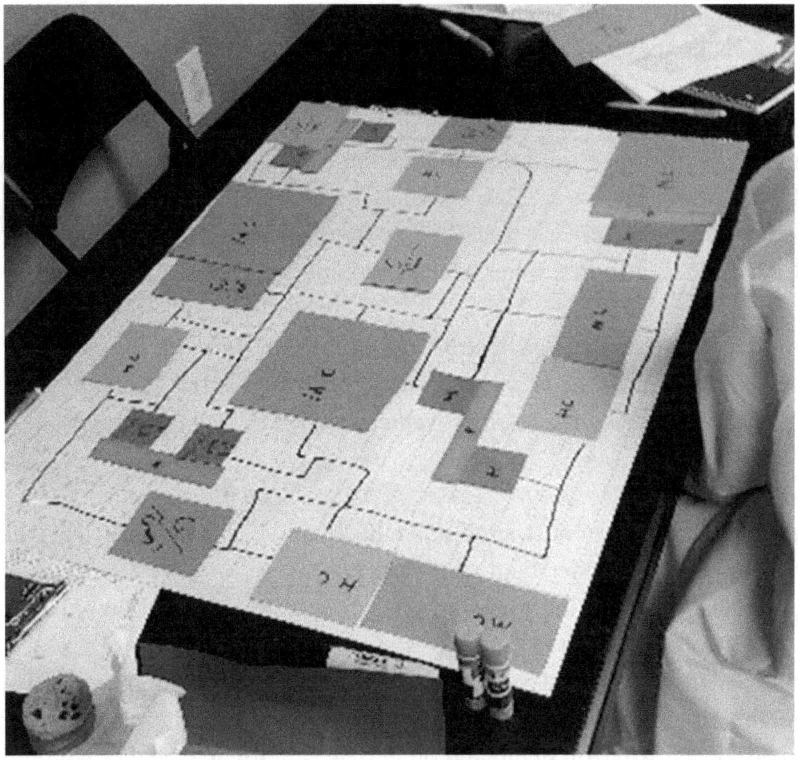

Procedure:

0-10min Review, what is civil engineering?
10-15--introduce design challenge
15-35 minutes--creating a block plan
35-50-- Present group plans
50-53—wrap up

Engage/Introduction:

Q: Review the Engineering Design Process; which part did they practice previous week? (EA: brainstorming, constrains, design and testing)

Introduce Civil Engineering--a field that applies the ED process to solving *infrastructure* and *civil* problems.

Teacher question (TQ): What kinds of infrastructure are there? What are the utilities, services, and public resources we use everyday? (Expected answers (EA): [likely to have difficulty answering at first without prompting] Bridges, airports, water, electricity, sewage, garbage pick up, highways, roads, stop signs, etc.)

Depending on your location, there may be infrastructure examples you can talk about—highway expansion, bridge building, property development, etc.

Explore:

Not in this lesson, but they could conduct some online searches about infrastructure in our everyday life and some of the challenges we have in keeping our infrastructure up to date.

Explain:

Introduce Civil Engineering--a field that applies the engineering design process to solving *infra-structure* and *civil* problems.

TQ: What kinds of infrastructure are there? What are the utilities, services, and public resources we use everyday? (EA: Bridges, airports, water, electricity, sewage, garbage pick up, highways, roads, stop signs, etc.)

Review or introduce constraints: ***Constraints** are things that limit my solution. For example, a constraint may be how much I'm allowed to spend, or how much time I have to complete the challenge. This time we're also going to learn about working together to solve design challenges.*

Elaboration:

Introduce design challenge, the constraints, and talk about what each role is and what they want from the design (make sure everyone understands their role)

The following is adapted from the Foresight Group description (http://www.fg-inc.net/civil-engineering-class/)

Design objective: With the new Amazon fulfillment center, Bessemer is growing! The city has decided to build a new neighborhood where the golf course is currently. Take a look at the map and where it is relative to your schools, the Amazon plant, and other places in town. The city would like you to design a city block layout that meets the needs of the community and stays under the $35million budget.

We are going to work in a team where everyone has different goals and priorities.

1. Have students form groups based on their assigned role (e.g., all Mayors together). Students should read about their assigned role and talk to others in the same role to be sure they understand what their values and priorities will be.
2. Students form groups with one person in each role. Tell students that they need to keep their role in mind and be sure the city block is designed with their priorities in mind.
    a. In these mixed groups, have students look at the size and cost of the different buildings in the worksheet.
    b. Each member of the group should share their goals and priorities while other members take notes in their worksheet.
    c. Make some initial decisions about which building is most important.
3. Hand out large sheets of paper and show students the prepared construction paper for each kind of building.
    a. Students should not immediately glue down their buildings until they have discussed their layout.
    b. Ask groups questions such as "will this school/business/etc. cause more traffic or noise?", "Where will there be the most traffic?", or "Can everyone get to work and school in the morning?"
    c. If students only place buildings, you may want to ask where roads will go.

   d. Remind students of their assigned roles and ask what choices were made that are consistent with their role's priorities. What choices are not consistent?

4. Students should use the worksheet to check their budget.
5. Have each group finalize their design by gluing down the buildings, naming their neighborhood and preparing to share their budget and design.
6. Each group should present their city block noting the decisions they made and what their final budget was.

Wrap up TQ: What were the constraints? (EA: money, location, negotiating with other perspectives). Were the constraints a problem? (EA: Yes, or they were good at negotiating)

Did everyone want the same thing for the block or did some members have different priorities? (EA: no, can discuss some of the negotiations that took place and how they decided on their plan)

Table 6. Example rubric for project

	Does not meet	Meets	Exceeds
Teamwork	- One or two individuals make most decisions about design	- Every team member provides input on the design	- Design based on extensive discussion or debate
Criteria	- Products do not explicitly align to or address stakeholder values	- Design choices are defended using one or more stakeholder values	- Reflects careful balancing of different stakeholder values
Constraints	- Project exceeds budget or teams did not complete a budget	- Project meets budget constraints	- Budget reflects multiple rounds of budgeting
Presentation	- Design is not described in detail with reference to design choices	- Describes the design choices they made	- Describes the design choices they made as well as the tradeoffs between different considerations (stakeholder, budget, etc.)
Creativity	- No evidence of creative solutions or design features	- Project shows creativity in how the buildings are arranged or by adding features of the block (roads, sidewalks, etc.)	- Introduces new resources/buildings to design in order to address values or introduces new constraints or criteria (pollution, social justice, animal rights, etc.)

Resources:

Activity developed by Foresight group: http://www.fg-inc.net/civil-engineering-class/
https://www.hofstra.edu/academics/colleges/seas/ctl/wisengineering/wisengineering_resources.html

ROLES ASSIGNED TO STUDENTS

(Text adapted from Foresight Group lesson)

Homeowners Association President

You have lived in the community for 20 years. You heard that a new developer is coming to town to transform the area of land next to your neighborhood. You come to the meeting to find out what the big plan is. You want to know what kind of stores are going into the area. Is the look of the development going to fit with the landscape of the area? How will the new development affect people who have been living in the community for years? You also heard that a homeless shelter might be built. What is that going to do to your property value?

Table 7. Stakeholder Preferences

Most important to you:	Least important to you:	Concerns
● Park ● Grocery Store ● Single family houses	● Mall ● Strip mall ● Homeless shelter	● Noise from schools or busy areas ● Traffic congestion ● Access to medical care ● Access to healthy food and schools

City Planner

You have worked for the City for 10 years. You have never worked with this developer before this project. You are concerned about boosting the city's economy, but you also have to take into account the needs of the community. Some city council members have recently been pushing to have more parks for children and families. The housing coalition has also been pressing for a new building for a homeless shelter and community center in the area.

Table 8. Stakeholder Preferences

Most important to you:	Least important to you:	Concerns
● Park ● Homeless Shelter ● Community Center	● Mall ● Strip mall ● Single family homes	● Homelessness ● Traffic congestion ● Runoff from concrete roads causing flooding ● Heat sinks caused by excessive concrete ● City services like Police and Fire

City Economic Developer

You have worked for the City for 15 years. You work closely with the developer, and are also interested in the revenue that the new development will generate. You can communicate the value of the development to the city planner to negotiate a plan that will generate a higher revenue for the community. You have the mindset that more commercial developments will provide more jobs to the city, and help eradicate joblessness and homelessness.

Table 9. Stakeholder Preferences

Most important to you:	Least important to you:	Concerns
● Medical Center ● Strip Mall ● Grocery Store	● Homeless Shelter ● Park ● Single family houses	● Noise from manufacturing ● Traffic congestion ● Access to medical care

Mayor

You have been Mayor for 2 years, and are coming up for re-election. You want to make everyone happy, particularly the people who are going to vote for you. The community members are concerned about jobs, that's what you care about. The community members are concerned about aesthetics, that's what you care about, too, all while making sure that the developer will want to come back if the city needs new developments in the future.

Table 10. Stakeholder Preferences

Most important to you:	**Least important to you:**	**Concerns**
● Single family houses ● Park ● School	● Mall ● Strip mall ● Grocery store	● Traffic congestion ● Access to medical care ● City services like Police and Fire

STUDENT WORKSHEET

Challenge

Bessemer is growing. Imagine that a developer (builder) wants to invest **$35 million** in a building project to establish a new city block to improve the city. They want to repurpose a nearby golf course (see map). You are working on a team of Civil Engineers who are planning the building project. You have to stay within the budget provided.

Review the different types of buildings you can choose to build on the budget page.

Stakeholders

Civil Engineers have to consider how new construction will affect people (including traffic), the local economy, the community as a whole, and the environment.

You have the technical expertise, but you need to know what is important to the community and what they want in the new development. You talk to each of these stakeholders to learn more.

Here are your notes from these interviews:

	Priorities
HomeOwners' Association President	
City Economic Developer	
City Planner	
Mayor	

GROUP WORK

Summarize Your Research

5 most important building projects	5 least important or most problematic buildings	Challenges and concerns

City Planning Budget: $35 million

Resources	Cost to build (in millions)	Number planned	Total Cost (Cost x Number)
Shopping center	5		
Emergency services	3		
Grocery store	3		
Pharmacy	1.5		
School	5		
Community center	2		
Medical center	4		
Homeless shelter	1		
Park	.5		
Single family house	.25		
Apartment complex	1.5		
TOTAL (35Mil max)			

1. Place your planned buildings on the grid. Indicate where there are key resources already—interstates, schools, major employers (draw on the edges of this block).
2. Draw in large and small roads you will need to get between all buildings. You can also draw in parking lots.
3. Consider:
   a. Where will a lot of people be traveling and causing traffic?
   b. Where will there be a lot of noise during the day? Night?
   c. What do your stakeholders want? What are they concerned about?

Chapter 4
Using Human-Centered Design to Partner for Continuous Program Improvement of STEM Programming

Ashlie Denton
Education Northwest, USA

Tong Zhang
Oregon MESA at Portland State University Foundation, USA

Kristin Moon
Portland Public Schools, USA

Jason Greenberg Motamedi
Education Northwest, USA

ABSTRACT

Oregon MESA provides students historically underrepresented in STEM fields with the skills, knowledge, and opportunities to develop their talents, explore technology-based careers, enter college, and compete successfully in the workforce. To provide project-based and participatory learning in its STEM out of school time program, MESA collaborated with teacher advisors and community members to co-develop and continuously improve its curriculum. Through this process MESA curriculum shifted from traditional inquiry-based programming that program leaders found to be limited and elitist toward an approach grounded in invention and human-centered design. This transition has meant that MESA is better able to build relevance and community partnerships that meet the needs of its students traditionally underrepresented in STEM. MESA has partnered with Education Northwest to evaluate the curriculum implementation and student and teacher Advisor outcomes. MESA has measurable impacts on students' sense of belonging, confidence and identities in STEM and invention, and ultimately graduation.

DOI: 10.4018/978-1-6684-7771-7.ch004

INTRODUCTION

Students who experience high-quality out-of-school time science, technology, engineering, and mathematics (STEM) opportunities in elementary, middle, and high school are more likely to follow STEM career pathways (Dorph et al., 2018; Sahin et al., 2017). This is especially true for students who have historically been underrepresented in STEM (Jackson et al., 2021). And these experiences are even more likely to develop students' STEM knowledge, skills, and engagement when they are authentic and culturally relevant (Baines et al., 2021; Jackson et al., 2021; Yerrick & Ridgeway, 2017).

So, what do we mean by *authentic*? Authentic STEM learning is hands-on, experiential, and includes a problem focus (Baines et al., 2021; Balemen & Keskin, 2018; Denton & Gandhi, 2020; Kasza & Slater, 2017; Stone, 2016; U.S. Department of Education, 2019). This type of project-based and participatory learning has been shown to better engage underrepresented students than traditional lecture and textbook methods (Haak et al., 2011). Furthermore, competitive and team-based projects make instructional material more enjoyable and engaging (Lemus et al., 2010) and improve communications and teamwork for underrepresented students (Kampe & Oppliger, 2011).

Additionally, authentic STEM experiences incorporate social and emotional learning (SEL) opportunities to increase students' positive identification with STEM fields and their ability to engage in collaborative problem solving (Garner et al., 2018; Jagers et al., 2021; Hoffman et al., 2020). For students to build positive STEM identities, learning experiences need to engage their intersectional racial, ethnic, and gender identities (Singer et al., 2020). The SEL experiences that foster a STEM identity can also help students develop a sense of belonging in STEM, which is crucial for building math and science efficacy and interest in adolescence (Hoffman et al., 2020). Additionally, project-based, community-oriented SEL opportunities focused on collaborative problem solving and self-management have been shown to improve outcomes for Black and Latinx youth from low- and middle-income families (Jagers et al., 2021).

Since 1985, Oregon MESA (Mathematics, Engineering, Science Achievement; MESA[1]) has been providing students with just this type of authentic STEM learning experience. MESA's mission is to provide students who have historically been underrepresented in STEM fields with the skills, knowledge, and opportunities to develop their talents, explore technology-based careers, enter college, and compete successfully in the workforce. The organization works to empower students in Oregon high schools and middle schools who are underrepresented in STEM with the knowledge, skills, and connections to thrive in college and career paths and potentially to see themselves as inventors.

MESA inspires students to pursue education and careers that incorporate invention through five key program areas.

- **MESA Schools Program.** This is the core of MESA's intervention. These weekly after-school sessions are led by a teacher Advisor in the partner school who engages students in hands-on science and engineering projects. The goal of the program is to help students develop academic skills, stay on track to graduate, and prepare for a postsecondary education. During the sessions, students work in teams to complete projects and prepare inventions for local and national competitions.
- **Local and National Competitions**. These science and technology competitions are integrated with the MESA curriculum, allowing students to creatively address real-world problems. Local competitions allow students to present their work to community members and peers from across the state. The winners of the local competition go on to compete at the MESA USA National Engineering Design Competition.

- **Family College and Career Nights.** Held at local schools, these evenings provide an opportunity for MESA families—particularly those from groups who have historically been underrepresented in STEM—to learn how to support their students along the pathway to a postsecondary education. The events are targeted to the entire family in order to promote intergenerational interest and involvement in MESA.
- **MESA Mentor Program.** Mentors are college and university students who work alongside the MESA teachers during the MESA Schools Program. Mentors serve as "near-peer" role models and provide support in multiple ways, from offering math and science content knowledge to sharing strategies for success in college and careers.
- **Career Awareness.** Field trips to local colleges, universities, and technology employers provide a sense of possibility for students who have typically never had exposure to higher education, STEM careers, or role models in these areas.

While all MESA activities support student engagement and sense of belonging in STEM, this chapter focuses on the activities associated with the MESA Schools Program and MESA local and national competitions. During the after-school program and competitions, all student work is grounded in human-centered design (HCD). HCD is a problem-solving approach that incorporates the human element in every stage of the design process (Zoltowski et al., 2012). HCD activities can help students develop key SEL skills (e.g., relationship skills) and 21st century skills (e.g., collaboration, communication, and creativity) through building human-centered and metacognitive approaches to invention, engineering, and science (O'Bryan et al., 2022).

Using HCD in early STEM learning experiences has been shown to engage historically underrepresented students in STEM, particularly girls (Gomoll et al., 2016). Teaching with HCD requires instructors to have a deep understanding of the process and all participants to be open to uncertainty and failure as part of the learning process (O'Bryan et al., 2022). Collaborative teamwork toward a shared goal is key to successful implementation of HCD strategies.

MESA Schools Program and MESA local and national competitions incorporate HCD through invention projects that are supported by teacher Advisors and mentors. MESA student project teams are small (two to four students), and each team member acts as a leader for the group during the school year. Students learn the basics of teamwork, creative thinking, invention, and HCD and are empowered to identify and solve real-world challenges in their communities. For example, one team may address visual impairment by designing a microcontroller-powered reading glove, while another team designs a robot that can help guide visually impaired students around the school. Students develop and test solutions in a safe, supportive environment—without an adult telling them what to do.

For the past 15 years, MESA has collaborated with teacher Advisors and community members to iteratively develop its curriculum. This effort is informed by regular feedback from students, families, teacher Advisors, and mentors. In 2009 MESA shifted from traditional inquiry-based, one-time projects to projects that address real-world, global problems through Design for the Other 90% (Smith & Cooper-Hewitt Museum, 2007). In 2014, with the support of a team of teacher Advisors and community members, MESA developed and piloted a curriculum that incorporates collaboration with actual clients in students' communities. This curriculum is based on HCD and incorporates the full cycle of invention education: problem identification, empathy, brainstorm, design, prototype, redesign, prototype, and entrepreneurship. After two years of implementation and iteration of the HCD curriculum, the MESA development team produced a modular Invention Toolkit with pacing guides. MESA USA adopted this approach nationally starting in 2018. Now, MESA teacher Advisors receive annual professional development training and provide regular feedback on the Invention Toolkit.

To evaluate implementation and student and teacher Advisor outcomes, MESA has partnered with Education Northwest, a research, evaluation, and technical assistance nonprofit whose mission is to use evidence to help partners solve educational challenges and improve learning. Through this partnership, Education Northwest has conducted formative and summative evaluations of the influence of participation in the MESA program on student and teacher Advisor outcomes.

Our formative evaluation of the program shows that most students find MESA to be collaborative, inclusive, and engaging, which helps to foster increased sense of belonging and confidence in STEM and invention skills (Denton & Gandhi, 2020). Teacher Advisors find MESA to be an important opportunity to connect with students in a deeper way than they can in the classroom. Additionally, using a quasi-experimental study design, we found that MESA students were significantly more likely than their peers who did not participate in the program to graduate from high school.

This chapter will describe the collaborations that have supported the development of MESA curriculum over time, as well as the student outcomes generated by participation in MESA. We describe how MESA worked with teacher Advisors and community partners to understand their perspectives and to build MESA curriculum in a way that best meets student needs, engaging in its own HCD process. Following the tenets of HCD, MESA has partnered with Education Northwest to collect feedback from students and teacher Advisors to continuously improve the experience and use of the MESA curriculum and to understand its influence on students.

BACKGROUND

In this section, we describe the iterative development of MESA curriculum over time, focusing primarily on 2009 to today. During this time, MESA experienced a shift in focus from traditional inquiry-based programming, which program leaders found limited and elitist, toward an approach grounded in invention and HCD. This transition has allowed MESA to provide programming that is more inclusive and relevant and better meets the needs of students historically underrepresented in STEM.

Identifying a Need

Oregon MESA activities have varied over the decades but have always focused on promoting skills and mindsets that prepare students to enter STEM fields. In the 1980s, MESA sessions typically involved one-time science and math activities, such as building mousetrap racers or doing the egg drop challenge. While these activities used traditional science inquiry approaches (Furtak, 2006), program leaders and partners gradually realized they were not adequately addressing the needs of the historically underrepresented students they served.

Through evaluative discussions about MESA curriculum, program leadership found that inquiry-driven projects have five drawbacks: they are often linear, begin with theory instead of function, can be solved with a single discipline, are not connected to the real world, and can be elitist. For example, inquiry-based science fairs were often targeted to White, middle-to-upper income youth engaging in college-level work.

To address these shortcomings, teacher Advisors and community partners sought an approach that would apply a growth mindset, engage students in functional and applicable projects, and give them an exciting and safe place to showcase their learning at the end of each year.

MESA began to identify invention-based education as a viable alternative to science inquiry. Starting in 2009, MESA engaged students in activities using team- and project-based participatory learning experiences, which are more engaging for historically underrepresented students (Haak et al., 2011; Kampe & Oppliger, 2011). These activities were unstructured and provided flexibility for teacher Advisors to choose their chapter's activities.

Design for the Other 90%

Initially, MESA's invention curriculum was a series of unstructured, "just for fun" engineering activities. However, program leaders began hearing from teacher Advisors that they wanted to engage students in engineering activities that addressed real-world problems and helped "make the world a better place." Therefore, MESA soon shifted its focus to Design for the Other 90% (Smith & Cooper-Hewitt Museum, 2007), an approach that intentionally focuses on the 90 percent of the world's population that does not have access to products and services that some students in the United States take for granted, including healthy food, clean water, and shelter. This approach provided students with a more authentic and relevant learning experience and created more consistency across the chapters, with teacher Advisors all leading similar activities in the fall of each year. These activities focused on sustainable lighting, water transportation, water filtration, and prosthetics.

Despite this consistency, there was no single strategy for identifying and sharing high-quality learning experiences across the more than 20 MESA chapters. To address this, a small team of teacher Advisors began to develop sharable sequenced lesson plans. A system of creation, use, and feedback emerged and continued to improve the lessons.

The MESA Design for the Other 90% project created four curricular units in which students collaborated to design, build, evaluate, and present a usable product that met the needs of a particular client. For example, students working on the water transportation curriculum designed and built water carriers that were physically and culturally appropriate for children in post-earthquake Haiti.

Findings from surveys, student focus groups, and Advisor feedback sessions indicated that participating students were highly motivated to help others and that developing solutions and products for real people was empowering. The feedback also revealed that students wanted even more relevance and even greater autonomy. Ultimately, this led to an approach in which students generated their own questions for investigation, providing an even stronger sense of ownership and connection to the work.

Scaling the Program

Based on positive feedback and student outcomes, program leaders began looking to scale the program to more schools in Oregon. First, they had to address gaps in the curriculum. With support from the Lemelson Foundation, MESA formed a curriculum development team composed of in-school and out-of-school educators from Portland Public Schools, Beaverton School District, and Parkrose School District. The team's goal was to develop a base set of materials that could be implemented with consistency, while also keeping the emphasis on student voice and autonomy and engagement with real clients.

Through discussion, the curriculum development team landed on HCD as a core strategy for engaging students and the community in real-world, problem-solving activities. The team compiled tools from the HCD framework that had shown the most success with middle and high school students and then created a guide to align the framework with Next Generation Science Standards. The team used four principles to guide their work:

- Leverage the students' strengths
- Acknowledge the importance of play in learning
- Value learning through failure
- Support students to develop a growth mindset

The team also formalized the "INVENT" acronym, which provided educators and students a way to track progress more easily through the HCD framework and formed the foundation of the Invention Toolkit.

The Invention Toolkit

The MESA Invention Toolkit is a resource for teachers and is focused on guiding students through the invention process. MESA published the first full draft of the Invention Toolkit in 2016. After further revisions, they published the final version of all toolkit units on OER Commons, a public digital library of open educational resources, in 2018.

The MESA team also brought in engineering and design professionals to better understand the skills students need to be successful in the rapidly evolving global economy and how organizations work collaboratively to create solutions. These conversations shaped the Invention Toolkit's focus on real-world careers and skills, including the 21st century skills of invention, such as understanding the need and user, brainstorming ideas, and collaborating with their peers.

The Invention Toolkit employs the HCD process to facilitate project-based invention education (Zoltowski et al., 2012; Figure 1). In the toolkit, the teacher is viewed as a facilitator of learning to help students work through a six-step process:

- **Interviewing and empathizing**: Interviews encourage students to empathize and view situations from another person's perspective. This allows students to develop innovative ideas, products, and solutions that are truly relevant, useful, and easy to use. When building for clients, students must be empathetic to the client, rather than sympathetic, and know that every person has an inherent dignity and should be treated with respect. Tools in this section include worksheets for guided interviews and sharing facts about a topic.
- **Naming and defining the problem**: After gaining a better understanding of client needs and requirements, students define the problem they want to solve. Students need a prompt that is narrow enough to guide their process but broad enough to keep them engaged and interested. For example, "Build a solar lantern for someone in a refugee camp" is a narrow, engineering-focused prompt that tells students exactly what to build. To foster innovation, the equivalent HCD prompt might be: "Create a device that someone in a refugee camp can use to walk to the bathroom after dark." In this example, students might build a solar light made of wearable technology, a lantern, or something that has a separate charging unit and does not look like a flashlight at all. Tools for this section include design briefs and priority grids.

- **Visioning and inspiring ideation**. Ideation involves generating possible solutions to the problem you previously defined. It is more than brainstorming because it also includes the evaluation of ideas. When working with students, we have found that it is essential to do this in a way that is equitable and that allows all voices and diverse perspectives to be heard. Tools in this section include brainstorming rules, materials for creating an ideation kit, and a decision matrix.
- **Experimenting and making a prototype**. Once students have finished the ideation process, they must narrow their solution to a design that meets a set of criteria and constraints. Then it is time to build and experiment with a device based on that design. In this phase, students need support with time and project management, prioritizing decisions, and understanding expectations around the status of a final prototype. Students use log books to develop goals for each build day and reflect on their success, priority grids and decision matrices to make building decisions, and example building materials and prototypes to envision the final product. Depending on the timeline of a class, program, or competition, the expectations for the completeness of the product can be adjusted. For example, students can develop a model, prototype, or product.
    - *A model* is a representation of what the finished device will look like and is used to help communicate a design idea. It is not always full size, and it lacks functionality.
    - *A prototype* is the first working iteration of the device. It is full size and has all the pieces. It lacks the finishing touches that would make it sleek and attractive to customers. The prototype is used to test the functionality of the product and make sure that it will fit the criteria and constraints of the product.
    - A ready-for-market *product* has usually gone through marketing and production. All it needs is to be distributed to its point of sale.
- **Engaging client feedback.** Where possible, it is important to allow students to gather feedback from their client about the usability and helpfulness of the final product. Does it meet their needs? Does it function as intended? Client feedback can provide relevant, actionable, and engaging learning opportunities for students.
- **Telling the world** is a process that engages students in basic marketing—how to tell a compelling story about their device so that people will be interested in it. The process encourages students to identify the challenges they faced and describe how they overcame them to create the working prototype. Presentations allow students to reflect on their learning process and demonstrate ownership of their talents and ideas. Students can present to the class or, ideally, at an event that may include families, school staff members, and volunteers from the community. Tools in this section include a presentation rubric.

Throughout the process, students are encouraged to **observe** their identities, values, and biases, as well as to **reflect** on their impact in the service of equity.

Each year, MESA develops pacing guides that help teachers navigate the toolkit. The guides are tailored to a specific theme and include lesson plans and resources related to the theme. In the example pacing guide below (table 1), Invention Toolkit components are interspersed between lessons on coding and using Arduinos. This pacing guide is used to prepare students to create an invention that incorporates the use of microcontrollers.

Figure 1. Human-centered design in the MESA invention toolkit

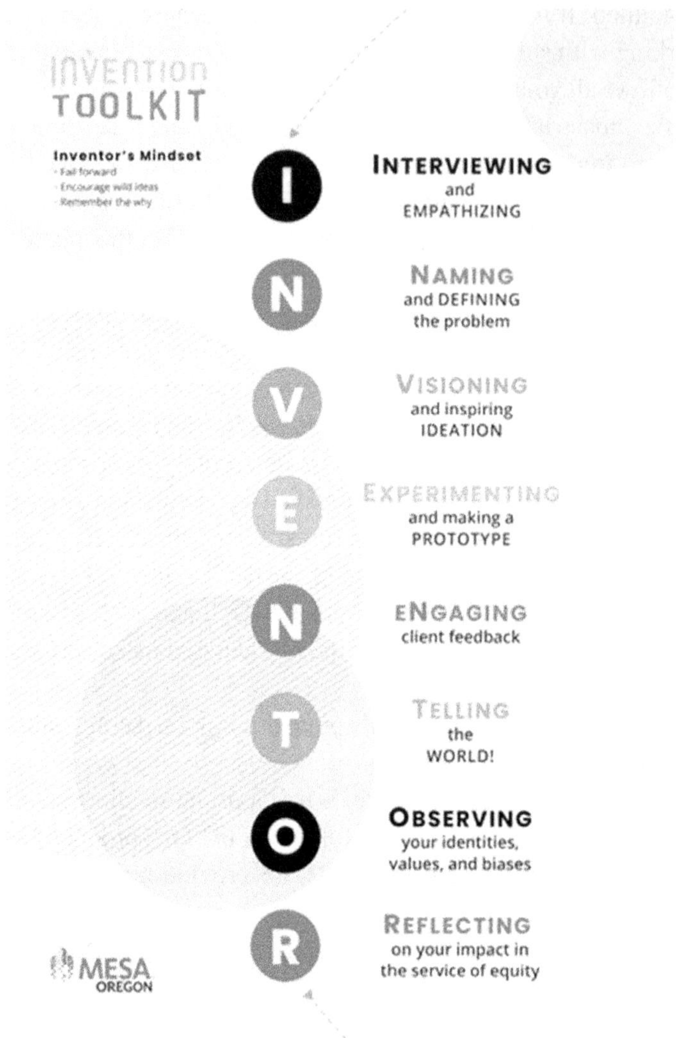

Engaging Community Partners

Invention education that is relevant and capable of engaging students from underrepresented groups requires community connection. To build community connections in a statewide program, Oregon MESA uses a regional center model to recruit and manage school and community partnerships, coordinate events, and support teacher Advisors. Regional coordinators in these centers can share both the knowledge and physical resources that teacher Advisors need to engage the invention curriculum in their school chapters.

At the statewide level, Oregon MESA staff members at Portland State University build strategic partnerships; engage stakeholders; develop funding; and provide curriculum, pedagogy, training, and evaluation. This reduces costs and allows regional centers to focus on leveraging relationships with schools, youth, families, and local partners.

Table 1. Example pacing guide using the MESA invention toolkit for a 22-session program

Week	Lesson Title	Activities
1	Introduction to MESA	Welcoming tradition, community agreements, scribblebots
2, 3	Introduction to Human-Centered Design	Teacher scenarios HCD activity; introduce National Engineering Design Challenge (NEDC)
4	Introduction to Computational Thinking	Introduce coding (binary/bits/bytes) as the basis for everything!); Binary Bracelets OR Capture the Flag (optional) OR Computational thinking cookies (algorithms)
5, 6	Introduction to Arduino*	Start Arduino lessons; basic circuitry and electrical parts Continue Arduino lessons (Blink and Sensors)
7, 8	Interviews/Empathy (Interviewing and Naming)	Arduino project debrief; form project teams and brainstorm clients; prepare client interview questions Conduct client interviews; define problem
9, 10	Ideate (Visioning)	Complete human centered-design brainstorm templates Complete design brief with project timeline
11	Design Brief Introduce Tinkercad - continuing Arduino	Introduce Tinkercad for Arduino by doing the one-button challenge (digital input using pushbutton)
12	Continuing Arduino	Continue Arduino lessons (self-paced) or project-specific Arduino coding
13	Advanced Arduino	Project-specific Arduino coding; transition to physical prototyping
14	Client Feedback Opportunity; Start Prototyping (Engaging and Experimenting)	Invite clients back to look at project ideas and provide feedback to groups; begin prototyping
15, 16	Prototype	Students work to prototype projects.
17	Mini-Presentations	Student groups present their projects; focus on what has been accomplished, next steps, and barriers to success; continue prototyping
18	Advanced Arduino	Address common issues/problem solve barriers to success/explore resources for help.
19, 20	Prototype	Final two weeks to complete prototype Invite client back to view final prototype and provide feedback
21, 22	Pitch (Telling the World)	Students complete pitch posters and film videos

Across the statewide network, MESA builds relationships with community-based organizations to support teacher Advisors with student recruitment. Through these partnerships, MESA can reach students who have been historically underrepresented in STEM in relevant and engaging ways. MESA also partners with culturally specific professional associations (e.g., National Society of Black Engineers, American Indian Science and Engineering Society, Society of Hispanic Professional Engineers, Society of Women Engineers), to recruit volunteers who reflect students' identities.

Continuous Improvement

MESA has continued to gather teacher Advisor feedback to improve the curriculum. Since 2018, the Invention Toolkit has also continued to add concepts from HCD. This has included a shift from sequenced lesson plans to developing tools that could be used throughout the school day to help students develop critical thinking skills.

The invention approach also helped MESA withstand some of the impacts of the COVID-19 pandemic. SEL is important for identity formation, which is a MESA priority and is deeply embedded in the Invention Toolkit. MESA is a very relational program—students join when they have a relationship with the teacher Advisor and stay when they have relationships with other students. Teachers are effective in recruiting new students to MESA and providing a connection back to the school.

Using the Youth Program Quality Assessment Tool (The Forum for Youth Investment, n.d.), MESA is engaging in continuous improvement of the Invention Toolkit. For example, program leaders are developing more training around emotional management, initiative, and responsibility. These are areas of SEL that are challenging for many students and are easily incorporated into MESA projects.

Implementing the Invention Toolkit

MESA uses a train-the-trainer model to provide teacher Advisors with the same experience in HCD they in turn provide to their students. Teacher Advisors participate in invention activities and may become stuck in the process and must rely on, learn from, and make connections with other teacher Advisors. These experiences help teacher Advisors understand the challenges their students may face and plan for how to support them.

During the summer, MESA program staff members use HCD principles in professional learning with teacher Advisors. Teachers attend a two-day Invention Institute where they practice using the Invention Toolkit. MESA staff members provide the training with the help of experienced teacher Advisors who can share more about their experience using the toolkit. Teacher Advisors also gain experience with the processes outlined in the toolkit. MESA brings in clients that are a good fit for the theme of the year (e.g., school nurses for the Health Heroes theme), and teacher Advisors interview the clients in the same way their students will interview clients during the year. By the end of the second day, teacher Advisors have had the opportunity to present a prototype to their colleagues and potentially to clients.

In the winter, MESA brings the teacher Advisors back together to support collaborative problem-solving around issues they are experiencing in their MESA chapters. The problems can range from technical issues to supporting students with soft skill development. MESA uses internal and external expertise to support teacher Advisors with overcoming these challenges. In addition, the winter train-the-trainer opportunity helps build relationships among teacher Advisors, which makes it easier to address questions and frustrations as they arise.

Feedback on MESA programming occurs throughout the year. MESA collaborates with a small group of teacher Advisors to test ideas before implementing them during professional learning days. Additionally, at the end of summer and winter sessions, teacher Advisors are given the opportunity to reflect on their experiences and opportunities for improvement. In late spring, MESA brings the teacher Advisors together for a feedback session that includes a conversation about the strengths and challenges of the materials, as well as ideas on how to improve or add to the tools in the Invention Toolkit. This feedback is reviewed and incorporated regularly, as MESA continues to revise its programming to best meet the needs of teacher Advisors, students, and the community.

Between 20 and 40 teacher Advisors participate in the toolkit training each year. This training is a requirement for operating a MESA program, and experienced teachers (those who have already taken the training at some point) are also invited to participate, often serving in a supporting role for the MESA trainers. Because the theme of the MESA challenge changes each year, even veteran teachers will gain additional insight and practice by interviewing the new "client" for the year.

Evaluating Implementation

Education Northwest has served as the external evaluator of the MESA program for nearly 20 years. In that time, the partnership has gradually incorporated HCD principles to co-design a system of evaluation that includes formative findings based on participants' experiences with the program and summative findings based on specific student outcome measures. The evaluation also incorporates a continuous improvement process, which is key to ensuring the program delivers high-quality outcomes for youth and makes it easier to communicate the value and impact of the program to district partners (Bryk et al., 2015; Penuel et al., 2011). The following sections describe these evaluation methods and outcomes.

MATERIALS AND METHODS

MESA uses evidence-based strategies to improve educational outcomes for all students, particularly students of color underrepresented in STEM. MESA relies on partnerships with teacher Advisors, schools, and community partners to recruit a diverse student population for its programming. For example, in the 2021–22 school year, most MESA students were from groups underrepresented in STEM, including underrepresented students of color (African American, Native American, Latino, multiracial, 59%) and students who identify as girls, gender diverse, nonbinary, or transgender (47%). Additionally, 78 percent of students were eligible for free and reduced-price lunch (Figure 2). This is representative of the composition of MESA in the years of the evaluations described below.

Summative Evaluation

With approval from its institutional review board, Education Northwest has conducted both formative and summative evaluations of student outcomes from participation in MESA. Education Northwest has conducted two rigorous summative studies examining the impact of MESA on student academic outcomes including course grades, assessment scores, and graduation rates. In 2016, Education Northwest asked the research question, "What is the impact of MESA's expansion into a new Oregon school district on participating students' test scores and grades?" Researchers found that MESA students had higher grades in science and math than their peers, even after starting out behind the district average. Low-income and underrepresented students of color showed the largest improvements (Greenberg Motamedi & Singh, 2016).

In 2020, to better understand whether these outcomes reflect the experiences of a broader selection of MESA students, Education Northwest conducted a more extensive impact evaluation of students in four Oregon school districts across 10 years. Again, researchers compared MESA students to a matched sample of Oregon students who did not participate in the program but were otherwise similar in all observable ways. Specifically, researchers compared their performance on the statewide grade 8 and high school (typically grade 11) English language arts (ELA), math, and science assessments, as well as high school graduation rates. Researchers found that MESA students had substantially higher rates of high school graduation than their matched peers who did not participate (Greenberg Motamedi et al., 2020).[2]

Figure 2. Most MESA students are from groups that have historically been underrepresented in STEM
Source: Oregon MESA 2022 Annual report: https://oregonmesa.org/wp-content/uploads/2022/12/Oregon-MESA_FY22-Annual-Report-Digital.pdf

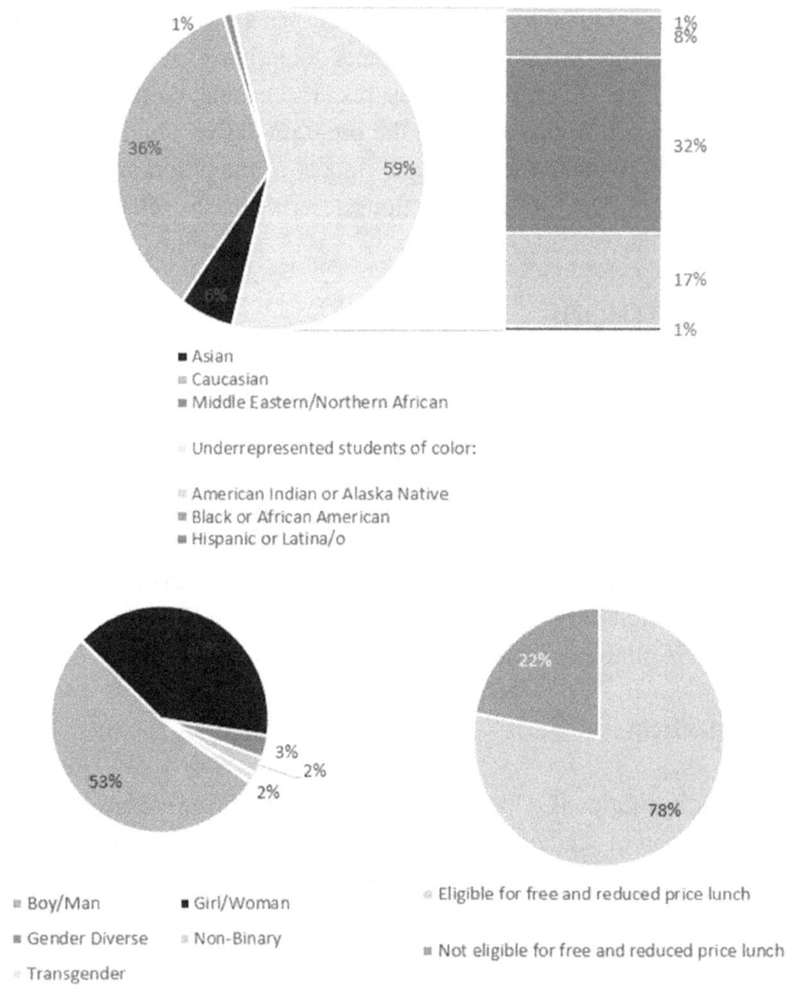

This study was based on data from 2008 to 2018, when MESA's curriculum was primarily focused on Design for the Other 90%. This kind of retrospective evaluation cannot address recent changes in the curriculum, such as the implementation of Invention Toolkit. Therefore, while the study provides important insights into student outcomes, an updated impact evaluation is needed.

Student Participants

To determine which Oregon students were MESA participants, Education Northwest researchers matched program records to Oregon Department of Education (ODE) administrative files using students' names, birthdates, and school districts. Researchers identified 431 students who participated in MESA in grades 6–8 during the period from 2009–10 to 2013–14 and who had the minimum necessary data to be included in the study, such as a state test score in grade 8 or high school.

MESA students in the study sample attended four Oregon school districts: Portland Public Schools, Parkrose School District, Hillsboro School District, and Beaverton School District. The students were members of three cohorts, as defined by their first year of grade 6, which was their first year of possible MESA participation. Depending on their cohort, these students took grade 8 and high school state tests, potentially graduated from high school, and matriculated to a postsecondary education institution in different years (Figure 3).

Figure 3. Description of study cohorts

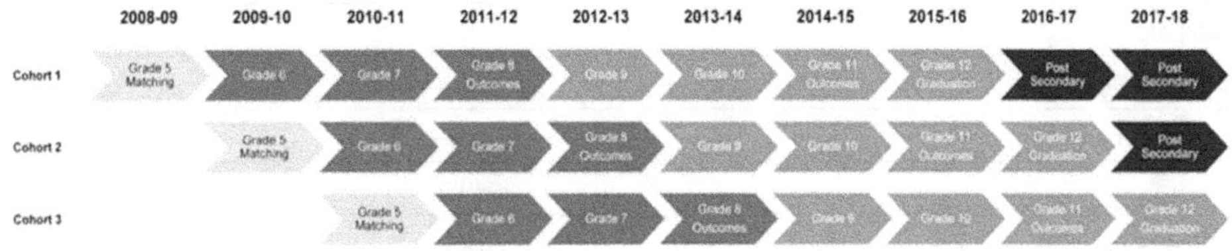

Comparison Students

Students in the comparison group did not participate in MESA but were otherwise similar in observable ways to program participants (Table 2). They were in the same grades and school districts and had similar academic achievement and demographic characteristics.

Education Northwest researchers matched MESA students to comparison group students in districts that offered MESA. Researchers used nine *covariates* (student-level data points that may predict student outcomes) to ensure the students were well matched: race/ethnicity; gender; grade level; state assessment results in ELA, math, and science in grade 5 (one year before students could participate in MESA); and eligibility for three federal programs: English learner services, special education services, and the free or reduced-price lunch program.

With these characteristics, Education Northwest researchers matched the 431 MESA students to one or more individuals in their district who were statistically similar to them from a comparison group of 11,839 non-MESA students. Researchers combined all MESA students into one group, regardless of when and for how many years they participated in MESA.[3]

MESA students had slightly higher assessment scores than the comparison group in grade 5. However, the differences were not statistically significant and were close enough to establish *baseline equivalence*, which means that MESA students and the comparison group students had characteristics that were comparable (*equivalent*) before the former participated in MESA (*at baseline*).

Table 2. Baseline results from fifth grade were well matched

	MESA students (*N*=431)	Comparison group students (*N*=11,839)
State test standard scores (OAKS)		
ELA	0.11	0.10
Math	0.26	0.23
Science	0.12	0.09
Program participation (ever)		
English learners	35%	29%
Federal lunch program	76%	65%
Special education	23%	15%
Demographic characteristics		
Female	49%	49%
Hispanic/Latino	32%	31%
Asian/Pacific Islander	15%	8%
African American	10%	5%
White	35%	50%
Other, including multiracial, Alaska Native, and American Indian	8%	5%

Source: Education Northwest analysis of MESA (2009–10 to 2013–14) and ODE data (2008–09 to 2017–18).

Summative Analysis

Education Northwest researchers used regression analysis to examine the relationship of MESA participation to student performance. Regression allowed Education Northwest researchers to estimate MESA's contribution to student learning based on students' performance on state standardized achievement tests or its impact on graduation. Regression also allowed researchers to compare student test scores or outcomes while accounting for other factors (such as prior achievement and demographic characteristics, as well as eligibility for special education services, English learner services, and the federal reduced-price lunch program).

Formative Evaluation

To understand student outcomes during the years since the summative evaluation, Education Northwest has continued its ongoing formative evaluation of MESA. Among other questions, researchers asked, "To what extent does MESA participation increase student outcomes in STEM and invention (e.g., engagement, knowledge, and confidence) and SEL (e.g., confidence, sense of belonging, and collaboration)?" Education Northwest has gathered student and teacher Advisor surveys since the implementation of the Invention Toolkit (2017–2022). These surveys were collaboratively created and informed by literature. Surveys are distributed to all participating students and teacher Advisors at the end of the year. All questions on the student survey are tested for statistically significant differences between:

- genders underrepresented in STEM (girls, nonbinary students, and other genders) and boys
- underrepresented racial and ethnic groups and White and Asian students
- MESA students attending middle schools and those attending high schools

Differences for these measures are reported only when they are statistically significant ($p \leq .05$) using the chi-square test and adjusting p-values for multiple comparisons.

RESULTS

Summative Evaluation Findings

Education Northwest researchers found that MESA students were significantly more likely than their peers who did not participate in the program to graduate from high school. Researchers did not find any other statistically significant differences between MESA students and their matched peers on the Oregon ELA, math, and science assessments.

On average, MESA students had higher science test scores than their matched peers in the comparison group (Table 3) in both grade 8 and high school. MESA students' ELA test scores were higher than the comparison group in grade 8, and the difference was nearly indistinguishable in high school. MESA students' math scores were lower than the comparison group in grade 8, and the difference was indistinguishable in high school. After accounting for prior achievement, program enrollment, and demographic characteristics, these differences were not statistically significant. This means the differences between MESA students and the comparison group could have been due to chance or normal population variation rather than program participation.

MESA students were more likely to graduate from high school than comparison group students. Education Northwest researchers found that 87 percent of MESA students graduated from high school in four years compared with 73 percent of the comparison group (Figure 4).

These differences were statistically significant ($p < .001$) after accounting for prior achievement, program enrollment, and demographic characteristics (Table 3). Of the 335 MESA students who graduated from high school, all but one did so in four years. In contrast, among the 8,457 comparison group students who graduated from high school, 756 (9%) needed more than four years. MESA students were significantly more likely to graduate from high school than comparison group students, regardless of how long it took.

Finally, Education Northwest researchers did not find any *interaction effects,* that is, statistically significant differences in how MESA interacts with graduation outcomes by students' gender, race/ethnicity, eligibility for the federal reduced-price lunch program, or English learner status. This means participating in MESA is likely to boost the likelihood of graduating from high school in four years for all students (Table 4).

Education Northwest researchers found that MESA students were 3.13 times more likely to graduate from high school in four years than comparison group students. Stated another way, we found that participating in MESA appears to increase the probability of graduating from high school in four years by 21 percentage points.

Table 3. Regression results for assessment scores

	Grade 8 ELA	High school ELA	Grade 8 math	High school math	Grade 8 science	High school science
MESA participation	0.05 (0.06)	0.00 (0.06)	0.05 (0.07)	0.03 (0.09)	0.10 (0.07)	0.07 (0.08)
Grade 5 achievement						
OAKS ELA	0.67*** (0.01)	0.59*** (0.02)				
OAKS math			0.71*** (0.02)	0.68*** (0.02)		
OAKS science					0.65*** (0.02)	0.59*** (0.03)
Program participation (ever)						
Federal lunch	-0.15*** (0.03)	-0.17*** (0.03)	-0.20*** (0.04)	-0.20*** (0.07)	-0.19*** (0.03)	-0.18*** (0.05)
Special education	-0.20*** (0.03)	-0.27*** (0.04)	-0.15*** (0.04)	-0.14** (0.05)	-0.24*** (0.03)	-0.16* (0.07)
English learner	-0.01 (0.04)	0.08* (0.04)	-0.04 (0.04)	-0.02 (0.04)	-0.07 (0.04)	0.03 (0.04)
Demographic characteristics						
Female	0.17*** (0.03)	0.09*** (0.03)	0.09*** (0.02)	0.04 (0.03)	-0.06* (0.03)	0.01 (0.03)
Hispanic/Latino	-0.14*** (0.04)	-0.11*** (0.03)	-0.14*** (0.04)	-0.12** (0.04)	-0.12*** (0.03)	-0.16** (0.06)
American Indian & Alaska Native	-0.12 (0.11)	-0.38** (0.12)	-0.19 (0.11)	-0.12 (0.09)	-0.19 (0.12)	-0.22 (0.12)
Asian & Pacific Islander	0.05 (0.05)	0.09 (0.05)	0.19** (0.06)	0.26*** (0.08)	0.10 (0.05)	0.10 (0.07)
Black/African American	-0.16*** (0.04)	-0.19** (0.07)	-0.17** (0.07)	-0.18* (0.08)	-0.27*** (0.04)	-0.40*** (0.09)
Multiracial	-0.07 (0.06)	0.00 (0.06)	-0.04 (0.06)	0.05 (0.07)	0.02 (0.03)	-0.02 (0.07)
Constant	0.17*** (0.04)	0.31*** (0.04)	0.23*** (0.05)	0.32*** (0.05)	0.26*** (0.04)	0.20** (0.06)
Variance components						
Level 2, constant	-1.85*** (0.11)	-1.91*** (0.10)	-1.16*** (0.09)	-1.772*** (0.109)	-1.51*** (0.08)	-1.02*** (0.11)
Level 2, residual	-0.47*** (0.02)	-0.42*** (0.01)	-0.42*** (0.02)	-0.335*** (0.0182)	-0.46*** (0.01)	-0.27*** (0.03)
Observations	10,827	8,474	10,808	8,339	9,255	6,140

*$p < 0.05$, **$p < 0.01$, ***$p < 0.001$

Note: Standard errors are shown in parentheses.

Source: Education Northwest analysis of MESA (2009–10 to 2013–14) and ODE data (2008–09 to 2017–18)

Figure 4. MESA students had a higher four-year graduation rate than their matched peers
Source: Education Northwest analysis of MESA (2009–10 to 2013–14) and ODE data (2008–09 to 2017–18).

Table 4. Regression results for high school graduation

	Four-year high school graduation	Ever graduated from high school
MESA participation		
Odds ratio	3.13*** (0.39)	1.76*** (0.25)
Grade 5 achievement		
OAKS ELA	1.16 (0.09)	1.16 (0.08)
OAKS math	1.34 (0.12)	1.39** (0.16)
OAKS science	1.04 (0.07)	1.07 (0.08)
Program participation (ever)		
Federal lunch	0.34 (0.08)	1.54*** (0.28)
Special education	0.85 (0.12)	0.85 (0.13)
English learner	1.68 (0.30)	1.54* (0.28)
Demographic characteristics		
Female	1.24 (0.12)	1.32 (0.13)
Hispanic/Latino	1.15 (0.19)	0.98 (0.16)
American Indian & Alaska Native	0.97 (0.37)	0.72 (0.27)
Asian & Pacific Islander	1.89 (0.50)	1.56 (0.43)
Black/African American	1.50 (0.28)	1.26 (0.20)
Multiracial	1.73 (0.32)	1.53 (0.30)
Constant	3.60 (0.78)	7.64*** (1.91)
Variance components		
Level 2, constant	0.23 (0.05)	0.18** (3.14)
Observations	10,689	10,689

*$p < 0.05$, **$p < 0.01$, ***$p < 0.001$

Source: Education Northwest analysis of MESA (2009–10 to 2013–14) and ODE data (2008–09 to 2017–18).

In comparison, being affected by poverty (defined as being eligible for the federal reduced-price lunch program) had a similar magnitude of effect on the likelihood of graduation. However, unlike MESA participation, this effect was negative—MESA participants were 21 percentage points more likely to graduate from high school in four years than comparison group students, and students affected by poverty were 20 percentage points less likely to graduate from high school in four years than their peers.

This outcome is notable among similar interventions. Education Northwest identified one intervention in the U.S. Department of Education's What Works Clearinghouse and one in a recently published review of after-school programs (Neild et al., 2019a, 2019b) that appear to have a statistically significant impact on high school graduation, but lower than MESA's impact. Similarly, Citizen Schools is an after-school program for low-income youth in grade 8 that includes career exposure, high school and college preparation, and academic enrichment. A quasi-experimental design study found that Citizen Schools participants were 1.70 times more likely to graduate from high school in four years than comparison group students – a good outcome, but still not the impact of MESA (Acaira et al., 2010). Finally, Early College is a dual enrollment program that supports underrepresented students as they take college courses while they are still in high school. An experimental study found that students participating in Early College were 1.48 times more likely to graduate from high school than students who were not selected to participate in the program (Berger et al., 2013). MESA's impact on graduation was larger than either of the comparison programs by increasing the probability of graduating from high school in four years by an odds ratio of 3.13.

Formative Evaluation Findings

In more recent years, Education Northwest has found that the use of the Invention Toolkit has resulted in STEM and SEL outcomes for students and teachers. In the most recent MESA survey, about three-quarters of students were from a group that has historically been underrepresented in STEM, yet 81 percent of students reported that MESA was a place where they felt they belonged. Overwhelmingly, they believed MESA is fun (86%) and that their confidence in using invention skills—problem solving, working on a team, product development—had improved (94%). Looking ahead to their futures, 80 percent affirmed that MESA taught them about careers in STEM (Denton & Gandhi, 2020).

For this chapter, Education Northwest researchers also reviewed the last five years of data on student and Advisor outcomes (2017–2022) since the piloting of the Invention Toolkit. During this time, researchers found that most students report high SEL outcomes, such as feeling a sense of belonging in MESA (90%) and finding MESA to be fun (95%; Figure 5).

As part of our analysis, Education Northwest researchers looked at differences for students that have historically been underrepresented in STEM (specifically underrepresented genders, underrepresented racial and ethnic groups). Although most MESA students come from groups that have historically been underrepresented in STEM, researchers found no significant differences among student groups on any measure.

Additionally, students report invention-specific outcomes. About three-quarters of students feel they built their confidence in invention skills through participation in MESA (Figure 6).

Additionally, many MESA students reported increased confidence in engineering, science, computer coding, and math skills (Figure 7). About 90 percent of students also reported learning more about careers in STEM.

Figure 5. Most MESA students report having opportunities to participate in MESA activities and finding MESA fun
Source: Education Northwest analysis of MESA student survey data from 2017–2022 (n=835).

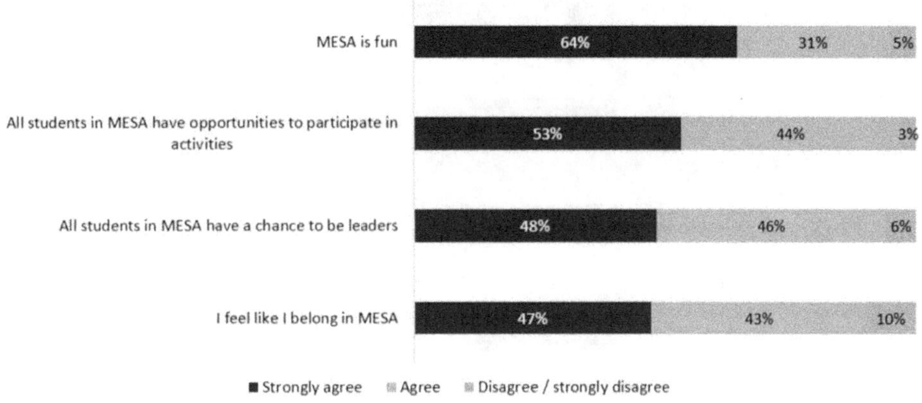

Figure 6. Many MESA students report increased confidence in invention skills
Source: Education Northwest analysis of MESA student survey data from 2017–2022 (n=835).

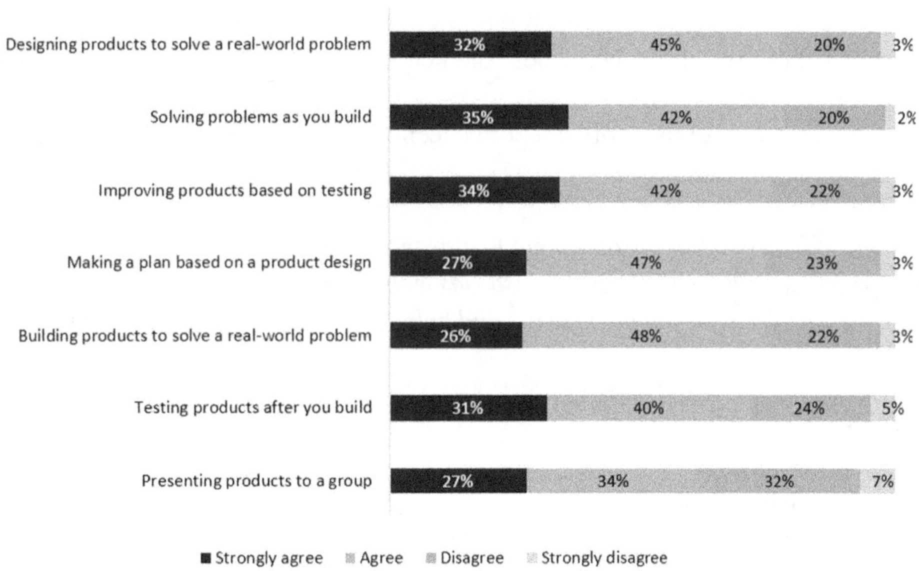

In open-ended survey responses, students often described an increased confidence as they worked through the invention process.

MESA [helped me build] more confidence to come up with a solution for an innovative project while collaborating with my group to build something that may have a good impact on someone's life. This is a good program for identifying the parts of the engineering process, while actually doing them. (MESA student)

Figure 7. Many MESA students report increased confidence in other STEM skills
Source: Education Northwest analysis of MESA student survey data from 2017–2022 (n=835).

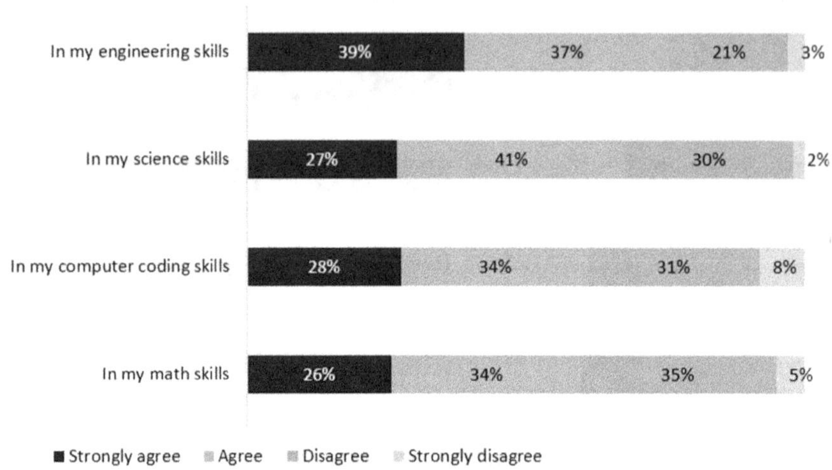

In open-ended questions, teacher Advisors were often enthusiastic about the growth mindset they were able to build on their students through the Invention Toolkit. "We build stuff, try stuff, and are ready to have stuff fail. And that's awesome! And hence, [my MESA students] just ROCK!" (MESA teacher Advisor).

More generally, teacher Advisors often reported feeling closer to their students through participation in MESA.

I'm closer to my MESA students because I get to play more of a coaching role, there's no grades, and I get to spend more time with them. These are also kids with STEM interests whom I avidly enjoy speaking to about their new projects, accomplishments, and hobbies. (MESA teacher Advisor)

DISCUSSION

MESA's hands-on, relevant, and problem-based approach to STEM reflects the most up-to-date research on how to engage students and improve outcomes (Dorph et al., 2018), particularly for students who have historically been underserved in STEM (Jackson et al., 2021). Additionally, MESA's authentic experiences incorporate SEL opportunities to increase students' identity, sense of belonging, and collaborative problem solving (Garner et al., 2018; Hoffman et al., 2020; Jagers et al., 2021). The HCD activities at the core of MESA's most recent curriculum can help students develop key SEL and 21st century skills (O'Bryan et al., 2022) and opportunities for STEM learning (Gomoll et al., 2016).

As a result, MESA is having a measurable impact on students' sense of belonging, confidence and identities in STEM and invention, and ultimately on graduation rates. There were no statistically significant differences in outcomes for students from groups who have historically been underrepresented in STEM. This suggests that the MESA curriculum, using real-world, team-based learning in a supportive and iterative environment, supports growth in STEM and SEL for all students, particularly those who have historically been underserved in STEM.

Moreover, the benefits of MESA's iterative curriculum development are demonstrable for all partners participating in the program. In addition to student outcomes generated by the curriculum, MESA administrators and teacher Advisors get consistent feedback and are better able to implement their programming to address the needs of students. Community partners were able to meaningfully participate in the development of the curriculum and continue to engage throughout students' invention process.

CONCLUSION

In partnership with teacher Advisors and local community members, MESA's co-development of the Invention Toolkit produced an approach to out-of-school-time STEM curriculum that provides students with the type of relevant, authentic, hands-on learning opportunities known to engage students who have historically been underrepresented in STEM. This approach is grounded in the tenets of HCD as it focuses heavily on gathering feedback from relevant partners (e.g., students, teacher Advisors, and community members) to support continuous improvement.

There are several pathways forward for this work. First, the Invention Toolkit can be used in in-school and out-of-school time settings to better engage students who have historically been underrepresented in STEM. To understand whether the outcomes experienced in Oregon MESA are more broadly generalizable, these tools could be used by practitioners in a variety of settings. This replication would be best supported with a robust formative evaluation and by building on the groundwork laid for a summative evaluation after several years.

More generally, the use of invention in STEM learning is not widely used but can create learning opportunities that are nonlinear, functional, relevant, interdisciplinary, connected to the real world, and open to all learners. This type of learning is clearly supported by the literature on STEM in general, but more work needs to be done to understand the specific ways that invention education supports the learning of students historically underrepresented in STEM beyond what is described in this chapter.

Additionally, there is room to replicate the underlying principles of MESA's approach to using HCD as a key strategy for program development. The implementation of any STEM program in partnership with K–12 requires learning and adaptation to fit local needs and contexts. In fact, contextualization and cultural responsiveness are some of the most challenging barriers to scaling successful out-of-school time STEM programs (Denton & Gandhi, 2020). By using an iterative approach that centers learning and empathy, programs can overcome some of these challenges.

Finally, the Invention Toolkit is relatively new. Replicating the analysis described in this chapter, ideally at regular intervals, will support the evaluation and continuous improvement of the toolkit's implementation and impact.

Supplemental Materials

Oregon MESA's *Invention Toolkit - Human Centered Design* curriculum is located here: https://www.oercommons.org/courseware/lesson/69204

REFERENCES

Acaira, E., Vile, J., & Reisner, E. R. (2010). *Citizen Schools: Achieving high school graduation: Citizen Schools' youth outcomes in Boston*. Policy Studies Associates, Inc. https://static1.squarespace.com/static/57542d1b0442628dccd81967/t/5898ae5f9f7456131b0b420d/1486401121825/PSA-Citizen-Schools-Youth-Outcomes-in-Boston.pdf

Baines, A. M., DeBarger, A., De Vivo, K., Warner, N., Santos, S., Brinkman, J., Udall, D., Zuckerbrod, N., Felsen, K., & Urban, R. (2021). *Key principles for project-based learning*. Lucas Education Research. https://www.lucasedresearch.org/wp-content/uploads/2021/02/Key-Principles-for-PBL-White-Paper.pdf

Balemen, N., & Keskin, M. Ö. (2018). The effectiveness of project-based learning on science education: A meta-analysis search. *International Online Journal of Education & Teaching*, *5*(4), 849–865.

Berger, A., Turk-Bicakci, L., Garet, M., Song, M., Knudson, J., Haxton, C., Zeiser, K., Hoshen, G., Ford, J., Stephan, J., Keating, K., & Cassidy, L. (2013). *Early college, early success: Early College High School initiative impact study*. American Institutes for Research. https://eric.ed.gov/?id=ED577243

Bryk, A. S., Gomez, L. M., Grunow, A., & LeMahieu, P. G. (2015). *Learning to improve: How America's schools can get better at getting better*. Harvard Education Press.

Denton, A., & Gandhi, E. (2020). *Scaling out-of-school time STEM programming: A national scan*. Overdeck Family Foundation and Education Northwest. https://educationnorthwest.org/sites/default/files/overdeck-ost-stem-report.pdf

Dorph, R., Bathgate, M. E., Schunn, C. D., & Cannady, M. A. (2018). When I grow up: The relationship of science learning activation to STEM career preferences. *International Journal of Science Education*, *40*(9), 1034–1057. doi:10.1080/09500693.2017.1360532

Furtak, E. M. (2006). The problem with answers: An exploration of guided scientific inquiry teaching. *Science Education*, *90*(3), 453–467. doi:10.1002ce.20130

Garner, P. W., Gabitova, N., Gupta, A., & Wood, T. (2018). Innovations in science education: Infusing social emotional principles into early STEM learning. *Cultural Studies of Science Education*, *13*(4), 889–903. doi:10.100711422-017-9826-0

Gomoll, A., Hmelo-Silver, C. E., Šabanović, S., & Francisco, M. (2016). Dragons, ladybugs, and softballs: Girls' STEM engagement with human-centered robotics. *Journal of Science Education and Technology*, *25*(6), 899–914. doi:10.100710956-016-9647-z

Greenberg Motamedi, J., Serrano, D., & Hanson, H. (2020). *Oregon MESA: Increasing the odds of high school graduation*. Education Northwest. https://files.eric.ed.gov/fulltext/ED606318.pdf

Greenberg Motamedi, J., & Singh, M. (2016). *Oregon MESA: Improving grades in science and mathematics*. Education Northwest. https://files.eric.ed.gov/fulltext/ED606319.pdf

Haak, D. C., HilleRisLambers, J., Pitre, E., & Freeman, S. (2011). Increased structure and active learning reduce the achievement gap in introductory biology. *Science*, *332*(6034), 1213–1216. doi:10.1126cience.1204820 PMID:21636776

Hoffman, A. J., McGuire, L., Rutland, A., Hartstone-Rose, A., Irvin, M. J., Winterbottom, M., Balkwill, F., Fields, G. E., & Mulvey, K. L. (2020). The relations and role of social competencies and belonging with math and science interest and efficacy for adolescents in informal stem programs. *Journal of Youth and Adolescence*, *50*(2), 314–323. doi:10.100710964-020-01302-1 PMID:32804294

Jackson, C., Mohr-Schroeder, M. J., Bush, S. B., Maiorca, C., Roberts, T., Yost, C., & Fowler, A. (2021). Equity-oriented conceptual framework for K-12 STEM literacy. *International Journal of STEM Education*, *8*(1), 1–16. doi:10.118640594-021-00294-z

Jagers, R. J., Skoog-Hoffman, A., Barthelus, B., & Schlund, J. (2021). Transformative social emotional learning: In pursuit of educational equity and excellence. *American Educator*, *45*(2), 12–17, 39.

Kampe, J., & Oppliger, D. E. (2011, June). *Effectiveness of team-based STEM project learning to recruit minority high school students to STEM. 2011 ASEE Annual Conference & Exposition*, Vancouver, BC. 10.18260/1-2--17820

Kasza, P., & Slater, T. F. (2017). A survey of best practices and key learning objectives for successful secondary school STEM academy settings. *Contemporary Issues in Education Research*, *10*(1), 53–66. doi:10.19030/cier.v10i1.9880

Lemus, J. D., Bishop, K., & Walters, H. (2010). QuikSCience: Effective linkage of competitive, cooperative, and service learning in science education. *American Secondary Education*, *38*(3), 40–61.

Neild, R. C., Wilson, S. J., & McClanahan, W. (2019a). *Afterschool programs: A review of evidence under the Every Student Succeeds Act.* Research for Action. https://www.researchforaction.org/publications/afterschool-programs-a-review-of-evidence-under-the-every-student-succeeds-act/

Neild, R. C., Wilson, S. J., & McClanahan, W. (2019b). *Afterschool evidence guide: A companion to Afterschool programs. A review of evidence under the Every Student Succeeds Act.* Research for Action. https://www.researchforaction.org/publications/afterschool-evidence-guide-a-companion-to-a-review-of-evidence-under-the-every-student-succeeds-act/

O'Bryan, L., Bohn, D., & Shehab, S. (2022). A Human-Centered Design Tool Kit for STEM-Based Capstone Courses. *Journal of College Science Teaching*, *51*(6), 23–28.

Penuel, W. R., Fishman, B. J., Haugan Cheng, B., & Sabelli, N. (2011). Organizing research and development at the intersection of learning, implementation, and design. *Educational Researcher*, *40*(7), 331–337. doi:10.3102/0013189X11421826

Sahin, A., Ekmekci, A., & Waxman, H. C. (2017). The relationships among high school STEM learning experiences, expectations, and mathematics and science efficacy and the likelihood of majoring in STEM in college. *International Journal of Science Education*, *39*(11), 1549–1572. doi:10.1080/09500693.2017.1341067

Singer, A., Montgomery, G., & Schmoll, S. (2020). How to foster the formation of STEM identity: Studying diversity in an authentic learning environment. *International Journal of STEM Education*, *7*(57), 1–12. doi:10.118640594-020-00254-z

Smith C. E. & Cooper-Hewitt Museum. (2007). *Design for the other 99%*. Smithsonian Cooper-Hewitt National Design Museum and Assouline Publishing.

Stone, B. (2016). Playing around in science: How self-directed inquiry benefits the whole child. *International Journal of the Whole Child, 1*(1), 1–10.

U.S. Department of Education. (2019). *Innovation spotlights: Nine dimensions for supporting powerful STEM learning with technology*. USDE. https://eric.ed.gov/?q=interdisciplinary+stem&ff1=dtySince_2017&ff2=pubInformation+Analyses&id=ED603418

Yerrick, R., & Ridgeway, M. (2017). Culturally responsive pedagogy, science literacy, and urban underrepresented science students. *International Perspectives on Inclusive Education, 11*, 87–103. doi:10.1108/S1479-363620170000011007

Zoltowski, C. B., Oakes, W. C., & Cardella, M. E. (2012). Students' Ways of Experiencing Human-Centered Design. *Journal of Engineering Education, 101*(1), 28–59. doi:10.1002/j.2168-9830.2012.tb00040.x

ENDNOTES

1 Oregon MESA is part of MESA USA and is housed at Portland State University. The activities described in this chapter apply exclusively to the Oregon MESA program unless otherwise indicated.

2 The data used for the summative evaluation covered years prior to the implementation of the Invention Toolkit due to data availability during the study and the long-term outcomes of focus. Analysis needs to be replicated to identify changes in summative student outcomes based on the use of the toolkit.

3 Some of these students participated in MESA for one year (71%), and others participated for two (21%) years or three years (8%).

Chapter 5
Supporting STEM Career Knowledge and Interest for Rural Grade Two Students Through a University–School Partnership

Stephanie C. Playton
Longwood University, USA

Fanni L. Coward
Texas Tech University, USA

Rebecca L. Hite
https://orcid.org/0000-0002-6275-3804
Texas Tech University, USA

Gina M. Childers
Texas Tech University, USA

ABSTRACT

STEM career information, hands-on content, and recruiting professionals to interact with students are challenging, especially within under-resourced areas. This chapter, within the STEM during the school day section, reports on a research-practitioner partnership (RPP) between a university and a rural school district. This collaboration resulted in a co-created curriculum to improve student (aged 7-8 years) knowledge of and interest in STEM careers. Among the 113 participants, 67 students participated in four modules with STEM content for up to five hours over 16 weeks. Observations, artifacts, and data from a STEM career metric were collected, including those who did not participate in the intervention (n=46). Analyses suggest that participants were more able to name STEM careers and make connections to content areas. Lessons learned are to monitor and pivot when students are struggling with technological challenges or lesson fatigue. Recommendations include devising ways to leverage an RPP and STEM professionals in the community to support STEM learning within elementary schools.

DOI: 10.4018/978-1-6684-7771-7.ch005

INTRODUCTION

This chapter describes a partnership between a University PK-12 STEM outreach center and a rural school district to grow grade 2 students' nascent knowledge (awareness) and interest in STEM learning and STEM careers in a diverse and rural school context. The described *researcher-practitioner partnership* (RPP) in this chapter engaged students in a curriculum on STEM related content which afforded them to interact with STEM experts in the field that connected to formal science content learning. The objective of this chapter is two-fold: 1) to describe the curriculum development process that transpired within the partnership such that other educators may replicate and implement the model for their own STEM partnerships; and 2) to demonstrate effectiveness of the current partnership's curriculum based on Social Cognitive Career Theory (SCCT) through the research conducted. The chapter's research was coordinated for both virtual and face to face classroom settings during the 2020-2021 school year and intended to address a need and a gap in rural communities for collaboration on best practices in both developing, measuring, and implementing STEM learning opportunities for lower elementary aged STEM students. To that end, the present research sought to address the following: to describe in what ways (if any) would an RPP-developed STEM curriculum intervention with STEM content and relevant career experts improve middle childhood aged learners' (i.e., ages 7-8 in grade 2) knowledge of and interest in STEM careers. Further, to what extent, if any, did participating grade 2 students report having STEM career aspirations?

BACKGROUND

Despite STEM jobs being on the rise in the U.S. (Bureau of Labor and Statistics [BLS], 2019; National Science and Technology Council [NSTC], 2018; Noonan, 2017), participation of all racial, ethnic, gender, and geographic groups in STEM careers falls short despite a growingly diverse U.S. population (Fouad & Santana, 2017). Federal financial resources like the Council of State Governments (2017) and the United States Department of Education (n.d.) have been actively funding STEM interventions, including activities to grow students' knowledge of STEM careers, to help close the U.S. STEM job participation gap. Historically, most of these efforts have been performed in informal and/or secondary education learning environments (van den Hurk et al., 2019). Presently, there is a shift towards funding STEM interventions at the elementary level (i.e., grades K-5) as research suggests students' perceptions of STEM form in middle childhood age (Kurz et al., 2015; National Academies of Sciences, Engineering, and Medicine, 2021; Playton et al., 2021; Schelegel, 2019), to address the ongoing STEM workforce need.

The impetus for this change from secondary to elementary level STEM interventions is to address the shortage in the workforce and was addressed in the United States' report and 5-year plan for the vision and goals towards STEM strategies. Interventions that take place among elementary aged students may provide a novel opportunity for youth to garner preliminary knowledge (awareness) of required skills and social and fiscal benefits of STEM careers (Bandura, 1998, 1999, 2005; Lent, 2013; Lent et al., 1994); herein called *STEM career knowledge*. Prior research suggests formal educational experiences that embed career information with the curriculum (STEM content and skills) had positively influenced upper elementary and middle grade students' knowledge of and interest in STEM careers (Colston et al., 2017; Kier, 2013; Kier & Blanchard, 2020; Kier et al., 2014; Li et al., 2010; Playton et al., 2021).

To craft such experiences, educators need support in creating learning opportunities in STEM, embedded with information on STEM careers. This support can be provided by STEM outreach institutions of higher education, who hold the knowledge and expertise to co-construct with school partners. Notably RPPs can support educational reforms and quality STEM experiences (McClure et al., 2017); therefore, this chapter presents the findings of a study in which an RPP co-created age-appropriate curriculum with an engaging instructional strategy (i.e., Just-In-Time Teaching) for middle childhood age (grade 2) students. Replete with assessments, the RPP evaluated changes in students' STEM knowledge and STEM interest, fostering their discovery of STEM career pathways concurrent with STEM content in four areas (i.e., matter, animal behavior and adaptations, weather, life cycles).

School District and University Partnership

Across the U.S., RPPs are informing the planning and development of classroom curriculum, pedagogies, and policies (Baker, 2003) to improve K-12 education (Coburn & Penuel, 2016). A University outreach entity, Longwood University's Institute for Teaching through Technology and Innovative Practices (IT-TIP), serves as a fiscal and administrative agent for a consortium of school districts in rural Virginia and partners with school districts for provision of STEM resources and local STEM experts to meet State mandates towards integrating STEM into the PK-12 classroom. To address this concern, ITTIP contacted a rural school district to gauge interest in partnering for the 2019-2020 school year to implement a program to grow grade 2 students' knowledge of STEM careers.

Theoretical Framework

A prominent framework for studying career interests is the Social Cognitive Career Theory (SCCT), a theory that stems from Albert Bandura's social cognitive theory. This framework emphasizes individual factors such as self-efficacy beliefs, outcome expectations, and personal goals, as well as factors among the interactions of people and their environments; as a result, SCCT is able to model one's career interests, career development, career choices, and career satisfaction (Bandura 1986, 1999; Lent, 2013; Lent et al., 1994). Hence, SCCT has been used across a diversity of populations (Lent, 2013), including rural (Ali & Menke, 2014; Kier, 2013) and historically underrepresented populations (e.g., racial, minority, gender) in STEM (Ali & Menke, 2014). When SCCT was initially conceptualized, the intent was to examine adolescence and adulthood career development (Lent et al., 1994). After the inception of SCCT, Lent (2013) advocated that early preventative applications can be utilized by providing diverse career considerations to children and adolescents, when options might appear narrow, providing a course of nurturing and strengthening their career-relevant interests over time. This expansion of SCCT suggests that one's interest in a career can be generated as a youth and is consistent with past research that having an early interest in STEM areas is a predictor for later learning and/or eventual career interests and choices (Blotnicky et al., 2018; DeBacker & Nelson, 1999; Jones et al., 2021; Nugent et al., 2015).

While there are multiple models of SCCT, the core constructs of SCCT found to influence one's career interest are: *self-efficacy, outcome expectations,* and *personal goals.* Self-efficacy can be defined as a set of beliefs and confidence levels regarding personal abilities that can be changed based on experiences and the negative or positive responses to those conditions. Lent (2013) described that self-efficacy can be strengthened or weakened through successes and failures and shapes one's beliefs about specific career domains, but self-efficacy does not solely determine one's career path. Lent contrasted outcome expectations as one's "personal beliefs about probable response outcomes" (p. 83), which affect one's perceptions of self-efficacy. For instance, a person might understand that a career as a lawyer may offer financial stability (i.e., positive outcome expectation) or value, but avoid academic and higher-education choices by doubting their success in speaking skills (i.e., low self-efficacy) or fears the negative reactions from social persuasions (like family and peers) engaging in that profession. In regard to *personal goals*, goals can help organize, direct, and sustain progression towards a person's behavior (Lent et al., 1994). Further, gaining or lacking progress towards these goals influence both self-efficacy and outcome expectations (Lent, 2013; Lent et al., 1994). Nevertheless, personal goals related to career development for middle childhood learners tend to be more provisionally oriented (without a plan), such as, *someday I could be an engineer*. Based on SCCT, providing a learning environment that can positively influence self-efficacy, outcome expectations, and personal goals can lead to STEM career knowledge and interest. Currently, most of the work done with SCCT examines career development in late adolescence and early adulthood (Lent & Brown, 2017), leaving a gap to explore potential career pathway development in middle childhood age. While there are five versions of SCCT utilized across research (Lent & Brown, 2017), this chapter would use the *Interest Model* with connections to their *Career Self-Management Model.*

SCCT Career Interest Model

SCCT suggests that career interest is nurtured or deterred. For example, Lent (2013) described that interest (or disinterest) may be determined through competency and valued outcomes and that interests can encourage goals and/or change self-efficacy and favorable outcomes through feedback. Furthermore, it is believed that any shifts in interests are significantly due to the SCCT constructs, self-efficacy and outcome expectations. However, career interest may also be influenced by social contexts or environmental conditions including one's race (Ali & Menke, 2014), gender (Inda et al., 2013), and socioeconomic status (Ball et al., 2017; Flores et al., 2017; Yerdelen et al., 2016). Provided that such social and environmental influences can affect the development of interests, they could also influence career choices in relationship to middle childhood learners.

When using the Interest Model to analyze the efficacy of STEM learning experiences on career interest, it is important to provide activities relevant to middle childhood learners (Lent et al., 1994) that are connected to content (Heisey & Kucan, 2010; Playton et al., 2021; Royce et al., 2012) and nurtured through positive value outcomes (Lent, 2013) that teachers, classmates, and parents often provide young learners. As social contexts are important to this model, it is beneficial to focus on these needs to help address the disparities that are currently represented in the STEM fields (Fouad & Santana, 2017). A more comprehensive model, the *SCCT Career Self-Management Model*, is warranted.

SCCT Career Self-Management

This model, by Lent and Brown (2013), focuses on the process of career development and how one adapts to career tasks throughout the life-span. This model leverages the same constructs of SCCT (i.e., self-efficacy, outcome expectations, personal goals) to understand the behavioral outcomes that support one's progress towards an educational or occupational objective. This model does this by including contextual and personality factors as well as the role of one's interests and abilities. Contextual factors could include access to resources or the feedback of others, while personality factors could include the willingness to have experiences or social interactions. The authors suggest this model does not directly focus so much on interest, but includes *adaptive behaviors* noted as "to be developmental presses (e.g., the need to make a decision), personal goals (e.g., career advancement intentions), or environmental considerations (e.g., coping with adverse work events) than by activity interest per se" (p. 563). The abilities factor encompasses the cognitive and interpersonal talents related to the age-appropriate adaptive career behaviors. For instance, ability could be the efforts one puts forth to succeed while enacting a certain behavior (e.g., interviewing for a job). Since this SCCT model focuses on the process of career development, it addresses the developmentally appropriate adaptive behaviors from young learners up through retirement age, providing a framework for understanding the relationship to middle childhood learners.

The Career Self-Management Model is vital to the present study as a means to study age-appropriate career behaviors through a developmental period. For middle childhood aged students, considered the *growth period*, age-appropriate behaviors include social learning, decision-making and the fostering of preliminary interests through opportunities (Lent & Brown, 2013). Concurrently, students also gain developmentally appropriate knowledge (like awareness) of diverse career considerations (Lent, 2013) through learning experiences that connect to classroom content (Heisey & Kucan, 2010; Royce et al., 2012). While the next stage of this model occurs during adolescent age, the *exploration period*, Lent and Brown (2013) recommend career behaviors should include exploring career paths and even acquiring experiences through volunteer or part-time work in the related career. Here is where career decisions begin to take shape in the form of choosing elective courses or training, and transitioning into the workforce; however, if students lack STEM career knowledge and viable options at middle childhood age, it would be difficult to base a decision on elective courses for career pathways as young teens, which underscores the importance and relevance of the present research study.

In sum, SCCT constructs with attributes of SCCT Career Interest model and the SCCT Career Self-Management model provide a blueprint from which to develop a STEM career intervention in the current study for middle childhood aged youth learning within the STEM classroom environment.

Literature Review

When developing curriculum for middle childhood learners, it is important to make explicit connections to STEM content that is age-appropriate (Colston et al., 2017; Heisey & Kucan, 2010; Playton et al., 2021; Royce et al., 2012), while providing an inclusive representation of STEM careers (Lydian Hill Foundation & Geena Davis Institute on Gender in Media [LHF/GDI], 2018; van Tuijl & van der Molen, 2016). Content-specific interest for middle childhood learners can be facilitated through use of compelling media (Royce et al., 2012), providing context to content (Heisey & Kucan, 2010), while avoiding stereotypical threats like representation of only one gender or race, that may dissuade female and minority youth (Deemer et al., 2014; LHF/GDI, 2018; van Tuijl & van der Molen, 2016). Providing an opportunity to learn about STEM careers through applicable content and context along with proper representation, while simultaneously connecting components of the SCCT and addressing age-appropriate adaptive behaviors, it is suggested that self-efficacy, outcome expectations, and goals could positively increase knowledge and interest in STEM careers by readdressing these issues within the curriculum.

Knowledge gains for middle childhood learners in STEM careers have been found beneficial to students when interactions with STEM experts are relatable (Colston et al., 2017), which also apply to high school (Angle et al, 2016) and middle school students (Blotnickey et al., 2018; Salonen et al., 2018; Yerdelen et al., 2016). Additionally, per van Tuijl and van der Molen (2016), a lack of knowledge of career requirements for these age ranges (i.e., K-12) suggests that curriculum should include elements of the educational background and skills needed for STEM careers.

While interest can be described as a motivational variable (Frenzel et al., 2010), it is also considered a latent variable that needs to be observed through multiple measures (Ainley, 2017) and sub-measures (Lent et al., 1994). For middle childhood learners, this means addressing situational factors of interest like technology and open-ended activities (Jung et al., 2019) or social influences, like their classroom peers and family members (Hernandez-Hernandez et al., 2016). In addition, situational factors can provide knowledge about STEM careers, like highlighting the verbal skills needed for STEM jobs (Kisselburgh et al., 2009). Pursuing jobs that demand high verbal skills is especially important for girls when choosing a career (Wang et al, 2013). In addition to making sure middle childhood students' STEM career experiences are facilitated through age-appropriate situational interest factors, content-specific interests and values can be addressed. Compelling media can be integrated (Royce et al., 2012) and provide context to content (Heisey & Kucan, 2010), while avoiding stereotypical threats (Deemer et al., 2014; LHF/GDI, 2018; van Tuijl & van der Molen, 2016) creating an opportunity to increase knowledge and facilitate STEM interest; furthermore, additional factors for middle childhood learners can include intrinsic values (Ball et al., 2017) that can be affected by parental expectations (Thomas & Strunk, 2017).

Curriculum

In order to develop a curriculum map and scope and sequence for a STEM career intervention, the current district and state guides for grade 2 STEM activities were analyzed. Glatthorn (1999) suggested that curriculum alignment begins with first analyzing the written curriculum for three types of outcomes. The first type of curriculum, *mastery curriculum*, is the curriculum that is based on state standards that are tested. In Virginia, there are Standards of Learning (SOL) that define essential learning for students at a certain grade level (Virginia Department of Education [VDOE], n.d.a.). In contrast, *organic curriculum* refers to the curriculum that is not directly tested. This curriculum includes general learning goals that encompass what schools hope to accomplish through elementary and secondary education experiences (Glatthorn, 1993). In Virginia, the organic curriculum can be compared to the State's initiatives of *Profile of a Virginia Graduate*, describing the knowledge, skills, experiences, and attributes that support students in being *life ready* upon graduation (VDOE, n.d.b.). Notably, the Profile of a Virginia Graduate framework focuses on developing students' citizenship through critical thinking, creative thinking, collaboration, and communication (per the Battelle for Kids (2019) Partnership for 21st Century Learning framework). Last, Glatthorn (1999) described *enrichment curriculum* as a curriculum that is not essential but may be of great interest to students. Glatthorn (1999) classified the enrichment curriculum as a judgment decision, meaning it provides an opportunity to connect classroom learning to real-world (and thus, interesting) applications and skills.

Virginia's current frameworks for K-12 education (e.g., standards, teaching framework and life ready experiences) suggest there can be career connections, but lack recommendations for specific activities that task students with exploring career connections in middle childhood years (i.e., ages 6-11). To redress this gap, the RPP sought to co-create a grade 2 curriculum intervention that leverages specific strategies to foster students' interest such to grow their knowledge of and interest in STEM careers.

After analysis of the current curriculum, further examination was enacted to ensure the updated curriculum for grade 2 aligned with the goals and objectives of the proposed intervention to state/local standards. A *Curriculum Map* (in *Supplemental Material*, Appendix A) was created for the RPP and based on the science curriculum which was iteratively aligned to meet teachers' classroom needs. The original curriculum map intended for in-person interventions to begin in Fall of 2020 and did not anticipate the changes in teaching modality to distance and hybrid learning facilitated by the COVID-19 pandemic. The school transitioned back from virtual to in-person instruction in January of 2021. Therefore, in the spring of 2021, iterations to the curriculum map within the RPP to review science content. Instruction implementation was intended to begin at the start of school year in the Fall of the 2019-2020; however, student participants and families had the choice of face-to-face instruction two-days a week with virtual instruction on Wednesdays, or five-days of virtual instruction. The instruction fluctuated to fully virtual in November 2020 and again in January 2021 due to their county's matrices of COVID-19 positive cases. This delayed the implementation until February 2021. In sum, the curriculum map and *digital communication to teachers* (in *Supplemental Materials*, Appendix B) reflects the theoretical framework selected to co-design a curriculum that leverages attributes of students' interest in STEM and STEM careers.

The RPP strived to retain STEM experts who represented diversity in gender and race so that all students had an opportunity to feel represented (see LHG/GDI, 2018; van Tuijl & van der Molen, 2016), which is reflected in Table 1. From February 2021- May 2021, participating grade 2 students had opportunities to interact with four scientists related to STEM content areas.

Table 1. Demographics of STEM career experts

Expertise	Location	Gender	Race
Physicist; Graduate Research Student	West Virginia University, Morgantown, WV	Female	White
Wildlife Research Biologist and Ecologist; Associate Professor	Mississippi State University, Starkville, MS	Male	White
Meteorologist; Local Television Station	Richmond, VA	Female	African American
Zoologist; Zoo	Norfolk, VA	Female(s)	White

Pedagogy

A teaching pedagogy that motivates students and fosters deeper learning is the Just-in-Time Teaching (JiTT) strategy (Andrea & Cupita, 2016; Novak, 2011). The basic cycle of the teaching strategy can be explained as 1) inclusion of pre-instruction questions for students to provoke interest, sometimes with limited background knowledge; and 2) the teacher reviewing students' answers to gain insight on how to integrate new learning into the upcoming lesson, helping to foster relevant learning (Novak, 2011). In JiTT, teachers, students, and even parents can become a *team of learners* as the cycle becomes a feedback loop for teaching and learning and often happens through both in-class and out-of-class experiences.

Vicarious experiences feed perceptions of self-efficacy (Bandura, 1998) by providing young learners the opportunity to talk with experts about their STEM fields/careers resulting in positive outcomes (Colston et al., 2017; Playton et al., 2021; Wilson et al., 2010). The RPP anticipated connecting self-efficacy to adaptive behavior for middle childhood learners by having students: 1) use decision-making skills to determine what questions would be asked to an expert during interactions with their class (i.e., teachers and students) and 2) practice their social skills interacting with the STEM expert to gain knowledge of what the job entails and how the student's classroom learning compliments the work done in that career (Lent & Brown, 2013). Other benefits for providing young learners with decision-making in learning could include creating a more open-ended learning environment (Jung et al., 2019). This kind of intervention could also include affordances for middle childhood learners, like in learning context (Heisey & Kucan, 2010; Royce et al., 2012; Playton et al., 2021) where students can connect their learning intrinsically. This attribute is important, because ideas on potential careers originate at a young age, signs of career disinterest have been identified in children as young as 6-8 years of age (Playton et al., 2023).

In sum, JiTT provides strategies for youth to develop behaviors that include social and decision-making skills, as a team of learners with their classroom teacher. Students not only need the opportunity to connect with STEM careers to build their own knowledge (Lent & Brown, 2013; Yerdelen et al., 2016) but also need to have those opportunities connected to a real-life learning context (Heisey & Kucan, 2010; Playton et al., 2021; Royce et al., 2012). Figure 1 represents an overall summary of the STEM curriculum (using SCCT) and pedagogy (JiTT) for the intervention.

Figure 1. Summary of STEM curriculum

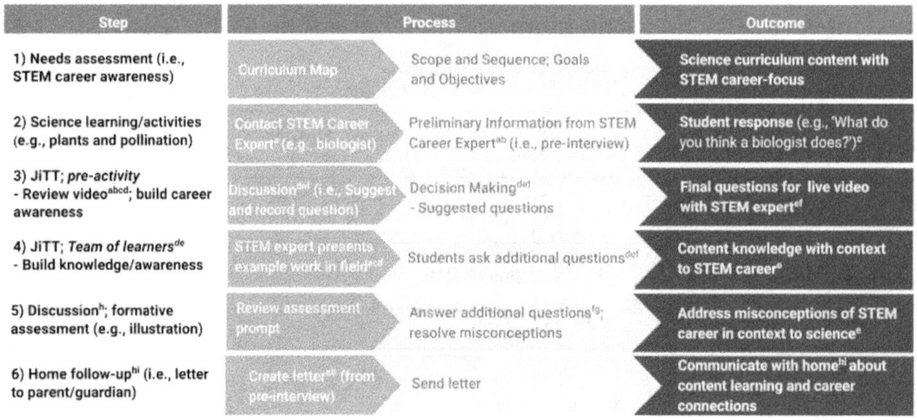

MATERIALS AND METHODS

Two elementary schools had participated in the research study; the first school contributed 67 students (as the treatment group) and a second school contributed 46 students (as the control group). Both schools were rurally located, served low-wealth students (i.e., approximately 91% received free or reduced lunch) and majority minority (63% and 60% identifying as non-white, respectively) populations. Participants were attending school either virtually or through blended learning experiences (face-to-face and virtual). This research study was approved by TTU in IRB 2020-99 on July 28, 2020 as a Category 1 exemption titled "A STEM Career Intervention for Primary Learners" and is a part of a larger set of studies on STEM career knowledge (Playton, 2021).

Co-Curriculum Design Process

Together, all partners, including university researchers, the school district's assistant superintendent, a community project manager, and two elementary school principals, developed an RPP agreement. The first step of the partnership was to conduct a needs assessment to be shared with community stakeholders regarding growing youths' STEM career knowledge. In order to develop the curriculum intervention, the assessment in a topic related to STEM career knowledge and interest was adapted from a community-engaged needs assessment (see Hite, 2020). A copy for use is provided in Supplemental Material: Appendix A, *Curriculum Needs Assessment for STEM Career Knowledge in Elementary Schools*. While administered in September of 2020, the curriculum assessment did not merit informed responses from community partners (specifically, teachers), as they themselves may not be knowledgeable of STEM careers (Javidi & Sheybani, 2017; Nadelson et al., 2009; van Tuijl & van der Molen, 2016).

Intervention

After each STEM expert agreed to participate, Grade 2 teachers posted an assignment using their online learning platform. Students were prompted to reflect on describing the specific STEM job they would learn more about during a live discussion. For instance, if students were learning about matter, they may be prompted to answer, 'What do you think a physicist does?'. The assignment required students to use a digital platform where students recorded a video response to a prompt, then were briefly introduced to a STEM professional through a pre-recorded video that provided background about a STEM career, as well as the types of skills needed for that career (in *Supplemental Materials*, Appendix C, *STEM Expert Pre-Interview*). The pre-recorded videos served as a way to provoke students' interests through media (see Jung et al., 2019; Royce et al., 2012). Since the intervention occurred during the 2020-2021 school year and was affected by the COVID-19 pandemic, students had a choice for remote or in-school learning at the time of the interventions. This meant that some students aggregated in small groups when in a classroom setting, or independently if remote learning, to respond to the prompt and watch the introduction video that included information about a STEM career expert's schooling, communal supports, and required skills, while explicitly relating content to the science curriculum. This information was important, because students often lack knowledge of the wide range of communal contributions (Diekman et al., 2010; Fuesting et al., 2017) and required skills (Angle et al., 2016; Kisselburgh et al., 2009; Salonen et al., 2018) that STEM careers encompass and embrace to foster knowledge and interest in STEM careers. Gender equity is a high priority for the STEM workforce (NSCT, 2018) and increasing knowledge of STEM career skills, like collaboration and communication (e.g., verbal skills), could encourage more females to pursue these fields (see Wang et al., 2013), so STEM career experts were asked to answer questions related to those skills. After viewing the introduction video, the students were asked to decide on a question to ask the STEM career expert during an upcoming live video conference. Next, the students were instructed to record their question through a video feedback platform. Because there was a reasonable number of participation in the voting and recorded questions (about 20-25 questions per session), all questions were vetted by the Researcher to be used during the video conference with the STEM professional, with priority to questions that could be asked by the student(s) themselves, synchronously with the virtual event. This pre-activity served as the 'warm-up' component of the JiTT strategy (Novak, 2011) and helped shape the live lesson with the STEM professional. The pre-activity also provided an opportunity for the students to be part of the decision making of what knowledge could be gained during the learning experience, an age-appropriate adaptive career behavior (see Lent & Brown, 2013). The integration of technology and open-endedness (e.g., decision making) in the activity also supported middle childhood learners' situational interests (Jung et al., 2019), making this pedagogically appropriate when trying to motivate (i.e., interest) students to learn. In addition to creating opportunities for students to make decisions about the interaction with a STEM career expert, the students also practiced social learning skills, another age-appropriate adaptive career behavior (Lent & Brown, 2013).

During video conferencing with a STEM career expert, students were either in their classroom or remote learning from home; however, the list of the students' questions were provided to the teacher and shared with students so they would know who would be asking questions during the video conference. This provided an opportunity for the teachers and students to become a team of learners, the second component of the JiTT strategy (Novak, 2011). During the live video conference, the STEM professionals presented to the class about their work and how it related to content they were learning in science and other subjects. Students would then take turns asking additional questions, providing an opportunity for students to practice their social skills while gaining knowledge about a STEM career professional, another age-appropriate behavior for middle childhood learners (Lent & Brown, 2017). Next, the JiTT lesson closed with the opportunity to gain some understanding of how the students expanded their knowledge about the STEM career and inform connections made to classroom learning objectives.

After interaction with the STEM career expert (i.e., video conference), students were prompted to draw a digital picture to illustrate a person actively engaged in the career that was explored, with relevance to the current science content (Appendix C). Student illustrations have been used as a relatively engaging assessment process for students that can be used to examine how middle childhood learners make sense out of ideas like social class (Howard et al., 2018), environmental perceptions and situations (Barraza, 1999), and conceptual understanding of science topics (Anderson et al., 2014), as a formative assessment tool. After the illustrations were complete, classes were encouraged to discuss the video conference and clarify any misconceptions or help answer any remaining questions. This was important, because teachers (van Tuijl & van der Molen, 2016) and classmates' views (Hernández -Hernández et al., 2016; van Tuijl & van der Molen, 2016) are sources of middle childhood learners social persuasion and self-efficacy, but it was unknown if it was made part of the learning experience for the in person and/or virtual students; however, there are other sources of social persuasion and self-efficacy for middle childhood-learners.

Parents and other relatives are also sources of social persuasion and self-efficacy for middle childhood learners (Hernández-Hernández et al., 2016; Thomas & Strunk, 2017; van Tuijl & van der Molen, 2016), including influencing expectancy value factors of STEM subjects (Thomas & Strunk, 2017). Parents may not be knowledgeable about STEM career fields (van Tuijl & van der Molen, 2016); therefore, a take-home resource guide was created to provide information on the STEM careers to be shared with parents and/or guardians (Appendix C) on students digital learning platform and for take-home.

Data Collection

To measure potential impact of the intervention, data were collected to describe curriculum efficacy via formative (qualitative) and summative (quantitative) metrics. Hence, the JiTT data collection occurred in the form of a descriptive case study (Yin, 2018). Qualitative data included artifacts of pre-knowledge (Data Collection A), observations of students' interactions with the STEM professional (Data Collection B) and student artifacts (e.g., digital pictures and text, Data Collection C). Quantitative data were collected through using a survey, the STEM-Future CIS, to provide descriptive statistics of students' nascent STEM career knowledge from before (pre) and after (post) the intervention.

Observations

Observations were conducted of students during the live video conference with a STEM expert using a video screen recording tool. Data was also collected by using a video screen recording tool regarding verbal interactions (e.g., questions and statements) directly between the STEM career expert and students asking questions. During that time, descriptives were also recorded as a means to document the setting students were interacting with during the learning experience (e.g., How many students appeared to be in a virtual location?) shown in Table 2. After observations of the interactions between students and STEM experts, data collection continued via student artifacts.

Table 2. Student participation by science topic and expertise during each live video call

Expertise	Science Context	# of Students	Questions	Descriptives		
				M/F/U	W/M/U	C/O/U
Physicist; Graduate Research Student	Forms of Matter	30	10	5/12/13	6/11/13	10/7/13
Wildlife Research Biologist and Ecologist; Associate Professor	Animal Behavior and Physical Adaptations	36	8	3/5/28	5/3/28	15/11/10
Meteorologist; Local Television Station	Weather	30	9	0/0/30	0/0/30	0/0/30
Zoologist; Zoo	Life Cycles	38	9	0/0/38	0/0/38	0/0/38

Note. Descriptives are described as the apparent observations during the live video calls where additional data was recorded and abbreviated as follows: Male (M), Female (F), White (W), Minority (M), Classroom (C), Other (O), Unknown (U).

Artifacts

Data was collected from student artifacts in several ways and additionally served as a pre-assessment and formative assessment for the teachers. First, before suggesting a question for the scientist or watching a video introducing the scientists, students were asked to answer the question, 'What do you think a (scientist job) does?'. Next, after watching the introduction video, students recorded video questions/ responses from an online platform where part of the data was collected and used to determine the video conference questions. Collection method of this pre-activity (i.e., scientist questions) changed as requested by the lead teacher. After the first intervention, grade 2 began skipping the pre-assessment (i.e., 'What do you think a (scientist job) does?') and typing their questions (i.e., formative assessment) into a collaborative spreadsheet for the remaining pre-activity instead of recording it online through video means. The second artifact that was collected included illustrations and accompanying text from student digital journal entries to help determine how the students translated their knowledge and interests (or disinterest) of the STEM career after the live interactions. Next, the quantitative data are described.

Survey

Students completed the validated STEM Future-CIS pre- and post-intervention (Playton, 2021), serving as the summative assessment. Data from the survey provided descriptive information on career aspirations, as well as STEM career knowledge (i.e., being able to name a STEM career). Interest was also measured across four constructs (science and math interest, technology interest, engineering interest, and future-self) and was analyzed through a Mann Whitney U with a two-sided level of significance ($\alpha = 0.05$) as the data met the assumptions of a non-parametric test.

Validity

Providing validity shows how the "trustworthiness, utility and dependability" of the research data and its interpretation (Zohrabi, 2013, p. 258). Qualitative data, across data sources, were analyzed to identify common themes and categories to substantiate findings (Creswell et al., 2003). Regarding quantitative data, the one-group pretest–posttest pre-experimental design has been criticized for not addressing threats to validity including events (history), time (maturation), and/or even sensitization to the treatment (Knapp, 2016); therefore, by collecting survey data from a second elementary school with similar demographics from the same district (that had not participated in the JiTT intervention) helps to mitigate this issue (Stanley & Campbell, 1980) of uncontrolled variables (e.g., history, maturation).

Trustworthiness and Reliability

Data collection must include measures of *reliability*. Reliability suggests that replicability could occur with identical results, which is difficult to do with qualitative data in a narrative form and more subjective, as opposed to quantifiable data in numerical form (Zohrabi, 2013). Once data were collected for the intervention, the qualitative data were analyzed and coded through emerging codes to help answer the research question. In order to obtain reliable coding of the qualitative data set for Data Collection A (i.e., student responses related to the career prompt) three researchers coded the interventions independently, with total (100%) agreement. Next, data on the questions students either predetermined or communicated during the live call was coded by two researchers (i.e., Data Collection B). Since initial coding did not have high agreement (i.e., 76%), a conference method was initiated. Then the codes were finalized by the author of the code book (Figure 2). The remaining data was addressed by two researchers and coded, but also did not have high agreement for Data Collection B (i.e.,78% and 74%). For instance, there was disagreement among the two coders where there was clearly a connection between the pre-video introduction (e.g., pictures of fishing) and the proposed questions (e.g., "How many fish have you caught?"). To measure intercoder reliability, percent agreement was calculated between coders on the data collected among all the interventions from the appropriate data set (i.e., Data Collection B) and a third researcher reconciled Data Collection B for 100% agreement. Data collection from student artifacts related to their digital picture and comments (Data Collection C) had three raters, meriting 92% agreement. Next, reliability statistics for the quantitative data are described. Notably, qualitative data (observations and digital artifacts) were intercoded as part of a larger data set (Playton, 2021), providing a way to report the reliability of multiple coders analyzing data (Creswell & Poth, 2018). The quantitative data included the reliability statistics, Cronbach's alpha, to determine levels of internal consistency of the sets of scale and test items (Field, 2013). The data on STEM career knowledge and aspirations were examined and coded as either a STEM or non-STEM job as defined by the BLS (n.d.) and included a STEM code if students listed subjects from the acronym (e.g., science, technology). Additionally, descriptive statistics were calculated to document STEM career knowledge and aspirations.

Figure 2. Code book

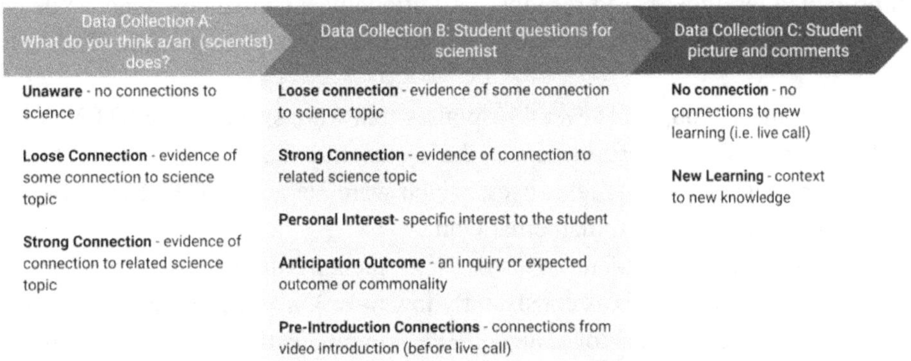

Data Analysis

To determine if there were differences in STEM career knowledge and interest between the treatment and control group, a Mann-Whitney U was selected (i.e., two independent groups, ordinal data, independence of observations) for the analysis of data (Field, 2013). The Mann-Whitney U was used to analyze each survey construct with a two-sided level of significance ($\alpha = 0.05$); however, since grade 2 did not have a control group for the pre-test, the treatment groups' pretest was used as a proxy for the control group pretest since neither grade 2 group, at the time of the pretest had received any treatment, and it is likely that the two groups were similar (albeit a limitation). To attribute students' growth in STEM career knowledge (by naming STEM careers introduced in the JiTT), a Chi-Square analysis was performed between treatment and control groups to determine if there was a difference between groups. Special attention was paid to the specificity (versus generality) of the STEM careers that students could name as visualization of their STEM career knowledge (per BLS, 2019).

RESULTS

Data Collection A intended to gain information on students' pre- knowledge of each STEM career. The beginning of the intervention included student video responses from the pre-introduction question (i.e., *What do you think a physicist does?*) and had a low response rate (*N*=5). When the participation results were discussed, the practitioners suggested that using (another) unfamiliar digital tool paired with teaching both virtual and face-to-face students added to the challenges of implementation, resulting in excluding the video responses from the remaining interventions as requested by the practitioners. This portion of the intervention and data collection was implemented previous to the students meeting only the physicist in context to their current science topic, *matter*, where most of the students (4 out of 5) made only general connections to science, or *loose connections*, as a career using knowledge directly related to (forms of) matter (e.g., '*I think a physicist predicts something and then tries it until it is right.*'). Only one student had made a strong connection to the science topic before reviewing the introduction video (i.e., '*I think a physicist is someone that.. ummm... a scientist that learns about matter.*'). This lack of knowledge in STEM career connections was paired with results from the STEM Future-CIS, in which responses from the treatment group's pretest (N=9) indicated that students were unable to name a STEM career when prompted for a text-based response. Additionally, a Chi-square analysis determined there was a significant difference between the control and treatment groups regarding STEM knowledge, X^2 $(1, N = 80) = 3.89$,

$p = .048$; the treatment group had several responses that were specific to a STEM job, whereas the majority of which were STEM careers introduced during the intervention (i.e., meteorologist, zoologist). The control group produced minimal responses, which were names of STEM careers in the acronym (i.e., scientist, technologist, engineer, mathematician).

Data Collection C related to student artifacts. After meeting the first two scientists, participation for this activity among the students was consistently low ($n = 3$, $n = 2$), reflecting numbers of students participating in the classroom, in person. During RPP communication, the teachers suggested using a platform that the students had used in the past to eliminate learning a new digital tool. Therefore, a new tool (Kahoot) was used to present the same questions and drawing activity for the third intervention with a gamification element and resulted in 12 responses. While the third implementation merited more responses, when it came to the fourth intervention the lead teacher asked to eliminate that activity since it appeared to fatigue the students after engaging with a scientist, and suggested it would be more beneficial for teachers to participate in an informal discussion with their students. Among the 17 responses, 59% were coded as making no connections to the current science topic, while 41% of codes suggested new learning. Figure 3 provides examples of student work where new learning or no connections were made from the scientist to current science learning.

In the survey, students were asked what they wanted to be when they grew up. Career aspirations from the pretest can only be reported for the treatment group, but half of them aspired to a job in a STEM career (e.g., doctor, astronaut, vet, nurse). After the intervention, a Chi-square analysis determined there was no difference between the control and treatment groups on STEM career aspirations, X^2 $(1, N = 80) = 1.97$, $p = .160$; however, several students in the treatment group reported having career aspirations specific to careers they had learned about during the intervention (i.e., zoologist, meteorologist). Next, grade 2 STEM career interest results from both the intervention and additional pre-, post-survey data are described.

While RPP participating teachers reported struggles of teaching and learning a new digital tool with virtual and face-to-face students with Data Collection A, they also reported students had difficulty with video recording a question previous to meeting the first scientist (for Data Collection B) using the same tool. In this data collection, students could suggest a question that would be included during the live call with the scientist(s), after they viewed an introduction video. Alternatively, students typed their questions into an online spreadsheet for the remaining activities. Of the 110 questions coded from this data, the majority of the questions related to personal interest (36%) and information shared during the pre-introduction video (27%), followed by anticipation outcome (18%), making strong connections to the science content (14%), and loose connections (5%). Figure 4 provides examples of student responses and the corresponding codes for Data Collection B.

Figure 3. Drawing assessment

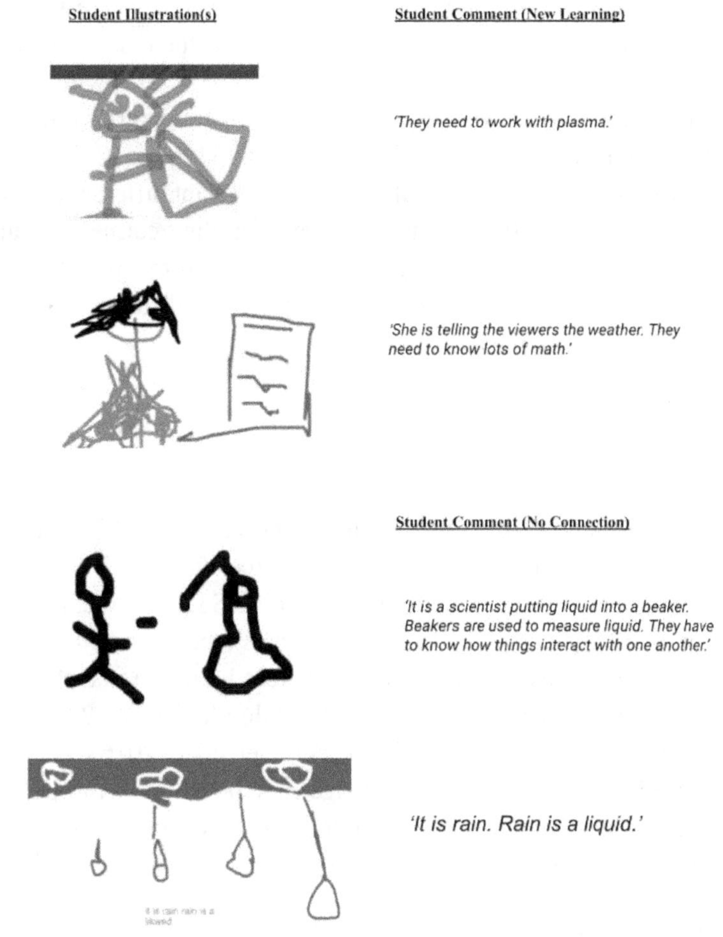

Figure 4. Examples and results of students coded questions

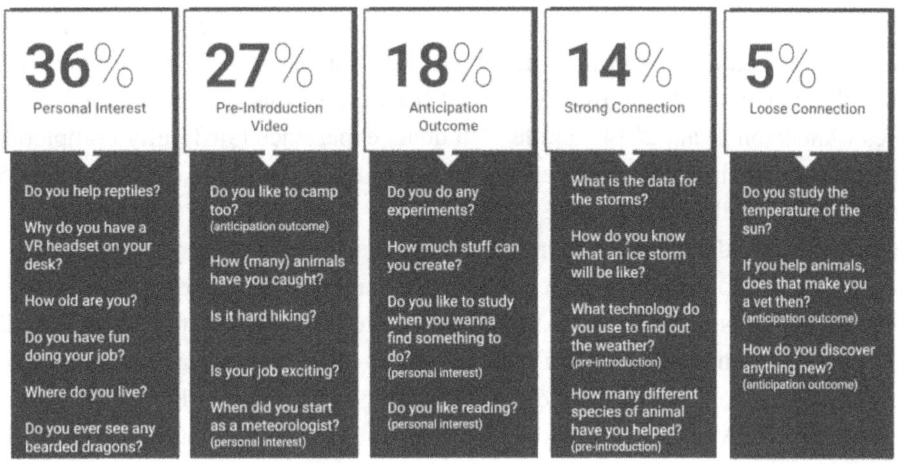

A Mann-Whitney U was used to compare data and determine if the two groups (control and treatment) were different after the treatment group received the intervention. The results suggested that Math and Science Interest levels in the treatment group (M = 55.44) differed significantly from the control group (M = 33.12) after the intervention, $U = 442.50$, $z = -4.13$, $p = .011$, with a medium effect ($\eta 2 = -0.45$). Next, Engineering Interest levels were examined in the treatment group (M = 50.83) and control group (M = 37.13), also finding a significant difference, $U = 627.00$, $z = -2.53$, $p = .011$, with a small effect ($\eta 2 = 0.03$). For the construct of Future Self, again a significant difference $U = 495.50$, $z = -3.67$, $p = .000$, with a medium effect ($\eta 2 = 0.40$), was found between the treatment group (M = 54.11) and control group (M = 34.27). Lastly, the Technology Interest level was examined for the treatment (M = 51.38) and control (M = 35.98) groups and also found a significant difference, $U = 574.00$, $z = -2.84$, $p = .005$, with a medium effect ($\eta 2 = 0.31$).

DISCUSSION

The outcome of this study evidenced that the JiTT intervention with STEM career experts for grade 2 students significantly improved their knowledge and interest in STEM careers. This growth was qualitatively evidenced by students being able to name specific STEM careers (post survey), describe their growing interests in STEM careers (observations, artifacts), and report having novel STEM career aspirations (in the post survey). These findings affirm current literature that there is a lack of STEM career knowledge within this age group (Angle et al., 2016; Blotnicky et al., 2018; Colston et al., 2017; Salonen et al., 2018; van Tuijl & van der Molen, 2016; Yerdelen et al., 2016). Furthermore, the present research extends literature by modeling how knowledge may be fostered through implementation of JiTT strategies, evidencing how students' career aspirations can be shaped with four interactions with scientists within a single school year.

Significant Knowledge Gains

Knowledge gains were found in the area of STEM careers suggesting a construction of new knowledge and connections to science content; this finding is supported by treatment group gains from pre-survey to post-survey (i.e., 0% to 23%) in being able to name a STEM job after the learning intervention and identify specific careers they had learned during the intervention. However, after meeting a scientist, participating students had a difficult time making connections as evidenced in their artifacts in Data Collection C. While drawings and illustrations have been used to analyze conceptual understanding of science topics (Anderson et al., 2014), grade 2 students experienced difficulty completing the digital drawing activity due to their underdeveloped fine motor skills (Cohen et al., 2021) or technology fatigue (Garfin, 2020), yet still yielded knowledge gains.

Furthermore, these findings affirm literature that suggests students as young as grade 2 benefitted from classroom interventions that connected the STEM content to a career context (Angle et al., 2016; Blotnicky et al., 2018; Colston et al., 2017; Salonen et al., 2018, Yerdelen et al., 2016). Additionally, the present study provides an understanding of how classroom experiences, which include speaking with a STEM expert and utilization of social and decision-making skills (Lent & Brown, 2017), are beneficial in increasing student STEM knowledge and fostering elementary STEM career interest (Playton, 2021; Playton et al., 2021). These findings support existing research studies that suggest the importance of recruiting diverse STEM professionals to engage with students (Deemer et al., 2014; LHF/GDI, 2018; van Tuijl & van der Molen, 2016). Additionally, the present study extends this literature by qualifying how to use JiTT strategies to provide students as young as grade 2 with learning experiences to broaden their STEM career knowledge and aspirations.

Improved Career Interest

Overall, grade 2 students had improved STEM career interest, which may be in part attributed to making connections across subject areas and skills (see Figure 3) within the JiTT intervention (i.e., pre-video, video call, letter home). In addition, this study revealed that participating students mostly posed questions to scientists that related to their own personal interest or from information they gained from the pre-introduction video. This finding affirms prior literature that students at this age benefit from learning in context (Schlegel et al., 2019; van Tuijl & van der Molen, 2016), building their own unique value of STEM careers (Hsu et al., 2017; Kisselburgh et al., 2009), and discerning community contributions (Feusting et al., 2017) of STEM professionals (Appendix C).

An interesting finding of this study regards how the intervention enabled students to make stronger connections to their science content after a pre-video and/or presentation from an expert by investigating their perceptions. For example, utterances with strong connections often related to the pre-introduction videos and anticipation outcomes often were related to personal interest (see Figure 4). However, the benefit of these perceptions and connections are related to SCCT modeling as personal interest is a variable in building other interests (i.e., STEM career). Now, discussion of the pre-, posttest data are summarized.

Limitations

There was a small sample ($N = 9$) of students from the treatment group that completed the pretest, but when using the treatment groups' pretest data as a proxy for a control, favorable outcomes were found. Key findings for grade 2 included how this intervention undisputedly influenced STEM career interest, despite the focus on only science careers. Students in grade 2 not only had increases in technology, engineering and math and science interest, but also in augmentation and their future self in STEM fields. These findings suggest an empirical benefit in developing interventions that support STEM career knowledge and interest by integrating the communities of STEM and a team of learners (i.e., teacher, parent, students).

There were also challenges; the participating Grade 2 students had difficulty making connections to their current science topic when asked what they knew about a STEM career; however, the data were limited to the first scientist, and career connections were not made explicit; drawing distinct and clear connections during instruction are warranted. Students were able to make stronger connections to content once they met their scientists in context to their science topic (i.e., Data Collection B). Furthermore, making connections from content to context is important (Angle et al., 2016; Colston et al., 2017) especially for grade 2 learners, and strengthens the assertion that youth learning is strongly influenced by social interactions (see Fox & Riconscente, 2008).

Recommendations

The success of RPP can be credited to frequent iterations between stakeholders, which were important to meet the needs of participating classrooms when implementing the co-designed STEM learning experience, further challenged with virtual and blended learning. Clear communication, lesson implementation roles and expectations, and scaffolding tools (e.g., video tutorials; Appendix B) facilitated meaningful classroom experiences to foster students' STEM knowledge and interest. Expansion of the RPP includes participation from additional elementary grade levels and school sites. Given the importance of including families and civic stakeholders in the STEM intervention, the RPP designed quarterly events hosted with STEM experts to serve the consortium of rural school communities to provide knowledge and experiences in a college setting for students and parents/guardians.

CONCLUSION

In conclusion, the schools, community, and outreach organizations (the RPP) were successful in their mutual effort to create and enact an intervention in which students had improved STEM career knowledge and interest, evidenced by students making connections to STEM content and naming relevant STEM careers, contributing to nascent STEM career interest and aspirations. In addition, SCCT constructs succeeded in modeling youths' personal interests and outcome expectations in STEM careers. Because this intervention was co-created for grade 2, this RPP was able to initiate and potentially expand students' career aspirations as they matriculate through grades 3 through 12. This success can be attributed to, in part, by leveraging the constructs of SCCT and utilizing curricular experiences that foster younger learners' interest (i.e., personal interests, outcome expectations) and relate content to context. By creating learning opportunities where cognitive competences can be cultivated and socially validated (Bandura, 1998), and include appropriate adaptive behaviors (Lent & Brown, 2017), we were able to broaden students' knowledge of and interest in STEM career fields.

The present research suggests that RPPs hold great utility in devising the cognitive capacities and developmental needs of students in this age range and as a partnership, were able to provide proper scaffolding techniques to assist with and assess student learning. Given that findings revealed gains in the post-survey after having interactions with four scientists, future research may wish to consider the effect/s of the number of scientists students interact with in regard to diminishing returns. Research could further explore the impacts of this type of JiTT intervention with different geographical locations (e.g., urban middle childhood students) to strengthen this study's internal and external validity. While gains in middle childhood aged students' knowledge and interest is not an immediate fix (i.e., of adolescent age ready to graduate and pursue career fields or college pathways), this study suggests middle childhood is an appropriate time to build STEM career knowledge and interest before students are of age that require choice in academic paths (in middle school).

REFERENCES

Ainley, M. (2017). Interest: Knowns, unknowns, and basic processes. In P. A. O'Keefe & J. M. Harackiewicz (Eds.), *The Science of Interest* (pp. 3–24). Springer International Publishing., doi:10.1007/978-3-319-55509-6_1

Ali, S. R., & Menke, K. A. (2014). Rural Latino youth career development: An application of social cognitive career theory. *The Career Development Quarterly*, *62*(2), 175–186. doi:10.1002/j.2161-0045.2014.00078.x

Anderson, J. L., Ellis, J. P., & Jones, A. M. (2014). Understanding early elementary children's conceptual knowledge of plant structure and function through drawings. *CBE Life Sciences Education*, *13*(3), 375–386. doi:10.1187/cbe.13-12-0230 PMID:25185222

Andrea, L., & Cupita, L. (2016). Just in time teaching: A strategy to encourage students' engagement. *HOW*, *23*(2), 89–105. doi:10.19183/how.23.2.163

Angle, J., Colston, N., French, D., Gustafson, J., O'Hara, S., & Shaw, E. (2016). Addressing the call to increase high school students' STEM awareness through a collaborative event hosted by science and education faculty: A how-to approach. *Science Educator*, *25*(1), 43–50.

Ball, C., Huang, K. T., Cotten, S. R., & Rikard, R. V. (2017). Pressurizing the STEM pipeline: An expectancy-value theory analysis of youths' STEM attitudes. *Journal of Science Education and Technology*, *26*(4), 372–382. doi:10.100710956-017-9685-1

Bandura, A. (1998). Self-Efficacy. In V. S. Ramachaudran (Ed.), *Encyclopedia of Human Behavior* (Vol. 4, pp. 71–81). Academic Press.

Bandura, A. (1999). Social cognitive theory: An agentic perspective. *Asian Journal of Social Psychology*, *2*(1), 21–41. doi:10.1111/1467-839X.00024

Bandura, A. (2005). The evolution of social cognitive theory. In K. G. Smith & M. A. Hitt (Eds.), *Great Minds in Management* (pp. 9–35). Oxford University Press. doi:10.5465/amr.2007.23467624

Barraza, L. (1999). Children's drawings about the environment. *Environmental Education Research*, *5*(1), 49–66. doi:10.1080/1350462990050103

Blotnicky, K. A., Franz-Odendaal, T., French, F., & Joy, P. (2018). A study of the correlation between STEM career knowledge, mathematics self-efficacy, career interests, and career activities on the likelihood of pursuing a STEM career among middle school students. *International Journal of STEM Education*, *5*(1), 1–15. doi:10.118640594-018-0118-3 PMID:30631712

Bureau of Labor and Statistics. (September 4, 2019). *Table 1.11: Employment in STEM occupations, 2018 and projected 2028.* United States Department of Labor. https://www.bls.gov/emp/tables/stem-employment.htm

Coburn, C. E., & Penuel, W. R. (2016). Research–practice partnerships in education: Outcomes, dynamics, and open questions. *Educational Researcher*, *45*(1), 48–54. doi:10.3102/0013189X16631750

Cohen, E. J., Bravi, R., & Minciacchi, D. (2021). Assessing the Development of Fine Motor Control in Elementary School Children Using Drawing and Tracing Tasks. *Perceptual and Motor Skills*, *128*(2), 605–624. doi:10.1177/0031512521990358 PMID:33496640

Colston, N., Thomas, J., Ley, M. T., Ivey, T., & Utley, J. (2017). Collaborating for early-age career awareness: A comparison of three instructional formats. *Journal of Engineering Education*, *106*(2), 326–344. doi:10.1002/jee.20166

Council of State Governments. (2017). *Federal funding for state employment and training programs covered by the workforce innovation and opportunity act*. Capitol Research.

Creswell, J., & Poth, C. (2018). *Qualitative inquiry and research design: Choosing among five approaches* (4th ed.). Sage.

Creswell, J. W., Plano Clark, V. L., Gutmann, M. L., & Hanson, W. E. (2003). Advanced mixed methods research designs. In A. Tashakkori & C. Teddlie (Eds.), *Handbook of mixed methods in social and behavioral research* (pp. 209–240). Sage.

DeBacker, T., & Nelson, R. (1999). Variations on an expectancy-value model of motivation inscience. *Contemporary Educational Psychology*, *24*(2), 71–94. doi:10.1006/ceps.1998.0984 PMID:10072310

Deemer, E. D., Thoman, D. B., Chase, J. P., & Smith, J. L. (2014). Feeling the threat: Stereotype threat as a contextual barrier to women's science career choice intentions. *Journal of Career Development*, *41*(2), 141–158. doi:10.1177/0894845313483003

Diekman, A. B., Brown, E. R., Johnston, A. M., & Clark, E. K. (2010). Seeking congruity between goals and roles: A new look at why women opt out of science, technology, engineering, and mathematics careers. *Psychological Science*, *21*(8), 1051–1057. doi:10.1177/0956797610377342 PMID:20631322

Field, A. (2013). *Discovering statistics using IBM SPSS statistics* (4th ed.). Sage Publications.

Flores, L. Y., Navarro, R. L., & Ali, S. R. (2017). The state of SCCT research in relation to social class: Future directions. *Journal of Career Assessment*, *25*(1), 6–23. doi:10.1177/1069072716658649

Fouad, N. A., & Santana, M. C. (2017). SCCT and underrepresented populations in STEM fields: Moving the needle. *Journal of Career Assessment*, *25*(1), 24–39. doi:10.1177/1069072716658324

Frenzel, A. C., Goetz, T., Pekrun, R., & Watt, H. M. G. (2010). Development of mathematics interest in adolescence: Influences of gender, family, and school context. *Journal of Research on Adolescence*, *20*(2), 507–537. doi:10.1111/j.1532-7795.2010.00645.x

Fuesting, M. A., Diekman, A. B., & Hudiburgh, L. (2017). From classroom to career: The unique role of communal processes in predicting interest in STEM careers. *Social Psychology of Education*, *20*(4), 875–896. doi:10.100711218-017-9398-6

Garfin, D. R. (2020). Technology as a coping tool during the coronavirus disease 2019 (COVID-19) pandemic: Implications and recommendations. *Stress and Health*, *36*(4), 555–559. doi:10.1002mi.2975 PMID:32762116

Glatthorn, A. A. (1993). Perspectives and imperatives - Outcomes-based education: Reform and the curriculum process. *Journal of Curriculum and Supervision*, *8*(4), 354–363.

Glatthorn, A. A. (1999). Curriculum alignment revisited. *Journal of Curriculum and Supervision*, *15*(1), 26–34.

Heisey, N., & Kucan, L. (2010). Introducing science concepts to middle childhood students through read-alouds: Interactions and multiple texts make the difference. *The Reading Teacher*, *63*(8), 666–676. doi:10.1598/RT.63.8.5

Hernández-Hernández, A. M., Viga-de Alva, D., Huerta-Quintanilla, R., Canto-Lugo, E., Laviada-Molina, H., & Molina-Segui, F. (2016). Friendship concept and community network structure among elementary school and university students. *PLoS One*, *11*(10), 1–17. doi:10.1371/journal.pone.0164886 PMID:27760171

Hite, R. (2020). *Curriculum needs assessment for formal settings* [EDCI 5321- Curriculum Theory: Design and Development]. Texas Tech University.

Howard, A., Swalwell, K., & Adler, K. (2018). Making class: Children's perceptions of social class through illustrations. *Teachers College Record*, *120*(7), 1–44. doi:10.1177/016146811812000704

Inda, M., Rodríguez, C., & Peña, J. V. (2013). Gender differences in applying social cognitive career theory in engineering students. *Journal of Vocational Behavior*, *83*(3), 346–355. doi:10.1016/j.jvb.2013.06.010

Javidi, G., & Sheybani, E. (2017). Empowering teachers to raise career awareness in computing: Lessons learned. Systemics. *Cybernetics and Informatics*, *15*(3), 10–15.

Jones, M. G., Chesnutt, K., Ennes, M., Mulvey, K. L., & Cayton, E. (2021). Understanding science career aspirations: Factors predicting future science task value. *Journal of Research in Science Teaching*, *58*(7), 937–955. doi:10.1002/tea.21687

Jung, Y. J., Zimmerman, H. T., & Land, S. M. (2019). Emerging and developing situational interest during children's tablet-mediated biology learning activities at a nature center. *Science Education*, *103*(4), 900–922. doi:10.1002ce.21514

Kier, M. W. (2013). *Examining the effects of a STEM career video intervention on the interests and STEM professional identities of rural, minority middle school students* [Doctoral dissertation, North Carolina State University].

Kier, M. W., & Blanchard, M. R. (2020). Eliciting students' voices through STEM career explorations. *International Journal of Science and Mathematics Education*, *19*(1), 151–169. doi:10.100710763-019-10042-z

Kier, M. W., Blanchard, M. R., Osborne, J. W., & Albert, J. L. (2014). The development of the STEM career interest survey (STEM-CIS). *Research in Science Education*, *44*(3), 461–481. doi:10.100711165-013-9389-3

Kisselburgh, L. G., Berkelaar, B. L., & Buzzanell, P. M. (2009). Discourse, gender, and the meaning of work rearticulating science, technology, and engineering careers through communicative lenses. *Annals of the International Communication Association*, *33*(1), 259–299. doi:10.1080/23808985.2009.11679089

Knapp, T. R. (2016). Why Is the one-group pretest–posttest design still used? *Clinical Nursing Research, 25*(5), 467–472. doi:10.1177/1054773816666280 PMID:27558917

Kurz, M. E., Yoder, S. E., & Zu, L. (2015). Effects of exposure on attitudes towards STEM interests. *Education, 136*(2), 229–241. doi:10.1643/0045-8511(2002)002

Lent, R. W. (2013). Social cognitive career theory. In S. Brown & R. Lent (Eds.), *Career development and counseling: Putting theory and research to work*. John Wiley & Sons.

Lent, R. W., & Brown, S. D. (2017). Social cognitive career theory in a diverse world: Guest editors' introduction. *Journal of Career Assessment, 25*(1), 3–5. doi:10.1177/1069072716657811

Lent, R. W., Brown, S. D., & Hackett, G. (1994). Toward a unifying social cognitive theory of career and academic interest, choice, and performance. *Journal of Vocational Behavior, 45*(1), 79–122. doi:10.1006/jvbe.1994.1027

Lydian Hill Foundation & Geena Davis Institute on Gender in Media. (2018). *Portray her: Representations of women STEM characters in media*. See Jane. https://seejane.org/wp-content/uploads/portray-her-full-report.pdf

McClure, E., Guernsey, L., Clements, D., Bales, S., Nichols, J., Kendall-Taylor, N., & Levine, M. (2017). How to integrate stem into early childhood education. *Science and Children, 55*(2), 8–10. doi:10.2505/4c17_055_02_8

Nadelson, L., Callahan, J., Pyke, P., Hay, A., & Schrader, C. (2009). A SySTEMic solution: Elementary teacher preparation in STEM expertise and engineering awareness. *American Society for Engineering Education, 106*(2), 157–168. doi:10.18260/1-2--5080

National Academies of Sciences, Engineering, and Medicine. (2021). *Science and Engineering in Preschool Through Elementary Grades*. National Academies Press. doi:10.17226/26215

National Science and Technology Council. (2018). *Charting a course for success: America's strategy for STEM education.* Whitehouse. https://www.whitehouse.gov/ostp

Noonan, R. (2017). *STEM jobs: 2017 update*. Office of the Chief Economist, Economics and Statistics Administration (ESA Issue Brief #02-17).

Novak, G. M. (2011). Just-in-time teaching. *New Directions for Teaching and Learning, 2011*(128), 63–73. doi:10.1002/tl.469

Nugent, G., Barker, B., Welch, G., Grandgenett, N., Wu, C., & Nelson, C. (2015). A model of factors contributing to STEM learning and career orientation. *International Journal of Science Education, 37*(7), 1067–1088. doi:10.1080/09500693.2015.1017863

Playton, S. C. (2021). *Primary STEM Career Interest and Awareness: Adaptation and Validation of a Survey for Measuring STEM Career Interventions with Primary Students*. [Embargoed doctoral dissertation, Texas Tech University].

Playton, S. C., Hite, R., & Childers, G. (2023). *Measuring STEM Career Awareness and Interest in Middle Childhood STEM Learners: Validation of the STEM Futures Survey.*

Playton, S. C., Hite, R., & Leach, P. K. (2021). A case study of a researcher-practitioner partnership in teaching STEM+C to rural elementary students. *Electronic Journal For Research In Science & Mathematics Education, 25*(2), 30–46.

Royce, C. A., Morgan, E., & Ansberry, K. (2012). *Teaching science through trade books*. NSTA Press.

Salonen, A., Kärkkäinen, S., & Keinonen, T. (2018). Career-related instruction promoting students' career awareness and interest towards science learning. *Chemistry Education Research and Practice, 19*(2), 474–483. doi:10.1039/C7RP00221A

Schlegel, R. J., Chu, S. L., Chen, K., Deuermeyer, E., Christy, A. G., & Quek, F. (2019). Making in the classroom: Longitudinal evidence of increases in self efficacy and STEM possible selves over time. *Computers & Education, 142*, 1–15. doi:10.1016/j.compedu.2019.103637

Stanley, J. C., & Campbell, D. T. (1980). *Experimental and Quasi-experimental Designs for Research*. Rand McNally.

Thomas, J. A., & Strunk, K. K. (2017). Expectancy-value and children's science achievement: Parents matter. *Journal of Research in Science Teaching, 54*(6), 693–712. doi:10.1002/tea.21382

United States Department of Education. (n.d.). *Education Innovation and Research*. USDE. https://oese.ed.gov/

van den Hurk, A., Meelissen, M., & van Langen, A. (2019). Interventions in education to prevent STEM pipeline leakage. *International Journal of Science Education, 41*(2), 150–164. doi:10.1080/09500693.2018.1540897

Virginia Department of Education. (n.d.a). *Profile of a Virginia graduate*. DoE. https://www.doe.virginia.gov/instruction/graduation/profile-grad/index.shtml

Virginia Department of Education. (n.d.b). *Standards of learning (SOL) and testing*. DoE. https://www.doe.virginia.gov/testing/index.shtml

Wang, M., Eccles, J. S., & Kenny, S. (2013). Not lack of ability but more choice: Individual and gender differences in choice of careers in science, technology, engineering, and mathematics. *Psychological Science, 24*(5), 770–775. doi:10.1177/0956797612458937 PMID:23508740

Wilson, J., Krakowsky, A. M., & Herget, C. J. (2010). Starting early: Increasing elementary (K-8) student science achievement with retired scientists and engineers. *IEEE Transactions on Education, 53*(1), 26–31. doi:10.1109/TE.2009.2022546

Yerdelen, S., Kahraman, N., & Taş, Y. (2016). Low socioeconomic status students' STEM career interest in relation to gender, grade level, and stem attitude. *Journal of Turkish Science Education, 13*, 59–74. doi:10.12973/tused.10171a

Yin, R. K. (2018). *Case study research and applications: Design and methods* (6th ed.). Sage.

Zohrabi, M. (2013). Mixed method research: Instruments, validity, reliability and reporting findings. *Theory and Practice in Language Studies, 3*(2), 254–262. doi:10.4304/tpls.3.2.254-262

APPENDIX A

Curriculum Needs Assessment for STEM Career Knowledge in Elementary Schools

The following, *Curriculum Needs Assessment for STEM Career Knowledge in Elementary Schools,* questions will be used to survey community partners including teachers, administrators, and parents, students or community members/stakeholder. The survey introduction will include the following information:

*The following survey is to gain knowledge regarding curriculum needs for STEM Career awareness in (District's Name). The survey questions will use the term "***STEM career awareness,***" which is defined as knowledge of science, technology, engineering or mathematics careers, and includes computer science careers. Please complete the survey and email XXXX@XXX if you have any questions or concerns.*

Depending on the stakeholder (e.g., classroom teacher, campus/district administrator, parent/student/community member or other), they will be prompted to complete the following six questions:

	Classroom teacher	Campus / district administrator	Parent / student / community member and/or other stakeholder
1	When considering your students' stem career awareness either in your classroom or school and the curriculum, what works?	When considering students' stem career awareness across your elementary schools, what works?	When considering students' stem career awareness across elementary schools in the district, what do you think is working?
2	What curriculum is most successful in your classroom?	What inputs (data) do you use to know what/how your curriculum is working across the elementary schools?	What sources do you use to understand what/how the curriculum is working across the elementary schools?
3	In your opinion, what needs to be improved in the curriculum to increase students' stem career awareness in your classroom or school?	In your opinion, what needs to be improved, in regard to curriculum, to increase students' stem career awareness across the elementary schools?	In your opinion, what needs to be improved, in regard to curriculum, to increase students' stem career awareness across the elementary schools?
4	What is your greatest professional development need in regard to stem career awareness curriculum?	What inputs (data) do you use to know what stem career curriculum should improve across the elementary schools?	What community-based resources do/would you use to evaluate if/how student stem career awareness has improved?
5	As you think about students' stem career interest, what long term planning (like in a plc for example) or ideas do you have for enacting curriculum reform?	As you think about students' stem career awareness across the elementary schools, what long term planning or ideas do you have for directing curriculum reform in your elementary schools?	As you think about students' stem career awareness across the elementary schools, what long term planning or ideas do you have for supporting curriculum reform in your elementary schools?
6	Anything else you want to add about stem career curriculum needs?	Anything else you want to add about stem career curriculum needs for elementary schools?	Anything else you want to add about stem career curriculum needs for elementary schools?

APPENDIX B

Curriculum Map

<table>
<tr><td colspan="2">Timeframe: November; Review/Implement in February</td></tr>
<tr><td colspan="2">Science SOL 2.3: The student will investigate that matter can exist in different phases.</td></tr>
<tr><td>Key Ideas
a) Matter has mass and takes up space;
b) Solids liquids and gasses have different characteristics; and
c) Heating and cooling can change the phases of matter.</td><td>Scientist
A Physicist whose research involves using certain techniques to measure the temperature and velocity of the plasma to understand the behavior of plasmas, since some plasma behavior is not yet understood. Specifically, it is important to understand outer-space-like plasmas, since it makes up over 99% of the visible universe.

Plasma applications include the process of making microchips that are used in many electronic devices, fluorescent lighting, and even medical applications.</td></tr>
<tr><td colspan="2">Timeframe: March</td></tr>
<tr><td colspan="2">Science SOL 2.5: The students will investigate and understand that living things are part of a system</td></tr>
<tr><td>Key Ideas
a) Plants and animals are interdependent with their living and nonliving surroundings;
b) An animal's habitat provides all of its basic needs; and
c) Habitats change over time based on many influences.</td><td>Scientist
A Wildlife Ecologist and Assistant Professor who studies how to keep wildlife and people safe, mostly through travel in planes and roads. Additionally, they research what wild pigs do and restore habitat for marsh birds with the help of drones and other technologies.</td></tr>
<tr><td colspan="2">Timeframe: April</td></tr>
<tr><td colspan="2">Science SOL 2.6: The students will investigate and understand that there are different types of weather on Earth</td></tr>
<tr><td>Key Ideas
a) Different types of weather have specific characteristics;
b)Measuring, recording and interpreting
c) Weather data allow for identification of weather patterns; and
d) Tracking weather allows us to prepare for the weather and storms.</td><td>Scientist
A Local TV Meteorologist who studies weather through instruments, interpreting data, and reporting it to the public to keep them safe and prepared for weather events.</td></tr>
<tr><td colspan="2">Timeframe: May</td></tr>
<tr><td colspan="2">Science SOL 2.4: The students will investigate and understand that plants and animals undergo a series of orderly changes as they grow and develop</td></tr>
<tr><td>Key Ideas
a) Animals have life cycles</td><td>Scientist
Two Zoologists who will discuss life cycle concepts of three guest animal ambassadors (e.g., butterfly, parrot, frog) for a zoo.</td></tr>
</table>

As soon as each scientist was secured and coordinated with the lead teacher and classroom schedules, an email would go out to communicate to teachers the expectations and plan. Here is a sample email communication:

We are confirmed for ***Friday, February 12 at 1-1:20pm****. I have updated all the 2nd grade lesson plans, as we discussed:*

TEACHER NAME 1
TEACHER NAME 2
TEACHER NAME 3
TEACHER NAME 4

Within each lesson plan (first page) you will find the video link introduction (i.e., Scientist, NAME) and Peardeck link (for each class) to be completed after meeting the scientist.

I look forward to this! Let me know if you have any questions.

Next, is an example of a digital document (view of **TEACHER NAME 1** lesson plan) created for each teacher's classroom as a resource related to the flow of the JiTT activity. All grade 2 classrooms were invited to one Zoom call to ask the scientists questions. Some of the items highlighted in yellow indicate actual dates and times of implementation. Items in blue underline indicate where there were URLs embedded into this digital communication.

TEACHER NAME, Grade 2 - Lesson Details
Contact: RESEARCHER NAME, EMAIL, PHONE NUMBER

<table>
<tr><td>Date:</td><td>Time: am/pm</td><td>Location: Zoom Link</td></tr>
<tr><td colspan="3">Teachers Name: TEACHER NAME Scientist: NAME (Physicist)</td></tr>
<tr><td colspan="2">Before Zoom with Scientist/s (DATE)</td><td>During/After Zoom with Scientist/s (DATE)</td></tr>
<tr><td colspan="2">1. Have students go to Flipgrid and record a video on what they think a physicist does: URL HERE

They will use their first name as their login. It is case sensitive, so they need a capital letter. This video link will help you with the directions for Flipgrid.

2. Have students visit another Flipgrid topic to watch an embedded video of the scientist we are going to meet. Students will post a question for the scientist (via Flipgrid video response): URL HERE

3, Students should go back and view their classmate's questions (Flipgrid link above). They will use the Like button to vote on the questions they would like asked during a video conference. Only 1-2 questions from each class will be chosen.</td><td>The following student(s) from your class will be prompted to ask their question to the scientist. Please make sure they know so they can be ready.

Name of Student/Question:
STUDENT NAME - Can you please come and be our student teacher?
STUDENT NAME - How do you discover anything new?
STUDENT NAME - How is your job doing?
4. After the scientist is done answering questions, students will be grouped back with their classroom teacher in either the classroom (in-person students) or online (teacher's own Zoom link), to complete the activity (i.e., Peardeck).

5. Students will go to the following Peardeck link: URL HERE

The video link will help you with the directions for Peardeck.
6. Students will respond to the prompts (i.e., text, drawing)
7. Teachers can review and discuss the experience with their students
a. What kinds of things did you learn from the physicist?
b. How does what we know about matter help her job?
c. Who thinks they would be interested in being a physicist? Why?
Note: A digital information sheet describing the scientist is provided to send home to parents.</td></tr>
</table>

TEACHER NAME, Grade 2 - Lesson Details
Contact: RESEARCHER NAME, EMAIL, PHONE NUMBER

<table>
<tr><td>Date:</td><td>Time: am/pm</td><td>Location: Zoom Link</td></tr>
<tr><td colspan="3">Teachers Name: TEACHER NAME Scientist: NAME (Wildlife Biologist)</td></tr>
<tr><td colspan="2">Before Zoom with Scientist/s (DATE)</td><td>During/After Zoom with Scientist/s (DATE)</td></tr>
<tr><td colspan="2">1. Have students write/answer the question, "What do you think a Wildlife Research Biologist does?" (then, email to RESEARCHER EMAIL)

2. Have students watch a short video the scientist we are going to meet:

Scientist, NAME (video)

3. Have students write a question for the scientist (then, email to RESEARCHER EMAIL).</td><td>The following student(s) from your class will be prompted to ask their question to the scientist. Please make sure they know so they can be ready.

Name of Student/Question: Check document

4. After the scientist is done answering questions, students will be grouped back with their classroom teacher in either the classroom (in-person students) or online (teacher's own Zoom link), to complete the activity (i.e., Peardeck).
5. Students will go to the following Peardeck link: URL HERE

The video link will help you with the directions for Peardeck.

6. Students will respond to the prompts (i.e., text, drawing)
7. Teachers can review and discuss the experience with their students -
a. What kinds of things did you learn from the biologist?
b. How does what we know about ecosystems help his job?
c. Who thinks they would be interested in being a biologist? Why?
Note: A digital information sheet describing the scientist is provided to send home to parents.</td></tr>
</table>

TEACHER NAME, Grade 2 - Lesson Details
Contact: RESEARCHER NAME, EMAIL, PHONE NUMBER

<table>
<tr><td>Date:</td><td>Time: am/pm</td><td>Location: Zoom Link</td></tr>
<tr><td colspan="3">Teachers Name: TEACHER NAME Scientist: NAME (Meteorologist)</td></tr>
<tr><td colspan="2">Before Zoom with Scientist/s (DATE)</td><td>During/After Zoom with Scientist/s (DATE)</td></tr>
<tr><td colspan="2">1. Have students reflect on, "What do you think a meteorologist does?"

2. Have students watch a short video of the scientist we are going to meet.

3. Have students write a question for the meteorologist on the Google Sheet.</td><td>The following student(s) from your class will be prompted to ask their question to the scientist. Please make sure they know so they can be ready.

Name of Student/Question: Check Google Sheet
4. After the scientist is done answering questions, students will be asked to complete the activity (i.e., Nearpod):
a. Students will go to the following Nearpod: URL HERE
i. Students will respond to the prompts (i.e., text, drawing)
5. Teachers can review and discuss the experience with their students -
a. What kinds of things did you learn from the meteorologist?
b. How does what we know about weather help their job?
c. Who thinks they would be interested in being a meteorologist? Why?
Note: A digital information sheet describing the scientist is provided to send home to parents.</td></tr>
</table>

TEACHER NAME, Grade 2 - Lesson Details
Contact: RESEARCHER NAME, EMAIL, PHONE NUMBER

<table>
<tr><td>Date:</td><td>Time: am/pm</td><td>Location: Zoom Link</td></tr>
<tr><td colspan="3">Teachers Name: TEACHER NAME Scientist: NAME (Zoologist / Zoo Educator)</td></tr>
<tr><td colspan="2">Before Zoom with Scientist/s (DATE)</td><td>During/After Zoom with Scientist/s (DATE)</td></tr>
<tr><td colspan="2">1. Have students write/answer the question, "What do you think a Zoo Educator does?" (then, email RESEARCHER EMAIL)

2. Have students write a question for the zoo educator on the Google Sheet.</td><td>The following student(s) from your class will be prompted to ask their question to the scientist. Please make sure they know so they can be ready.

Name of Student/Question: Check Google Sheet (link to be provided by TEACHER NAME)

3. After the scientist is done answering questions, students will be asked to complete the activity (i.e. Peardeck):
a. Students will go to the following Peardeck link: URL HERE

The video link will help you with the directions for Peardeck.
b. Students will respond to the prompts (i.e., text, drawing)

4. Teachers can review and discuss the experience with their students -
a. What kinds of things did you learn from the zoo educator?
b. How does what we know about life cycles help their job?
c. Who thinks they would be interested in being a zoo educator? Why?</td></tr>
</table>

APPENDIX C

Three Documents

STEM Expert Pre-Interview

Pre-Interview Correspondence (*Example*)

Thank you so much for agreeing to video conference with grade 2 from (school name and location). Our students are looking forward to asking you some questions and would love to first know a little more about you and your career. Before we ask you some pre-interview questions, here is a little background on why we are interested in learning about your career!

Background

Our students in (grade level) are currently learning about weather! As part of their weather unit, they are learning about weather instruments to measure and record data to interpret and identify patterns, the different types of weather and their specific characteristics. Having this knowledge can help track weather events and allows time to prepare for things like hurricanes or floods.

Questionnaire for STEM Expert

This questionnaire is so that we can properly introduce you and your career to the students in order to guide their questions for the upcoming video conference. It has been put into an online survey form to collect your answers, but the questions are below so that you know what to expect before clicking the link: https://forms.gle/yPft8i7YoubCrgdD8

1. What is your name?
   Feel free to tell us a little about yourself! (e.g., family, hobbies, travel)
2. Where do you work?
   Where is it located? (e.g., New York City, Southern California, London)
3. How long have you been in this profession?
4. What kinds of things does your profession require?
   What kind of schooling and/or field experience did you need to qualify for this profession? (e.g., a 4-year engineering degree)
   "Soft skills" ? (e.g., communication, critical thinking, collaboration, creativity)
   https://www.thebalancecareers.com/list-of-soft-skills-2063770
   "Hard skills" ? (e.g., technology, software/programs, mathematics, data)·
   https://www.thebalancecareers.com/what-are-hard-skills-2060829_
5. Describe how your profession supports communal needs.
6. What is something that drew you to this profession? (e.g., a personal interest, goal, use of technology

STEM Career Expert

Student Name:____

Today we met a (STEM Career). Draw a picture below of what you think a (STEM Career) looks like while working and using knowledge about (science content):

Write a sentence or two about your picture:

____

____

____

What other skills does a (STEM career) need?

____

Parent Resource/Letter

(Date)

Dear Parents and/or Guardians:

Your child has been learning about matter in the science classroom and how it exists in phases, with varying characteristics. To help them understand how what they are learning is relevant to STEM career fields, they did a video conference with (Name), a physicist working on research at (Workplace) in (Location). Here are a few things we learned about Dr. (Name) and their job as a physicist through a pre-interview:

Tell us a little about yourself.

My hobbies include doing yoga, painting, and I love taking walks!

How long have you been in this profession?

This is my second year as a physicist, research assistant.

What kind of schooling and/or field experience did you need to be a physicist?

You usually need to be a graduate student to be a research assistant, so a bachelor's degree in physics or engineering is needed.

What type of soft skills does your job require (e.g., communication, critical thinking, collaboration, creativity)?

Communication, collaboration, creativity, and keeping a positive attitude.

What type of hard skills does your job require (e.g., technology, software/programs, mathematics, data)?

Operating advanced devices (e.g. lasers, oscilloscopes), physics, coding, and mathematics.

How does being a physicist support the community?

Research at (Workplace) supports communal needs in many different ways. One example is when the physics department gets funding from the government to conduct research! This research helps other scientists, engineers, and researchers with making decisions about their own work. Conducting research also helps people that are not scientists understand things that have not been able to be understood or even observed before!

What is something that drew you to this profession?

I've always loved math and science!

According to the *Bureau of Labor and Statistics*, the U.S. national salary average for a **Physicist** is:

$122,220

After talking with Dr. (Name), ask your child(ren) if they can describe how a physicist needs to know some of the same things they are learning about in school!

Thank you,
(Teacher)

Chapter 6

Development of a Low-Cost Affordable Inquiry-Based System-Thinking STEM Prospectus:

Food, Water, and Energy Learning Module Workbooks

Tonya Maria Gerald-Goins
North Carolina Central University, USA

Porche' L. Spence
North Carolina Agricultural and Technical State University, USA

Gregory D. Goins
North Carolina Agriculture and Technical State University, USA

ABSTRACT

The Food, Energy, and Water Learning Module (FEWLM) workbooks are an effort by URM Ph.D. STEM faculty to decrease the leaky STEM pipeline and 'pay it forward' via the creation of high-quality out of school learning experiences for 'at-risk' URM 3rd – 8th grade students less competitive in STEM. Curriculum is based on Next Generation Science Standards, with a system thinking approach that provides hands-on, self-driven and team-building activities with a goal to increase scientific exposure and encourage more 'at-risk' URM youth to enter and remain in the STEM pipeline. Analysis of survey data of FEWLM workbooks use by over 50 elementary and middle school youths implemented at area public after-school programs, and a two-week summer STEM enrichment camp, suggest increases in scientific exposure and awareness about environmental sustainability and human health issues suggest FEWLM made a positive impression by increasing 1) their interest in STEM, 2) their perception of STEM, and 3) confidence in their ability to participate in STEM.

DOI: 10.4018/978-1-6684-7771-7.ch006

INTRODUCTION

Our efforts at the Diverse and Integrative STEM Continua Using Socio-environmental Systems In and Out of Neighborhoods or *Discussion Network* attempt to improve access to high-quality science, technology, engineering, and mathematics (STEM) learning experiences for 'at risk', underrepresented minority (URM) students in K-12th grades who traditionally have been underserved. The DISCUSSION Network Core Team envisioned achieving this goal by increasing the number of prepared, motivated underrepresented students by disseminating evidence-based interventions as low-cost inquiry-based science learning materials that emphasize scientific literacy, knowledge retention, and quantitative skills. The DISCUSSION Network takes a socio-environmental system approach to promote a common view of natural resources for sustaining communities. We present an integrative, systems-thinking approach for teaching science while informing youth about the connection between the environment and human health, through a student-centered, inquiry-based curriculum focusing on the Food, Energy, Water (FEW) nexus (Mohtar & Daher, 2012). Therefore, the DISCUSSION Core Team identified, developed, and implemented a FEW Learning Module (FEWLM), which is a scaffolded, low-cost, inquiry-based learning curriculum built on the Next Generation Science Standards (NGSS; National Research Council, 2013; NGSS Lead States, 2013) designed for formal educational programs, after school programs, summer camps, homeschool instructors and any other STEM-related outlets (Table 1).

Table 1. Table of FEWLM modeling and real-world applications with Bloom's taxonomy

Module Number	FEWLM Module	Modeling Applications and STEM Features	Real-World Environmental Applications	Bloom's Taxonomy
1	Scientific Inquiry with Decimals and Fractions through the Colorful World of m&m's Workbook OR Exploring Scientific Inquiry and Mathematical Thinking with Skittles Workbook	Math Applications Basic Statistics Sample Replication Data Collection Data Analysis	Scientific Inquiry	Understand Apply
2	My Sprouting Bean Seeds Scientific Workbook	Seed Germination Biological Organisms	Food Insecurity	Understand Apply Analyze
3	Let's Take a Biochemical Journey Through Osmosis with a Naked Hen Egg Workbook	Cell Structure Cell Membrane Osmosis	Health disparities	Understand Apply Analyze
4	The Beneficial Health Impacts of Fruits and Vegetables	Cell Membrane Biochemistry	Health Antioxidants pH	Understand Apply Analyze
5	Exploring the Link Between Land Management, Plant Growth and Water Quality Using a Mini-Garden Model	Plant Growth Water Quality Land Management	Water Pollution Food Insecurity	Understand Apply Analyze Synthesis

(Spence et al, 2021)

Through the development and dissemination of an engaging inquiry-based curriculum which emphasizes the scientific process, quantitative literacy skills and environmental sustainability focused on the FEW Nexus, our DISCUSSION Network contributes to broadening participation in STEM by using fun, integrative, system-thinking, system-modeling STEM activities that intervene at the elementary and middle school critical juncture of 3rd–8th grade to engage 'at risk' underrepresented and underserved youth in scientific inquiry-based learning while providing a culturally relevant experience that encourages them to become interested in STEM careers. This chapter contains information about 1) The reason for the development of DISCUSSION and the DISCUSSION Network; 2) the reason for the development of FEWLM; 2) The implementation of and student perspectives about FEWLM impact; and 4) successes, challenges, and limitations of FEWLM as a resource that can assist educators whether used in formal or informal educational settings. The results of this chapter were generated in various outreach programs.

BACKGROUND

The Food, Energy, and Water Learning Module (FEWLM) workbooks are serendipitous outcomes of a common goal of several URM Ph.D. holding tenured and tenure-track STEM faculty members to decrease the leaky STEM pipeline and 'pay it forward.' Thus, leading to the creation of high-quality, hands-on, sustainable STEM learning activities that encourage URM and traditionally underserved 3rd - 8th grade students that post-secondary education and a career in STEM is possible. URM includes any racial group whose percentage of the population in a field, such as STEM, is lower than their percentage of the population in the country. These racial groups in STEM, historically include African American, Hispanic Americans, Native Americans, Alaskan Eskimos and Pacific Islanders. Now, women and those with disabilities are also included as URM in STEM. The underserved are defined as those that have limited or no access to resources; the disenfranchised, such as those found in central urban and rural areas.

FEWLM developed because of an effort led by a team of five African American Ph.D. STEM professors in Mathematics, Physics, Chemistry, Soil Sciences, and Crop Sciences, (Core Team) at two historically black colleges and universities (HBCUs), North Carolina Agricultural and Technical State University (NCAT) in Greensboro, NC, and North Carolina Central University (NCCU) in Durham, NC, with a common agenda to bridge critical junctures contributing to the diversity of the STEM workforce by taking part of a national effort to broaden URM participation and underserved K-12 students in STEM that was initially funded by the National Science Foundation's *Inclusion across the Nation of Communities of Learners of Underrepresented Discoverers in Engineering and Science* (NSF INCLUDES). NSF INCLUDES (n.d.) is a comprehensive national initiative designed to enhance U.S. leadership in STEM discoveries and innovations by focusing on broadening participation of URM and underserved in these fields at the national scale. Our collaborative effort to compete for an NSF INCLUDES grant was based on our individual passions to turn our broadening participation on the undergraduate and graduate levels to the K-12 levels. In addition, the geographical location of NCAT and NCCU, which are centrally located in a region where the Brookings report described as among the ten metro areas where poverty is growing fastest, could benefit from our combined 75+ years of using the Tinto model to retain undergraduate URM STEM undergraduate majors throughout our respective careers.

The initial effort by NCAT and NCCU URM faculty members was a pilot project that developed into the DISCUSSION. The purpose of the DISCUSSION was to build and sustain partnerships with organizations within URM and underserved communities, with a common focus to encourage, persuade, as well as broaden participation of URM and underserved K-12 students to pursue STEM as a realistic postsecondary major and career occupation. Elementary and secondary education in mathematics and science are the foundations for entry into postsecondary STEM majors and STEM-related occupations as suggested by a National Science Board (NSB) report, *The State of U.S. Science and Engineering* by Burke et al. (2022). Recent data generated and reported by the NSB indicates that the United States ranks higher in science literacy (7th out of 37 Organization for Economic Co-operation and Development [OECD] countries) than in mathematics literacy (25th of 37 OECD countries). Moreover, the average U.S. science score was greater than the OECD average and since 2006, has improved by 13 points. What is discouraging is the average U.S. mathematics score in 2018 was lower than the OECD average and has not changed significantly since 2003. It is imperative for our nation's development of sustainable communities (Goudie, 2009), for the U.S.A.'s protection, welfare, economic effectiveness, and continued international leadership (National Academy of Sciences, Engineering, and Medicine [NASEM], 2011; President's Council of Advisors on Science and Technology [PCAST], 2012; NSB, 2022), to provide educational experiences for all of its citizens that will produce a diverse, capable, and inventive STEM labor force. However, for decades, URM, with the labels for these groups changing to include Persons Excluded from science because of Ethnicity and Race (P.E.E.R.) or racially minoritized individuals (Hinton et al., 2020), in STEM fields (NSB, 2019). The lack of diversity is perilous since the U.S. Census Bureau report by Vespa et al. (2020) regarding the proposed 2060 U.S. demographic composition, suggested that 66% of children are projected to be a race other than non-Hispanic White. Although the percentage of URMs receiving undergraduate and graduate degrees in STEM is slowly increasing, the proportion of URMs attaining STEM degrees and employment in the STEM workforce, especially in mathematics, statistics, and physical sciences, is consistently low (NSB, 2019). Long-standing social and economic barriers in our society have contributed to the continued low percentage of URMs in STEM fields and careers (de Graaf et al., 2005; Musters et al., 1998), with STEM 'pipeline leaks' are exhibited as early as the K-12 levels, (NSB, 2019). Moreover, the complex interactions of socio-economic factors affecting any relegated community hamper students with poverty, compounded by under-resourced public education, that can lead to a loss or lack of interest in education, let alone, STEM, with a under-preparedness for college, thereby severely limiting the number of necessary URM students that can pursue STEM degrees thus preventing potential increases in the number of minorities as STEM professionals (PCAST, 2012).

Across the United States, a sizable percentage of minority youth are disadvantaged because their lives are constrained by one or more socio-economic barriers, such as, single-parent households, parents with low educational levels, unemployed parent(s), frequent residential changes or homelessness, food insecurity, household obligations, household poverty and impoverished community, and attending low-performing public schools. These are risk factors that have been linked to poor academic performance of all students (Robbins et al., 2012). Often P.E.E.R. youth from these 'at risk' backgrounds are unaware of opportunities to participate in STEM experiences early in their education, and therefore, many do not perceive STEM as relevant to their lives, let alone, as a possible career path to follow. P.E.E.R. youth lacking an early and positive connection to STEM reach middle school already having developed a negative perspective of STEM (Goodwin, 2016), and their motivation to learn this subject diminished (Dweck, 2007). However, there is evidence which suggests that URM youth become more motivated and gain more interest in STEM when they are exposed to culturally relevant science learning experiences that demonstrate a connection of STEM to their culture and community (Estrada et al., 2011). Furthermore, encouraging these youths that they are smart and that putting in extra time and effort will lead them to high achievements, thus encouraging them to look beyond their circumstances and change to believe they are smart, which increases their motivation to learn and helps them develop the skills to seek solutions to problems, encourage coping strategies to master setbacks, and defeat negative stereotypes (Dweck, 2007).

While gathering information to support our effort, we recognized the need to focus on 'at risk' students (Figure 1). At-risk students are the less-competitive STEM students whose learning gap increases from the highly competitive STEM students between 5th and 6th grade. Therefore, the DISCUSSION proposed that with the assistance of mentoring and the use of well-thought-out STEM-related classroom activities, the learning gap between highly competitive STEM and less competitive STEM students would decrease, thus increasing the number of URM and underserved students in the leaky STEM pipeline.

Figure 1. FEWLM influence on science interest and career aspirations for (afterschool program) and middle school youth (summer camp). Results from pre-survey and post-survey of elementary and middle school youth participants attitude toward science before and after use of FEWLM workbooks. The numbers above the bars represent the percent response of participants to the three inquires.
(Spence et. al., 2021)

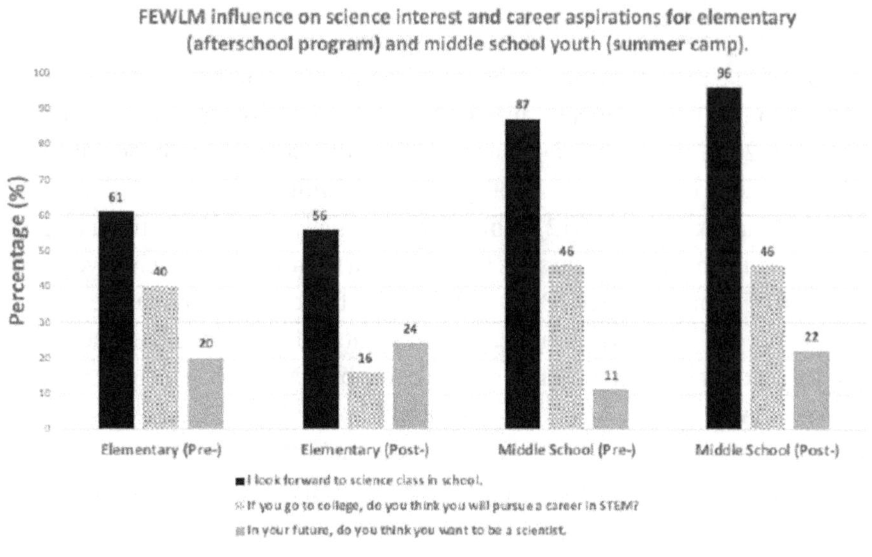

STEM-Related Activities to Help Bridge the Learning Gap

During the development of DISCUSSION, the NSF Advisory Committee for Environmental Research and Education (AC-ERE) established that complex ecosystems are an essential focus for STEM education because they contain important ideas about national standards and provide an integrating context across several science domains (NSF, 2015). We chose to 1) provide sustainability in education through ecological and environmental awareness and responsible behavior around the food, energy, water, and health nexus; (2) provide knowledge about food systems and nutrition, and healthy eating, especially consumption of fresh fruits and vegetables by the adoption and use of two (2) interrelated theoretical frameworks that unified our approach to research, instruction, and assessment in complex ecosystem contexts and socio-environmental justice. The two interrelated theoretical frameworks include 1) System Thinking Hierarchical (STH; (Assaraf & Orion, 2005, 2010) and 2) Structure, Behavior and Function (SBF) frameworks, which were originally developed to describe engineered systems, but have proven to be flexible as tools for facilitating systems thinking in diverse contexts. The STH model identifies critical skills inherent for thinking and reasoning about systems, such as: (1) the ability to organize system components, processes, and interactions within a framework of relationships; (2) the ability to recognize hidden dimensions of the system (i.e., understand natural phenomena through patterns and interrelationships not seen on the surface); and (3) the ability to think temporally (e.g., retrospection and prediction) and understand that current system functions are a result of past interactions, while future events may be a result of present interactions. SBF decomposes systems into their elemental constituents, including system components (structures), relationships between components (behaviors), and the holistic role or output of a system (function). SBF provides a unified framework for instruction and assessment that employs models and modeling to learn about biological systems. SBF has been used extensively in biology curricula at Michigan State University for both life science majors and non-majors (Dauer & Long, 2015). SBF-based assessments have potential to document progression in students' thinking about biological systems and identify critical patterns in student reasoning that can inform subsequent teaching (Goel et al., 2009).

There are numerous imbalances in the natural systems that provide our food, energy, and water (FEW), such as increasing environmental, economic, and political pressures stemming from population growth, climate instability, heterogeneous distribution of natural resources, habitat destruction and land use (Mohtar & Daher, 2012). It is essential that our society develop a knowledgeable citizenry so we can devise the best approaches to integrate natural and human-built environments that supports a growing demand for FEW (Mohtar & Daher, 2012), while conservatively and efficiently managing irreplaceable natural resources, (i.e., freshwater, fertile soils for agricultural crop and medicinal vegetation generation) as well as, regulation of climate change (Daily et al., 1997). Developing a populace that is STEM-literate begins with educating our youth. Innovative, student-centered, evidence-based curricula that engage youth early and sustain their interest in STEM should contribute also to broadening participation of URM students in STEM (PCAST, 2012). Also, the *Vision and Change Report* suggest the use of inquiry-based learning modules for formal and informal STEM educational settings that piques the curiosity in STEM (American Association for the Advancement of Science [AAAS], 2011) that will develop into an appreciation of the interconnective ecological processes of nature, as opposed to the separated, scientific concepts of math, physics, chemistry, and biology (Goodwin, 2016).

Use of an experiential, scientific method-based curriculum for youth to build core competencies in science, including hypothesis development, experiment development and testing, data generation and interpretation, as well as self-reflection of the learning experience, enhances STEM learning (AAAS, 2011; Goodwin, 2016). This has been shown especially to be the case for elementary school youth who are members of various demographic groups, classified by socio-economic status, gender, achievement level, and English proficiency levels (Cuevas et al., 2005). For instance, Cuevas et al. (2005) find that employing an inquiry-based intervention significantly improves the ability for elementary school youth to conduct scientific inquiry and more specifically implement procedures, record information, and make conclusions. While Marx et al. (2004) affirms that an inquiry-based and technology infused curriculum can help middle school students who are in underperforming urban schools to learn science, several other studies also provide strong evidence of the positive impacts of using an inquiry-based curriculum to improve academic achievement at the elementary and middle grade levels (Abdi, 2014; Şimşek & Kabapınar, 2010). However, studies have shown major challenges associated with dissemination and implementation of empirically proven curriculum (Estrada et al., 2011; Harper & Griffin, 2011) designed to narrow the achievement gap and increase STEM literacy for minority children at the elementary and middle school levels (Lent et al., 2005; NASEM., 2011). Also, the COVID Pandemic revealed the breadth of the digital divide gap for students who fall into the categories, after the United Nations declaring in 2016 that technology is an essential and basic need for everyday life.

To implement the DISCUSSION Network, two members of the Core Team began building relationships with partners such as the non-profit organization, Alamance-Burlington Closing the Achievement GAP, Inc., located within the Alamance-Burlington County Public School System (ABSS) and John Avery Boys and Girls Club of Greater Durham (BGCGD), the first African American Boys and Girls Club in the United States, as well as several other non-profit and faith-based community organizations to create the DISCUSSION Network. The DISCUSSION Network has and continues to develop and promote a spirit of socio-environmental comprehensiveness with the current partnerships and continues to add more partners whose direct or indirect focus is to encourage and broaden the participation of URM and underserved K-12 students in post-secondary STEM majors and STEM career occupations.

DISCUSSION NETWORK PARTNERS

John Avery Boys and Girls Club

The DISCUSSION established a connection with the John Avery Boys and Girls Club of Greater Durham (BGCGD), the first African American Boys and Girls Club in the United States. The Boys and Girls Club is a private, nonprofit organization that provides an afterschool program with specialized activities for youth during the traditional school year, primarily to a population of students meeting the criteria of URM in STEM and underserved 'at-risk' youth (Boys and Girls Club of Durham and Orange Counties, 2017). The BGCGD's after school program occurs every weekday during the traditional school calendar year between 2:30 pm and 6:30 pm. The afterschool program offers several concurrent activities that focus on leadership development, academic success, computer literacy and coding, fitness and stress management, health and life skills, creativity, and culture awareness (Boys and Girls Club of Durham and Orange Counties, 2017). The BGCGD youth attend public, public charter and private schools located in Durham County, NC. Furthermore, the BGCGD was located approximately two miles from NCCU's campus; the facility has since moved to another location (Spence et al., 2021).

The sessions at the BGCGD after school program were conducted by the three African American female Core Team members from NCAT and NCCU with years of teaching STEM students from diverse socioeconomic and educational backgrounds. In addition, these professors have also taught numerous non-STEM majors, thus understanding the need to modify the delivery of fundamental STEM-based concepts by providing connections to the real-world for all students. The three female African American STEM professors have motivated STEM and non-STEM undergraduate students to learn content. This knowledge and ability of the faculty enabled them to effectively teach STEM using the FEWLM curricula. FEWLM workbooks allow students to experience the activities while their interaction with the URM scientists provides a tangible opportunity for students to see a likely future career. A Ph.D. is not needed to use these materials, even though they were developed by Ph.D. scientists. The FEWLM workbooks are designed to be clear and easy-to-follow so that elementary and middle school teachers, parents and STEM outreach instructors can guide students through the activities. Older students might complete some activities on their own with minimal supervision.

FEWLM was an ideal curriculum for the elementary school youths attending the BGCGD, because the DISCUSSION Network goals align perfectly with the BGCGD's mission. DISCUSSION faculty from NCCU visited the 3rd–5th grade class at the BGCGD afterschool over a 4-week period (2018) to implement the FEWLM curricula three days a week (Monday, Tuesday, and Wednesday) in two 2-week intervals. Each session began around 4 pm for 2 hours. The BGCGD sessions involving FEWLM consisted of the following first session schedule: 1) workbooks handed out to participating students, 2) review of the workbook activity by the lead instructor, 3) distribution of activity relevant materials, and 4) a demonstration of the proposed activity by the lead instructor followed by a trial run by the students. The second session involved a review of the workbook and completion of the activity by the students.

Alamance-Burlington County Public School System (ABSS) and the NCAT/NCCU "HydroPHonics" STEM Enrichment Summer Camp

The DISCUSSION Network established a connection with a non-profit organization, Alamance-Burlington Closing the Achievement GAP, Inc., located within the Alamance-Burlington County Public School System (ABSS). The mission of the Alamance-Burlington Closing the Achievement GAP (2016, para. 2) Inc., is to "promote raising the achievement levels of all students and closing the achievement gap of ethnic minority and lower socioeconomic students." The DISCUSSION Network collaborated with Alamance-Burlington Closing the Achievement Gap, Inc. through ABSS to organize and implement a cost-free two-week *HydroPHonics* STEM Enrichment Summer Camp to increase scientific literacy and critical thinking skills using FEWLM. The selected URM ABSS student participants were rising 6th graders to rising 9th graders. One of the members of the Alamance-Burlington Closing the Achievement GAP, Inc. organization serves as the ABSS Director of Outreach Services and was instrumental in organizing the application process, transportation, and summer camp counselors (certified ABSS STEM teachers). The DISCUSSION Network provided University faculty (NCAT and NCCU), undergraduate counselors, professional development workshops, facilities, consumables, and sustenance.

The two-week HydroPHonics Summer Camp was led by a URM male and a URM female Core Team members from NCAT and NCCU on the NCAT campus. Three workbooks were introduced and used, *Exploring Scientific Inquiry and Mathematical Thinking with Skittles*, *My Sprouting Bean Seed*, and *Let's Take a Biochemical Journey through Osmosis with a Naked Hen Egg*. The HydroPHonics Summer Camp lasted two weeks and consisted of 40 URM middle school students that ranged from rising 6th to rising 9th graders; 80% male and 20% female. There were five URM counselors (two African American female educators from the Alamance–Burlington School System, one African American female NCAT graduate student and two NCAT STEM undergraduates, both African American STEM majors; one female and one male who was also a member of the football team. The students arrived on campus by 9:00 a.m. and assembled in an auditorium of Barnes Hall on NCAT's campus. Lead instructors reviewed the previous day's activities before presenting the current day's schedule. Students were then sent to two adjoining teaching labs at 9:30 a.m. Each student participant was required to wear safety goggles and a disposable lab coat; a lab safety course was given to all counselors and participants. Each student participant was given a workbook, which was reviewed by one of the lead instructors, and any required materials. Student participants could then work according to the instructions given in the workbook. Students were also building a hydroponic system and a vivarium at each lab table. Lunch was scheduled from noon to 1:30 p.m. in the Williams Cafeteria on NCAT campus. The walk from Barnes Hall to Williams Cafeteria allowed students an opportunity to get fresh air, physical activity, and an informal tour of NCAT's campus. After lunch, students returned to Barnes Hall to work on any incomplete tasks from the morning. Everyone assembled in Barnes Hall Auditorium for discussions and to watch short relevant videos. Some afternoons, students presented a brief daily overview of their work, or they listened to a STEM professor from NCAT. Students departed NCAT at 3:00 p.m. every day.

MATERIALS AND METHODS

Assessing Women and Men in Engineering Pre-College Outreach STEM Activity Survey

A modified version of the *Assessing Women and Men in Engineering* (AWE, 2010) pre-college outreach STEM activity surveys was revised by the DISCUSSION evaluators and administered via paper form to the camp participants. Unique identifier of the camp participants was based on student initials and birthdates. These surveys were designed to be flexible for researchers to add and remove specific questions provided by AWE without compromising the validity of the survey. The pre-college outreach STEM activity pre and post upper elementary school surveys were administered to the 3rd-5th grade youth at the beginning and end of the 4-week BGCGD NCCU DISCUSSION STEM club. The pre-college outreach STEM activity for middle school science surveys were administered at the beginning and end of the two-week NCAT/NCCU DISCUSSION Network HydroPHonics STEM Enrichment Summer Camp. The AWE pre-college outreach STEM activity assessments (NSF Grant # HRD–0120642 and HRD–0607081) are validated tools designed to understand to what extent formal or informal educational activities are meeting programmatic goals and to survey youth participants about 1) attitudes regarding science, 2) educational background and experiences and 3) career ambitions (Figure 2; AWE, 2008, 2010). To supplement the AWE surveys, a Student Background Information Sheet and Science Interest Questionnaire were given to elementary school youth at the BGCGD prior to beginning FEWLM (Spence et al., 2021).

Please note that the data in Figure 2 was recorded and updated so that scoring is consistent for both groups of students. Likert-type responses from the middle school youth survey were re-coded as Yes (slightly, moderately, a great deal) and No (not at all) to be consistent with the elementary school youth survey.

Surveyed Participant Groups

A total of 55 youths was engaged in science inquiry using the FEWLM curricula: 27 elementary age participants (3rd–5th graders) at the BGCGD after school program, NCCU DISCUSSION STEM Club between March and April 2018 and 38 middle school age (rising 6th– rising 9th graders) participants in a two-week NCAT/NCCU DISCUSSION HydroPHonics STEM Enrichment Summer Camp (Spence et. al., 2021). Each participant fit the description of a URM in STEM; however, youth who self-identified as African Americans represented the largest demographic percentage at the BGCGD after school program

Figure 2. Network's common progression agenda built on structural, behavioral, and functional descriptions of a socio-environmental system (Musters, de Graaf et al. 1998). Capturing average 6th - 8th grade URM students the leaky STEM pipeline occurs during (magnifying glass).

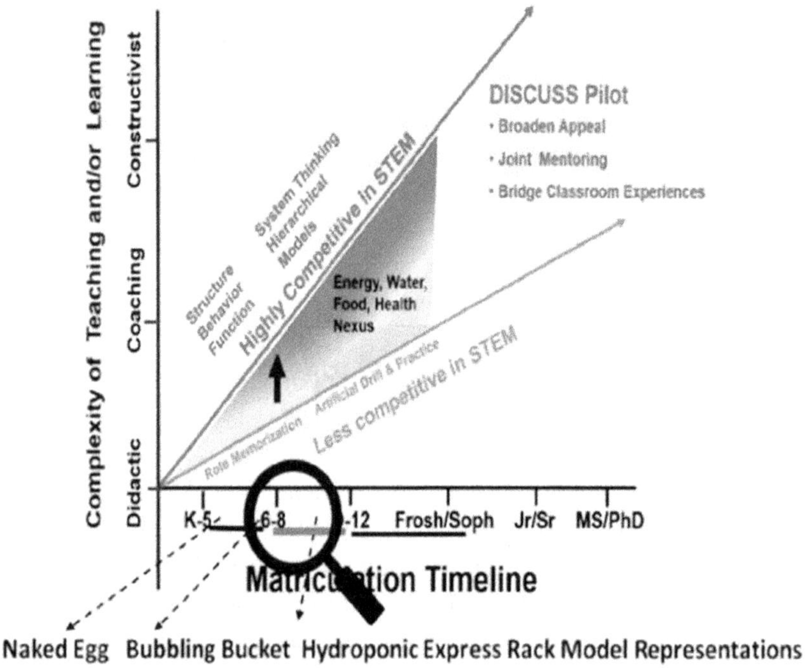

(96%) and the NCAT/NCCU DISCUSSION HydroPHonics STEM Enrichment Summer Camp (50%). Furthermore, most of the youth participants self-identified their gender as male (60%). The elementary school youth participants attended public charter schools (67%) in Durham County, NC, while the middle school youths attended public schools (85%) in Alamance-Burlington, NC. A sizable percentage of the youth participants came from families with parents who were college graduates (71% after school program and 83% summer camp participants). This study was reviewed in accordance with federal regulations governing human subjects research as an expedited review study (NCAT IRB #17- 0109).

FEWLM Curriculum Development

A DISCUSSION Core Team member with an elementary school age child at the time, noticed the use of worksheets that describe science activities and contain predetermined answers; but this is not science. Science involves the joy of discovery and comprehension of our world. The FEWLM workbooks provide students with the opportunity to explore, practice, and apply the scientific method while learning the interconnectedness of FEW. The data collected during these activities are not pre-determined and will vary each time experiments are conducted. Likewise, the participants will learn, practice, and apply skills such as observing, recording, graphing, analyzing and interpreting the data to test their own hypothesis and formulate a conclusion. In addition, the idea of affordable programming was the guiding principle in identifying, designing, and developing FEWLM.

It was conceptualized by the DISCUSSION Network Core Team to address the issue of access through lowering socio-economic barriers that limit participation in STEM for minority communities. FEWLM is an assembly of age-appropriate, culturally sensitive STEM curricula workbooks immersed with environmental conservation, sustainability, and human health foci. The curricula development started with evidenced-based science learning materials we modified to be hands-on, self-driven and team-building activities and then compiled into workbooks (Ahmed et al., 2021; Carlsen et al., 2010; Khoo et al., 2017; Nolte, 2008; Science Buddies, 2012; Spence et al., 2020). The curricula emphasize the application of scientific inquiry, quantitative literacy, conceptual understanding, and data analysis and interpretation. These FEWLM workbooks were developed by members of the DISCUSSION Core Team. The content was designed to provide an educational experience for elementary and middle school youth to engage in the scientific method and comprehend science, not as isolated facts, but as collective concepts that interlink ecological systems with human well-being. Through FEWLM, the youth are encouraged to act like scientists by working individually and/or collaboratively to observe, record, graph, analyze and interpret data to test a hypothesis and formulate a conclusion. Self-reflective, open-ended questions are provided to encourage students to think about what they learned while completing the workbook (Spence et. al., 2021).

This curriculum is a reactive plan that benefits many 'at risk' elementary and middle school youth who might otherwise have been disregarded as potential STEM professionals by exposing them to "real world" problem solving using the scientific method. Furthermore, emphasizing problem-solving approaches to environmental issues inspires interest, motivation, and value for STEM learning for students and improves STEM educational options for teachers (Bandura & Locke, 2003; Musters et al., 1998; Trauth-Nare, 2015). Using the NGSS as a standard guideline, the set of FEWLM workbooks collectively invoke use of the eight (8) science and engineering practices: 1) Asking questions and defining problems; 2) Developing and using models; 3) Planning and carrying out investigations; 4) Analyzing and interpreting data; 5) Using math and computational thinking; 6) Constructing an explanation and designing a solution; 7) Engaging in an argument stemming from evidence; and 8) Obtaining, evaluating, and communicating information (Spence et al., 2021). An inquiry framework uses mathematics and systems-thinking processes of science and engineering to prepare K-12 students from diverse socio-economic and ethnic backgrounds to develop an in-depth understanding of content as well as develop key skills such as written communication, collaboration, inquiry, problem solving and flexibility that will serve them throughout life (Januszyk et al., 2016). Moreover, note that the activities are gauged on the six levels of cognitive learning according to the revised version of Bloom's Taxonomy (Table 1).

The suggested total cost for the complete set of FEWLM workbooks is less than $300 for 30 students (see supplemental materials); the cost varies with the number of students. The cost to implement the individual workbooks varies between $26 and $100. The items needed to conduct the targeted activity can be purchased at the local Dollar Tree, Walmart, Lowes, pet store (e.g., PetSmart) or grocery store for reasonable prices, unlike science kits provided by other companies and organizations. The suggested store and pre-COVID price for each item required to complete a specific FEWLM activity is provided with the instructor's information at the beginning of each individual workbook. The overall cost will decline significantly during subsequent years because several items can be reused. Suggestions for reducing the total FEWLM cost consist of (1) including items on the classroom supply list at the beginning of academic year, (2) asking parents to donate items during the academic year, (3) purchasing school supplies in August during the 'back to school' sale, or (4) ordering the supplies online and shipping them to the respective store to avoid shipping cost. The time required to complete each workbook varies from two hours to 19 days (for two and a half weeks) depending on the activity (see Spence et al., 2021).

Food Energy and Water Learning Module (FEWLM) Workbooks

The completion of the FEWLM workbooks in a sequential order provides a scaffolded learning experience for 3rd – 8th graders. Scaffolding the curriculum allows for the repetition to learn, practice, and apply the scientific method (D'Costa & Schlueter, 2013), and allows students to recognize a connection between the environment and human health as well as nurturing numerical literacy (Spence et. al., 2021). These workbooks are designed to be appropriate for youth 3rd through 8th grade. Instructors can choose to use all the workbooks or select specific workbooks that align with their grade level curriculum. After downloading the FEWLM workbooks from the QUBES website (see supplemental materials), printed as loose sheets initially, they can be modified to a specific grade level by the removal of information too advanced for a particular age group or not printed prior to assembling, followed by stapling the worksheets to create a workbook. Part I consists of individual activities that are primarily designed to engage elementary school youth. Advanced versions, such as Part II of each workbook, are designed to provide middle school youth with a more independent and collaborative learning experience. Instructions for teachers to effectively implement these activities as well as the elementary and middle school NGSS learning objectives are provided with each workbook. The FEWLM workbooks were designed using graphics available at teacherspayteachers.com. A detailed description of each workbook can be found at the QUBES site (see supplemental materials); a brief descriptions of its contents are:

- ***Scientific Inquiry with Decimals and Fractions through the Colorful World of m&m's® Workbook***. During this two-part activity, elementary and middle school youth learn the scientific process and engage in mathematical thinking while enjoying an edible treat. Moreover, they learn to record, analyze, and interpret data gathered from tallying and sorting plain chocolate m&m's® by color characteristics. Students learn how to use their data to create bar graphs, which is a valuable tool for interpreting and communicating results. The overall goal is to promote mathematical thinking by determining whether the percentage of colors in each bag of m&m's® is identical. Part I, focuses on learning how to tally and sort data into categories followed by rudimentary mathematics such as calculating ratios and percentages, in fraction and decimal form. Part II, a continuation of part I, aims to teach elementary statistics such as calculation of the mean, median, and mode as well as introduction to sample replication and variability. The required time to complete the *Scientific Inquiry with Decimals and Fractions through the Colorful World of m&m's®* workbook is estimated to be two hours (per Spence et al., 2021).
- ***Exploring Scientific Inquiry and Mathematical Thinking with Skittles Workbook.*** The Skittles activity is an exact replica of the m&m's® activity and an excellent alternative for students who are allergic to nuts and chocolate.
- ***My Sprouting Bean Seed Workbook***. This two-part inquiry-based activity is an adaptation of the "Pocket Seed Viewer" provided by *ask a biologist* (https://askabiologist.asu.edu/content/pocket-seed-viewer) that is designed to allow students to practice the scientific process. During this activity, students learn about the seed germination process, the history of black, pinto, and lima beans as a staple food for Central, North, and South American indigenous cultures, respectively, and the human health and environmental benefits of including beans in their diet. During Part I, the youth

observe and compare the root and shoot growth rate of black, pinto, and lima beans in a pocket seed viewer over a seven-day period. During Part II, the middle school youth continue to learn the importance of sample replication and variability via working collaboratively as a team of five "student researchers" by combining their individual data to calculate average, median, and mode for the growth length of root and shoot for each bean. Students use line graphs to illustrate the average growth rate of root and shoot over the seven-day period for each bean. (Spence et al, 2021).

- ***Let's Take a Biochemical Journey Through Osmosis with a Naked Hen Egg Workbook***. This is a classic hypothesis-driven, inquiry-based experiment that introduces fundamental biological and chemical laws and theories because of understanding osmosis and the effects of hypotonic, isotonic, and hypertonic conditions on a naked hen egg (i.e., cell model). This hand-held cell model helps students learn how common foods most people eat may affect the cell and eventually, the whole body. They discover the negative impacts of unhealthy eating habits, e.g., too much sugar or salt in their diet, can impact their cellular function and learn how too much of these molecules and compounds, sugar and table salt , respectively, are associated to human diseases such as diabetes, obesity, and high-blood pressure; diseases known to plague minority communities (United States Department of Health and Human Services, 2021). During Part I, the youth observed osmosis in the naked egg cell model using various salt concentrations while expanding their knowledge about scientific inquiry by way of identifying the independent, dependent, experimental, and standardized variables. Students practice illustrating their data using both bar and line graphs and learn the appropriate graph to use for illustrating their data. Part II of the activity provides the middle school youth with the opportunity to apply the scientific concepts learned during Part I to design and implement their own experiment and investigate the effects of one independent variable (i.e., sugar, vinegar, corn syrup, rubbing alcohol, corn- starch or baking soda) on a naked hen egg. (Spence et al., 2021).
- ***Beneficial Health Impacts of Fruits and Vegetables Workbook.*** This hands-on, inquiry-based learning activity reaffirms the importance and benefits of eating fruits and vegetables. Students learn how red cabbage juice can be used as a pH indicator solution as well as molecules within red cabbage and other vegetables or fruits that provide the natural defenses of antioxidants against cellular oxidative stress by detecting the strength of antioxidants against free radicals, such as hydrogen peroxide. During this activity, the youth collect qualitative data by documenting the changes in color of red cabbage solution for various fruits and vegetables. They continue to expand their knowledge about scientific inquiry by identifying independent, dependent, experimental, and standardized variables; they learn the difference between qualitative and quantitative data and practice when to illustrate their data using a bar graph. (Spence et al., 2021).
- ***Exploring the Link between Land Management, Plant Growth and Water Quality Using a Mini-Garden Model Workbook.*** Horticultural plants, grown specifically for human use, offer diversity to human diets and enhance the human living environment. This hands-on, multi-week activity allows students to continue practicing the scientific method while learning about the physical and chemical properties of soil, gardening practices, and the importance of water quality. This is a fun and practical activity for middle school youth to learn about techniques for growing food, the influence of water quality, and the implications they have on land management. During this activity, students will collaborate as a research team to create mini-garden models with the aim of investigating the practice of applying organic materials, such as turf grass clippings, leaves, and other natural materials on plant growth as well as nitrogen leaching and water movement through

a soil profile. Moreover, students will learn how the carbon to nitrogen (C:N) ratios of organic materials influence the rate of nitrogen availability for plant uptake. This activity is designed for students to collect plant, soil, and water quality data daily for up to 19 days (about 2 and a half weeks). At the end of the experiment, students will calculate the mean growth lengths, plot line graphs, analyze their results, and write down their explanations, discussions, and conclusions in their workbook. Additionally, this activity will enable students to self-reflect on what they learn. Spence et al., 2021).

Research Question

With the review of the literature as it relates to "at risk" student groups, the DISCUSSION Network Core Team members developed opportunities for the targeted "at risk" URM 3rd – 8th grade students to be exposed to mentorship and inquiry-based STEM activities. The questions are as follows: What type of impact do the FEWLM workbooks have on the students exposed to them and is the impact favorable or not?

RESULTS

Impact of FEWLM on Interest in Science and a Scientific Career

50% of the elementary school youth at the BGCGD after school program aspired to attend college after high school and 67% of the youth wanted to enter non-science professions as an adult. Even though a small percentage of the youth aspired to have a STEM career (Spence et al., 2021), many self-reported they enjoyed science (56%) and agreed with the statements "I am good at science" (86%), "I am good at math" (75%), and "I am good at engineering" (60%). Despite their confidence in performing science, only 31% of the youth declared science and 52% of the youth declared math was their favorite subject in school. Many believe they are creative (86%) and enjoy learning how things work. When asked "have you carried out a scientific experiment in your classes at school," 56% of the elementary school youths responded "yes" and 8% responded "I don't know" (See Spence et al., 2021).

Elementary School Youth

Spence et al. (2021) reported elementary school age youth self-reported that family interest in science, family encouragement to study science, and family enthusiasm about a science career was low. Informal science learning experiences, such as visiting museums and exhibits, seemed to help pique their interest in science; however, these experiences did not increase the likelihood of considering a career in science. Despite the fact their teachers made science interesting and encouraged them to learn STEM, most of these youths had truly little interest in becoming a scientist. Overall, they enjoyed learning science and their attitude about scientists was positive. These youths understood the importance of science in the real world, recognized that scientists help make people's lives better (72%), and stated they would enjoy a job helping to protect the environment (71%), yet only a few aspired to become a scientist (Figure 1). Furthermore, unpublished draw-a-scientist data suggests that many of the elementary youth associate scientists with working with chemicals in the lab.

After engaging with the FEWLM workbooks, the elementary youth self-reported that FEWLM helped them understand science better, motivated them to want to work harder in school, and increased their confidence in doing science. Furthermore, the survey captured a slight decrease in elementary school youth looking forward to attending science class and their interest in pursuing a science related career; however, their interest in becoming a scientist slightly increased. Only 38% reported they wanted to try to redo the FEWLM activities at home. Overall, the elementary school youth in the BGCGD after school NCCU DISCUSSION STEM Club (78%) stated they would recommend the FEWLM curricula to their friends.

Middle School Youth

All the middle school age youth participating in the NCAT/ NCCU DISCUSSION HydroPHonics STEM Enrichment Summer Camp aspired to attend college after graduating from high school. Among these middle school youths, 46% think they will pursue a career in a science related field; however, only 11% think they want to become a scientist. Several middle school youths self-reported they receive good grades in science (55%) and math (60%), yet many agree that science is a difficult subject (48%), especially when it involves math (48%). Even though the middle school youth had truly little interest in becoming a scientist, many look forward to attending science class (87%).

After engaging with the FEWLM curricula at the NCAT/NCCU DISCUSSION HydroPHonics STEM Enrichment Summer Camp, the middle school youth participants reported that FEWLM helped them understand science better, motivated them to want to work harder in school, and increased their confidence in doing science. There was an increase in the percentage of middle school youth looking forward to attending science class (+9%) and thinking about becoming a scientist (+4%). Moreover, the middle school youth reported FEWLM increased their interest to study science in college, helped them to better understand their career goals, made them think about what they will do after graduating from high school, and increased their confidence in their ability to participate in science projects and activities (Figure 1). Overall, most participants attending the NCAT/NCCU DISCUSSION HydroPHonics Summer Enrichment Camp (94%) would recommend the FEWLM curricula to their friends, and several were interested in trying the FEWLM activities at home (77%).

DISCUSSION

A fortuitous outcome for the DISCUSSION Network Core Team was to identify, develop and implement the scaffolded low-cost, inquiry based FEWLM workbooks that provides a culturally relevant learning experience for 'at risk' underrepresented elementary and middle school youth while encouraging them to become interested in STEM careers (see supplemental materials). For example, once students create the naked egg, students measure the mass of the egg after sitting in hypertonic salt or sugar solutions over time. The effects of the solution on the mass of the egg are correlative of a 'cell' in a person diagnosed with hypertension or diabetes, two metabolic disease states that disproportionately affect members of URM ethnic groups.

FEWLM promotes learning science by performing the scientific process to conduct research experiments, which aligns with the Next Generation Science Standards. FEWLM workbooks are designed for these youths to collaborate self-sufficiently as a group of 'scientists in training,' which teaches team-building skills. Furthermore, FEWLM focuses on solving real world problems using an interdisciplinary approach with an emphasis on data interpretation and promotes self-reflection as learning mechanisms. Data collected from the AWE survey instrument suggest FEWLM is effective for teaching and positively promoting scientific literacy and quantitative skills to broaden the participation in STEM across socio-economic barriers for elementary and middle school youth. This outcome is correlative with the outcome of outreach programs regardless of demographic or country. (Ghazzawi et al., 2021; Markle et al., 2022). These low-cost, inquiry-based learning modules provide a scaffolded curriculum for youth to learn, practice, and apply scientific inquiry while learning concepts associated with the FEW nexus (Table 1). Through FEWLM, students learn that scientific concepts associated with the food, energy and water nexus are not isolated facts but interconnected and interdependent systems with complex solutions that require systems thinking. Evidence from the elementary and middle school youth indicates FEWLM made a positive impression on the youth and increased students' interest in STEM, their perception of its value to their lives, and confidence in their ability to participate in STEM. The youth have some understanding of STEM fields. In underrepresented communities, 'traditional scientific fields' most familiar are medical doctors, chemists, biologists, and engineers. The word 'scientist' is broad and can represent many different fields. During our sessions, the youth would discuss their career ambitions and talk about specific scientific professions. Less familiar scientific careers are soil scientists, ecologists, astronomers, or astrophysicists, etc., because they do not often see people in these fields in their communities. However, a possible caveat to the results that was not addressed by the survey was the impact of being taught the FEWLM Workbooks by DISCUSSION Network Core Team members who are African American. Studies on the impact of teachers' students can identify with resounding clarity that representation for students before 3rd grade has a profound impact on whether the students graduate from high school, let alone pursue a college degree. Therefore, the impact of FEWLM workbooks may generate different results if the material is presented by an educator who is not representative of the class population they are instructing.

The goal of this project was to provide a low-cost, high-quality science curriculum that has relevance and is practical to the lives of URM students in K-12 grades. This has been achieved by explicitly embedding concepts in learning materials that stimulate thinking about solutions to human and environmental health issues, and threats to conservation and Earth ecosystem balance. Although the target group for this curriculum is URM students, it does have value and wide appeal beyond just this group (Spence et al., 2021). Recently, a DISCUSSION Network Core Team member, used the "*Beneficial Health Impacts of Fruits and Vegetables Workbook*" to an afterschool program that consisted of a diverse racial, gender and socioeconomic group of students at a middle school in Durham, NC. The adults who volunteered were also a diverse group who were members of the outreach component of the Durham Museum of Life and Science as well as scientists from companies in the nearby Research Triangle Park (RTP) and educators at the middle school. Anecdotally, the feedback concerning whether the workbooks impacted the students was incredibly positive. Moreover, the volunteers enthusiastically declared they learned some fundamental concepts they did not previously know. For example, the *H* in pH is the concentration of Hydrogen ions in a solution.

Advantages and Disadvantages of Implementing FEWLM

There are advantages and disadvantages for instructors who adopt FEWLM to teach scientific inquiry. FEWLM provides affordable and student-centered learning for elementary and middle school youth to learn that science is a process and supported with data and other forms of evidence. These workbooks are designed for instructors who teach science inquiry in both formal and informal educational settings. There are no predetermined answers. Every time the FEWLM workbooks are implemented, a new data set is generated. The cost for adopting these workbooks is less than $300 for a class size of 30 students, with the cost for executing these modules during subsequent years becoming lower because several items on the materials lists can be reused. Moreover, the cost to use all 6 of the FEWLM workbooks in a class setting with a lower number of students would be even more affordable.

The FEWLM budget does not account for the cost of printing each student their own workbook. These workbooks are lengthy, with the number of pages for a given activity ranging from 14 to 31 and requiring each student to have their own copy to complete the activity. For instructors at institutions with printing restrictions, the length of the workbook could be a deterrent for selecting to use FEWLM. As a substitute for printing individual workbooks, instructors can print one copy for each group, and the students can complete the activity in composition notebooks. If instructors plan, 100-sheet composition notebooks, which are quite reasonable in cost during a back to school sale and can be included on the parent's school list at no additional cost for instructors. Furthermore, instructors can use the composition notebook as an opportunity to teach elementary and middle school youth the purpose and concept of using laboratory notebooks as a scientific tool.

Most school systems have constraints regarding instructional time and topics to cover during an academic year. Third through eighth graders enrolled in public schools typically receive approximately 33 hours of school per week (Hoyer & Sparks, 2017). On average, third graders in public school spend 2.9 instructional hours per week learning science while the science instructional time for eighth graders is 4.3 hours. Private elementary and middle schools spend less time teaching science than public schools (Hoyer & Sparks, 2017). Therefore, it could be challenging for teachers to complete all the FEWLM workbooks because the length of time required to complete all the activities could interfere with instructional time dedicated to other subjects, such as reading. Nevertheless, teachers could use FEWLM as a mechanism to indirectly practice reading, writing, and mathematics while increasing the instructional time allotted for science (Spence et al., 2021).

Study Limitations

The BGCGD after school program and NCAT/NCCU HydroPHonics summer camp were not designed to fully measure the comprehension of the subject matter; therefore, a pre/post-assessment was not administered to measure learning gains. The AWE survey instruments implemented did not included open-end questions to allow the youth to explain their answers. Furthermore, due to competing activities during the BGCGD after school program and flexible pick-up times, the number of youths completing the FEWLM workbooks varied each day making it difficult to keep accurate attendance records. As a result, it was challenging to efficiently evaluate the comprehension of the subject matter using the associated grading rubrics. Even though the summer camp was structured to be a day-long program and the participants were able to dedicate more time completing the activities, the workbooks were not collected and evaluated to determine comprehension of the subject matter because the middle school youth were encouraged to take their workbooks home and continue the experiments. They were not instructed to return their workbooks to evaluate comprehension and learning outcomes. Attendance was not taken at the NCAT/NCCU HydroPHonics summer camp.

CONCLUSION

In summary, FEWLM is an affordable approach to teaching STEM to 'at-risk' URM students categorized as less competitive in STEM. Evidence suggests a collaborative effort by community organizations that target 'at-risk' URM 3rd–8th graders can use the low-cost interventions of FEWLM workbooks to develop a sense of connection, a change in attitude towards STEM, and the utility goals and values generated to benefit 'at-risk' 3rd–8th grade URMs. Although encouraging, the evidence presented in this study is limited in generalizability, so further research is needed to replicate and extend it to other outreach settings. Through affordably programmed science content and curricular materials developed, the DISCUSSION Network is connecting STEM learning to the immediate, relevant, and personalized life experiences of students from a range of socio-economic backgrounds, resulting in increases in science interest and literacy, and this approach may be an effective stimulus for driving social change that improves the participation of URM in STEM.

Supplemental Materials

The elementary and middle school Next Generation Science Standards learning objectives for each FEWLM workbook is available for download as supplementary materials. Educators interested in adopting FEWLM can access it at the QUBES DISCUSSION website: https://qubeshub.org/publications/2287/1.

REFERENCES

Abdi, A. (2014). The effect of inquiry-based learning method on students' academic achievement in science course. *Universal Journal of Educational Research*, *2*(1), 37–41. doi:10.13189/ujer.2014.020104

Ahmed, M. W., Anderson, Y., Gerald-Goins, T., Hollowell, G. P., Salim, E. T., Sangutei, T., Simpson, B., Spence, P., Whittington, D., & White, S. L. (2021). Promoting STEM-literacy by designing decision-driven interdisciplinary courses for non-science majors. *Journal of STEM Education: Innovations and Research*, *21*(3), 28–34.

Alamance-Burlington School System. (2016). *7/11/16 Closing the achievement GAP*. Alamance-Burlington School System.

American Association for the Advancement of Science. (2011). *Vision and change in undergraduate biology education: A call to action.* AAAS.

Assaraf, O. B.-Z., & Orion, N. (2005). Development of system thinking skills in the context of earth system education. *Journal of Research in Science Teaching*, *42*(5), 518–560. doi:10.1002/tea.20061

Assaraf, O. B.-Z., & Orion, N. (2010). System Thinking Skills at the Elementary School Level. *Journal of Research in Science Teaching*, *47*(5), 540–563.

Assessing Women & Men in Engineering. (2010). *Pre-college out-reach STEM activity upper elementary school surveys.* The Pennsylvania State University and University of Missouri. http://aweonline.org/online- surveys.html

Assessing Women & Men in Engineering. (2008). Pre-college out- reach STEM activity pre-college middle school science surveys. The Pennsylvania State University and University of Missouri. http://aweon- line.org/onlinesurveys.html

Bandura, A., & Locke, E. A. (2003). Negative self-efficacy and goal effects revisited. *The Journal of Applied Psychology*, *88*(1), 87–99. doi:10.1037/0021-9010.88.1.87 PMID:12675397

Boys and Girls Club of Durham and Orange Counties. (2017). *2017 Impact report*. Boys and Girls Clubs of America.

Burke, A., Okrent, A., & Hale, K. (2022, January 18). *The State of U.S. Science and Engineering 2022*. National Science Board. https://ncses.nsf.gov/pubs/nsb20221

Carlsen, M. H., Halvorsen, B. L., Holte, K., Bøhn, S. K., Dragland, S., Sampson, L., Willey, C., Senoo, H., Umezono, Y., Sanada, C., Barikmo, I., Berhe, N., Willett, W. C., Phillips, K. M., Jacobs, D. R. Jr, & Blomhoff, R. (2010). The total antioxidant content of more than 3100 foods, beverages, spices, herbs, and supplements used worldwide. *Nutrition Journal*, *9*(1), 3. doi:10.1186/1475-2891-9-3 PMID:20096093

Cuevas, P., Lee, O., Hart, J., & Deaktor, R. (2005). Improving science inquiry with elementary students of diverse backgrounds. *Journal of Research in Science Teaching*, *42*(3), 337–357. doi:10.1002/tea.20053

D'Costa, A. R. S., & Mark, A. (2013). Scaffolded Instruction Im- proves Student Understanding of the Scientific Method & Experimental Design. *BioOne*, *75*(1), 18–28.

Daily, G. C., Alexander, S., Ehrlich, P. R., Goulder, L., Lubchenco, J., Matson, P. A., Mooney, H. A., Postel, S., Schneider, S. H., Tilman, D., & Woodwell, G. M. (1997). *Ecosystem services: Benefits supplied to human societies by natural ecosystems*. Ecological Society of America.

de Graaf, H. J., Musters, C. J. M., Noordervliet, M. A. W., & ter Keurs, W. J. (2005). An instrument for interactive setting development goals of a socio-environmental system. *Ecological Economics*, *52*(1), 19–29. doi:10.1016/j.ecolecon.2004.05.006

Estrada, M., Woodcock, A., Hernandez, P. R., & Schultz, P. W. (2011). Toward a model of social influence that explains minority student integration into the scientific community. *Journal of Educational Psychology*, *103*(1), 206–222. doi:10.1037/a0020743 PMID:21552374

Ghazzawi, D., Pattison, D., & Horn, C. L. (2021). Persistence of Underrepresented Minorities in STEM Fields: Are Summer Bridge Programs Sufficient? *Frontiers in Education*, *6*, 630529. Advance online publication. doi:10.3389/feduc.2021.630529

Goel, A. K., Rugaber, S., & Vattam, S. (2009). Structure, behavior, and function of complex systems: The structure, behavior, and function modeling language. *Artificial Intelligence for Engineering Design, Analysis and Manufacturing*, *23*(1), 23–35. doi:10.1017/S0890060409000080

Goodwin, T. (2016). Educating for ecological literacy. *The American Biology Teacher*, *78*(4), 287–291. doi:10.1525/abt.2016.78.4.287

Goudie, C. A. (2009). Institutionalizing ecological literacy: A critical component of the new "green" economy. *Business Perspectives*, *19*(3), 16–21.

Harper, S. R., & Griffin, K. A. (2011). Opportunity beyond affirmative action: How low-income and working class Black male achievers access highly selective, high-cost colleges and universities. *Harvard Journal of African American Public Policy*, *17*(1), 43–60.

Hinton, A. O., Termini, C. M., Spencer, E. C., Rutaganira, F. U. N., Chery, D., Roby, R., Vue, Z., Pack, A. D., Lillian, J. B., Garza-Lopez, E., Marshall, A. G., Lewis, S. C., Shuler, H. D., Taylor, B. L., & McReynolds, M. R. (2020). Patching the Leaks: Revitalizing and Reimagining the STEM Pipeline. *Cell*, *183*(3), 568–575. doi:10.1016/j.cell.2020.09.029 PMID:33125882

Hoyer, K. M., & Sparks, D. (2017). *Instructional time for third and eighth graders in public and private schools: School year 2011-12*. Department of Education.

Januszyk, R., Miller, E. C., & Lee, O. (2016). Addressing student diversity and equity: The next generation science standards are leading a new wave of reform. *Science Scope*, *39*(8), 16–19. https://www.jstor.org/stable/43827310. doi:10.2505/4s16_039_08_16

Khoo, H. E., Azlan, A., Tang, S. T., & Lim, S. M. (2017). Anthocyanidins and anthocyanins: Colored pigments as food, pharmaceutical ingredients, and the potential health benefits. *Food & Nutrition Research*, *61*(1), 1361779–1361779. doi:10.1080/16546628.2017.1361779 PMID:28970777

Lent, R. W., Brown, S. D., Sheu, H.-B., Schmidt, J., Brenner, B. R., Gloster, C. S., Wilkins, G., Schmidt, L. C., Lyons, H., & Treistman, D. (2005). Social cognitive predictors of academic interests and goals in engineering: Utility for women and students at historically black universities. *Journal of Counseling Psychology*, *52*(1), 84–92. doi:10.1037/0022-0167.52.1.84

Markle, R. S., Williams, T. M., Williams, K. J., deGravelles, K. H., Bagayoko, D., & Warner, I. M. (2022). Supporting Historically Underrepresented Groups in STEM Higher Education: The Promise of Structured Mentoring Networks. *Frontiers in Education*, *7*, 674669. doi:10.3389/feduc.2022.674669

Mohtar, R. H., & Daher, B. (2012). Water, energy, and food: The ultimate nexus. In Encyclopedia of Agricultural, Food, and Biological Engineering (2nd ed., pp. 1-5). CRC Press, Taylor, and Francis Group. https://doi.org/ doi:0.1081/E- EAFE2-120048376

Musters, C. J. M., de Graaf, H. J., & ter Keurs, W. J. (1998). Defining socio-environmental systems for sustainable development. *Ecological Economics*, *26*(3), 243–258. doi:10.1016/S0921-8009(97)00104-3

National Academy of Sciences. National Academy of Engineering, & Institute of Medicine. (2011). *Expanding under- represented minority participation: America's science and technology talent at the crossroads*. The National Academies Press. https://doi.org/doi:10.17226/12984

National Center for Science and Engineering Statistics. (2019). *Women, minorities, and persons with disabilities in science and engineering: 2019*. N. S. Foundation. https:// www.nsf.gov/statistics/wmpd

National Research Council. (2013). *Next Generation Science Standards: For States, By States*. The National Academies Press. doi:10.17226/18290

National Science Foundation (2015). *America's Future: Environmental Research and Education for A Thriving Century. A 10 Year Outlook: A Report by the NSF Advisory Committee for Environmental Research and Education*. NSF.

National Science Foundation. (n.d.). Inclusion across the Nation of Communities of Learners of Underrepresented Discoverers in Engineering and Science. *NSF INCLUDES*. https://www.nsf.gov/news/special_reports/big_ideas/includes.jsp

Next Gen Science. (2013). *Next Generation Science Standards: For States, By States*. The National Academies Press. https://www.nextgenscience.org/

Nolte, K. (2008). *Red Cabbage*. University of Arizona. https:// cals.arizona.edu/fps/sites/cals.arizona.edu.fps/files/cotw/ Red_Cabbage.pdf

President's Council of Advisors on Science and Technology (U.S.). (2012). *PCAST engage to excel: Producing one million additional college graduates with degrees in science, technology, engineering, and mathematics*. President's Council of Advisors on Science and Technology (PCAST).

Robbins, T., Stagman, S., & Smith, S. (2012). *Young children at risk: National and State Prevalence of Risk Factors Fact Sheet*. National Center for Children in Poverty: Mailman School of Public Health Columbia University. https:// www.nccp.org/publication/young-children-at-risk/

Science Buddies. (2012). *Cabbage Chemistry--Finding Acids and Bases*. Scientific American. https:// www.scientificameri- can.com/article/bring-science-home-cabbage-chemistry/

Şimşek, P., & Kabapınar, F. (2010). The effects of inquiry-based learning on elementary students' conceptual understanding of matter, scientific process skills and science attitudes. *Procedia - Social and Behavioral Sciences, 2*(2), 1190-1194. https://doi.org/https://doi.org/10.1016/j.sb- spro.2010.03.170

Spence, P. L., Gerald-Goins, T., & Goins, G. (2021). *Food, Energy and Water Learning Modules (FEWLM) Workbooks*. The DISCUSSION Network, QUBES Educational Resources. https://qubeshub.org/publications/2287/?v=1

Spence, P. L., Phillips, R. S., McAllister, A. R., White, S. L., & Hollowell, G. P. (2020). Student-scientist curriculum: integrating inquiry-based research experiences and professional development activities into an introductory biology laboratory course. *Journal of Microbiology & Biology Education, 21*(3). https://doi.org/doi:10.1128/jmbe. v21i3.2225

Trauth-Nare, A. (2015). Influence of an intensive, field-based life science course on preservice teachers' self-efficacy for environmental science teaching. *Journal of Science Teacher Education, 26*(5), 497–519. doi:10.100710972-015-9434-3

United States Department of Health and Human Services. (2021). *Office of Minority Health Home*. United States Department of Health and Human Services. https://minorityhealth.hhs. gov/

Vespa, J., Armstrong, D., & Medina, L. (2020). *Demographic Turning Points for the United States: Population Projections for 2020 to 2060*. USCB. https://www.census.gov/library/publications/2020/demo/p25-1144.html

Section 2
STEM in the Community

Chapter 7
Identity-Centered STEM Curricula for Black Girls:
An Intersectional Intervention for Black Girl Joy

Nicole M. Joseph
Vanderbilt University, USA

Christy L. Erving
University of Texas at Austin, USA

Kimberlyn A. Ellis
Vanderbilt University Medical Center, USA

ABSTRACT

This chapter presents data from Black Girls Becoming Summer Research Institute (BGB), a two-week summer residential program for rising seventh and eighth grade Black girls at Vanderbilt University. The program was intentionally designed with Black girlhood celebration and joy as its centripetal force—all instructors, chaperones, volunteers, and even the photographer were Black women. The girls spent two weeks engaging in STEAM workshops that also supported their socio-emotional development while developing academic literacies. Using intersectionality and Black girlhood as lenses for understanding the various learning experiences, the authors present narratives that reclaim educational spaces in Black girls' own images. In so doing, this chapter serves as one model for higher education administrators seeking to expand their STEM footprint within their community as well as researchers interested in the research impacts on student learning outcomes from STEM-related programming.

DOI: 10.4018/978-1-6684-7771-7.ch007

INTRODUCTION

Black girls seldom have educational spaces that honor their full humanities (Brown, 2013). In fact, a large body of research shows that schools and schooling are places and processes that dehumanize, demean, marginalize, and violate Black girls (Carter Andrews et al., 2019; Halliday, 2019; Morris, 2016). Thomas (2021) pointed out that society and schools force stereotypes onto Black girls and their names are "erased and replaced with labels for metaphorical boxes of stereotypes: loud, angry, emotional, and disruptive" (p. 9). Elevating such harmful research against Black girls is one way of talking back and disrupting such consequential education research; but Milner (2021) points out that we should not stay there; that there is work to be done. Therefore, designing and conducting education research that matters for public consumption and practice, and ensures equitable processes and results, is necessary.

This chapter advances knowledge about the Black Girls Becoming (BGB) Summer Research Institute, a novel STEAM experience for rising 7th and 8th grade Black girls. Specifically, we delineate the design principles and components of the BGB's curriculum and experience to provide technical assistance to other university faculty seeking to expand their STEM outreach footprint within their communities. We also share findings from the inaugural research study from summer 2022. We do this as part of our argument that any STEM outreach opportunities focused on marginalized students should be coupled with social science research. It is critical to not only assess the effectiveness of the program, but also to add to the literature on marginalized students' STEM learning opportunities in informal educational spaces. Emerging research suggests informal learning spaces can be insightful and imaginative locations for disrupting domination, push out, and white supremacy in STEM so that students' STEM interest, agency, sense of belonging, and positive discipline identities can shine (Collins, 2018; King & Pringle, 2019; Lane & Id-Dean, 2023). One important approach to conducting such research is through an intersectional lens. Intersectionality affords understanding into broader systems and structures of oppression that often get in the way of marginalized learners' robust learning and achievement (Crenshaw, 1990). We organize the chapter by first providing a positionality statement of the authors, followed by a section that unpacks the design principles of BGB. We then discuss the materials and methods used to bring to fruition a successful program, followed by a section on the BGB's longitudinal research study. Finally, we include a brief discussion about the public and scholarly impacts on the girls' learning and life outcomes from BGB.

BACKGROUND

We are a group of intergenerational Black Girl Cartographers, meaning we have deep concern and advocacy of Black girls' and women's "health, lives, well-being, and ways of being" (Butler, 2018, p. 33) across the lifespan. We self-identify as Black researchers and members of the Intersectional Study of Black Women and Girls in Society (Vanderbilt University, 2023a) leadership team, a Trans-Institutional Programs (TIPs) grant initiative made up of all Black women faculty and doctoral students focused on forms of resistance and barriers Black women and girls face in learning spaces (Vanderbilt University, 2023b). Black Girls Becoming is one part of the TIPs initiative, and the curriculum and study were conceived through this larger initiative funded through Vanderbilt University. Thus, our own biases and assumptions are that all Black girls deserve boundless possibilities in educational spaces whereby their intersectional identities are affirmed, their ways of being and knowing themselves and the world are valued, and their adolescent development into adulthood is a solidarity project with their teachers, families, and communities. Although prior research findings influenced the design principles of BGB, our partiality shaped the research questions we asked and guided our analyses. This is reflected in the interpretations about the structural barriers and forms of resilience Black girls face in the society and schools, as well as what transformative intersectional interventions might look like to dismantle disparities and enhance advancement.

Researchers have shown that there is no such thing as purely objective research, thus we aim for strong objectivity (Harding, 1992), in which the authors substantiate who they are and their assumptions. At the same time, we bring a standpoint (Harding, 1992) as Black women, living in Black bodies, experiencing our own racialized and gendered education in the United States. Nicole Joseph attended K-12 schools that were racially and ethnically diverse. She was in advanced courses across her education trajectory which came with a vetting process, so teachers held high expectations for her. She did not realize the structural inequalities in her schools until high school and became aware of other Black students' limited access to a high-quality education. She eventually became a mathematics teacher and made it her mission to use ambitious and equitable instruction and advocacy for Black students, especially Black girls, to have better chances for life success. Christy Erving attended a Black-affirming elementary school which also consisted of a predominantly Black student and teacher population. With the encouragement of her elementary school teachers, she applied for and was accepted into a middle school academy and, subsequently, a high school magnet, both with selective admissions processes. As such, in grades 7-12, she found herself in a more racially and economically diverse environment of students and teachers. During this time, she often reflected on the stark contrast in opportunities and resources between her local school district and the middle and high school academy/magnet schools she attended. Upon completion of high school, she attended predominantly white institutions of higher education, but simultaneously maintained a commitment to Black students through participation in Black student and Black women-led organizations throughout college and graduate school. She has been a faculty member at three different predominantly white institutions of higher education in the U.S. South and is committed to mentoring and elevating the voices of Black and other minoritized students in PWI higher education contexts. Kimberlyn Ellis attended a middle-class and primarily white K-12 education system that simultaneously upheld white supremacy and patriarchy in the classroom while refusing to acknowledge minority identities and experiences. While Kimberlyn had a small sample of teachers that nourished her talents and were significant in her development and journey of self-actualization, many teachers knowingly and unknowingly embedded racist and discriminatory comments, practices, and pedagogies in their direct interactions with Kimberlyn and the classroom curricula. An inexplicit, and sometimes explicit, motto of educators in Kimberlyn's K-12 education system was 'We don't see color.' This proved to be a common scapegoat for ignoring both the needs of their minority students and refusing to admonish the racist caricature of Native Americans that served as their school mascot. Ironically, Kimberlyn also 'didn't see color,' as she did not have a single Black teacher, male or female, until she graduated from high school and attended an HBCU for college. As a doctoral student, Kimberlyn unapologetically centers Black populations, especially Black women, in her graduate research, volunteerism, and activism.

We acknowledge that we live with and embody tensions. On the one hand, education in the United States is broken and does not care about Black girls' lives. On the other hand, we know that Black girls will show up for class on Monday, so we work to support them, their families, and the teachers that teach them. We see BGB's work as an ongoing quest for Black girls' learning to focus on deeply understanding themselves as beautiful, worthy, and critically conscious thinkers and knowledge producers as they become young adults with gifts and talents to give to themselves, their communities, and the world.

Black Girls Becoming Summer Research Institute: Design Principles

BGC is a summer residential living and learning community experience model held at Vanderbilt University. It is a fully sponsored program funded by grants and is a part of a larger initiative that draws on the intellectual resources in sociology, law, STEM, education, African American and Diaspora Studies, religious studies, and health. This initiative has created a transdisciplinary hub of research, discovery, and teaching activities centered on elevating and understanding structural barriers and forms of resilience Black women and girls experience across various social contexts in society, and how intersectional interventions might be created to expand opportunities and increase pathways to success. BGB was not a traditional partnership with a local school district because we have found that sometimes districts may have different agendas limiting the partnership. For example, districts may not have wanted a single focus on just Black girls, but girls of color. BGB used social media to attract and recruit girls into the program. We will say more about recruitment later.

The set of values that acted as a compass for BGB included Black girlhood and the Five Cs Model of Positive Youth Development (PYD). BGB is one type of *intersectional intervention* because it was designed with more inclusive constructs than traditional STEM curricula. For example, the curriculum centered the narratives of Black girls and increased representation within the classroom; all 39 instructors, chaperones, volunteers, and staff identified as Black women, as research suggests that representation in the classroom has a strong impact on Black girls (Gershenson et al., 2018). The program hosts 20 Black girls from the local Nashville area. The girls live in the dorms and eat in the cafeteria while learning about themselves and their peer-sisters through classes in science, critical consciousness, the arts (visual, physical, and abstract), research fundamentals, financial literacy, and evening centering. The purpose was to design a learning environment where Black girlhood is centered and elevated without the "white gaze;" the white gaze means the expectation of assimilation and constant comparison to white girls in every way, from forms of femininity to academic intelligence (Carter Andrews et al., 2019; Morris, 2015; Ray, 2022). Research shows that Black girls face negative issues in schools and classrooms that hurt their development, identities, and achievement in large part because Black girls are often mischaracterized and misunderstood through stereotypes; many educators expect Black girls to look and behave like white girls (Gholson, 2016; Gholson & Martin, 2019; Morris, 2019). But because education scholars seldom study Black girls through an intersectional lens, issues they face fly under the radar; after all, what gets promoted in social media and national news is that Black women are among the most educated group in the United States and have been receiving college degrees for years (Davis, 2020). However, when a closer look is considered, extreme vulnerabilities are present and persistent. For example, national data show that Black girls in the K-12 system are 5.5 times more likely to be suspended than white girls, and more likely to receive multiple suspensions than any other gender or race of students (Onyeka-Crawford et al., 2017). In the District of Columbia, where Black girls represent 73 percent of girls enrolled in school, but 94 percent of all girls suspended, they are an astounding 17.8 times more likely to be suspended than white girls (Onyeka-Crawford et al., 2017). If critical masses of Black girls are out of school, their learning time, and opportunities to learn are significantly decreased. However, some recent research on the impacts of COVID on adolescent Black girls showed that much learning was happening. For instance, in their study that interviewed a total of 25 participants whose ages ranged from 9–18 years,

Crooks et al. (2022) found that the majority of the girls shared feelings of empowerment and self-love as consequences of the pandemic that may not have occurred without the opportunity to isolate themselves from others during quarantine. Other positive areas of life outcomes included weight loss, focusing on their mental health, and becoming more critically self-aware and reflective.

The racialized/gendered discipline disparities data are driven by how Black girls are adultified and criminalized in schools. Adultification bias is a form of racial prejudice where children of minority groups, typically Black children, are treated by adults as being more mature than they actually are. When adults, particularly educators, view Black girls in these ways, Black girls receive harsher treatment and discipline; clemency is seldom given to Black girls in comparison to white and other girls of color (Morris, 2016). Research has shown that because of this adultification, teachers and general people in society think that Black girls need less protection, less nurturing, less comfort, and less support; this bias against Black girls starts at age five (Epstein et al., 2017). Thus, there is a critical need for youth empowerment programs designed specifically for Black girls to thrive in informal spaces because schools have been found to be inhospitable places for them to learn, develop strong academic identities, and achieve at high levels (Carter Andrews et al., 2019).

The Five Cs Model of PYD was also an important component to the design of the BBGP. This is a framework for youth that outlines five psychological, behavioral, and social characteristics that indicate youth are thriving and, when high levels of the Five Cs are developed, ultimately this can result in contributing to self, community, and society. Based on almost three decades of developmental science, the Five Cs Model of PYD is the most empirically supported framework today (Bowers et al., 2010). The model focuses on the positive characteristics that enable adolescents to lead productive and healthy lives and contribute to the world in ways that benefit themselves, their families and communities, and civil society. The five principles of this framework are connection, confidence, character, competence, and contribution.

- *Connections*: BGB provides opportunities for the girls to make connections through the curriculum, the staff and faculty, each other, their families, and the broader college community. Evidence from our survey data showed that the girls felt safe and a sense of belonging because we built into the curriculum social learning activities emphasizing bonding and making meaningful connections.
- *Confidence*: Our surveys also showed that across the two weeks, the girls felt a sense of self-worth because they were not penalized for being Black girls—they were encouraged to keep their phones, create TikTok videos of their experiences at BGB to provide the girls freedom of expression and unpoliced girlhood elevating their voices in advocating for themselves.
- *Character:* Character development is also emphasized in BGB; the girls supported each other in taking responsibility and had multiple opportunities to show their individuality and independence. For example, in the critical consequences course, we had four girls 'come out' in community with the instructor and their peers as LBGTQ. All the girls showed empathy and care for their fellow sisters who were brave enough to take such an important step in their identity development.
- *Competence*: BGB mines for Black girls' competence; not just academically, but all the many ways that one can show what they know and can do. Schools seldom give Black girls these opportunities, but BGB opens space for Black girls to shine and see their brilliance. STEM outreach collaborations and opportunities need to understand that 'extending brilliance' beyond hegemonic

descriptions is critical for Black girls' identities, motivations, and belonging (King & Pringle, 2019).

- *Contribute:* Finally, BGB helps the girls to contribute and lead in a variety of ways. For example, at the closing luncheon the girls facilitated the program, and each girl presented their research in poster style to an audience of not just their parents, but several university upper administrators including the Deans of colleges and senior administration.

Overall BGB engages adolescent Black girls along with their families so that they are empowered to reach their full potential. These design principles are significant for increasing the likelihood of transformative and liberatory learning spaces.

MATERIALS AND METHODS

This section includes details about recruitment for BGB including the chaperones and instructors. We also describe and highlight the curriculum and schedule of activities.

Recruitment

Any Black girl who lived in the local and metro area and was a rising 7th or 8th grade student could apply for BGB. We advertised on social media platforms and within a week, we had 55 applications for 20 spots, which indicated that there is a great demand for such programming. The girls completed a comprehensive application that included race and gender demographic data as well as open-ended questions about their school and life experiences. We provide 10 examples of questions included on the application:

1. How do you like to learn (i.e., working alone, listening to the teacher, working with large groups, working with small groups, etc.)?
2. Would you like to attend college? Why or why not?
3. What do you think are the obstacles in your school environment that get in the way of you reaching your fullest potential?
4. Do you feel valued in your school? Why or why not?
5. What career(s) would you like to pursue and why?
6. What has been your greatest accomplishment so far?
7. What is your greatest strength or best quality?
8. What is an area in your life where you want to grow or improve?
9. What does being a Black girl in society mean to you?
10. When you hear the word sisterhood, what is the first thing that comes to mind?

Some of the responses to "What does being a Black girl in society mean to you?" are provided; all names are pseudonyms.

- Being a Black girl in society to me means celebrating who I am, embracing my curly hair, my skin, and my creative ideas. It also means being a leader to others, showing love to people, and making opportunities for other Black women. (Alisha)

- As a Black girl in society, I feel I have to work harder than my non-black peers. I have to make sure I let my light shine and show how great of a Black girl I am. Not that I'm a threat to society. (Lauren)
- It means that everything for you will be harder or more difficult. (Eliza)
- To me I see it as having to work harder than the next person and having to be confident in my abilities. (Katy)

These responses show variation of Black girls' understanding of the challenges they will likely face in life in addition to the importance of celebrating their personhood even while being racialized and gendered in society and schools.

We used a rubric to evaluate and rank the applications (see Supplemental Materials) after which we invited a selection of applicants to do 30-minute Zoom interviews where the girls discussed more about why they wanted to participate in BGB. It was important to select Black girls who do not often get these types of STEM opportunities, rather than Black girls who have plenty of social capital, girls whose parents are in STEM or had a STEM higher education, or other privileged contexts. An analysis of the girls' responses to the interview questions revealed a salient theme about the importance of affinity groups. One participant stated, "I want to experience a program where all the people look like me." A couple of the girls expressed in their interviews how they felt isolated and excluded many times because they were the only one or a few Black students in their school or classrooms. All the girls were excited to participate in a summer program that was being attended by only Black girls.

Teachers, teaching assistants, and chaperones also completed an application and participated in interviews that focused on their commitment to Black girls, curriculum design and pedagogies. We recruited teachers from our own networks. For example, the science team included four Black tenured and tenure-track faculty women in chemistry, basic sciences, and biomedical sciences at Vanderbilt University. The financial literacy instructor was a successful Black woman entrepreneur who had previously worked on Wall Street; thus, part of her financial literacy curriculum included teaching about investments. Teaching assistants and chaperones were also undergraduate and graduate students at Vanderbilt University with a range of experience working with adolescent Black girls and having passion and commitments for Black girls' positive life outcomes. Chaperones were also rigorously vetted as these individuals were responsible for the care of the girls overnight. In addition to their zoom interviews, they went through extensive training for protection of minors. We also had volunteers, videographers, and a photographer—all of whom were Black women. All 39 staff and faculty members passed a background check and were trained on the Protection of Minors training. BGB was lauded by Vanderbilt University's risk management office as one of the best summer groups with which they had ever worked.

Curriculum and Activities

The girls had a rigorous schedule that started at 7:00 a.m. and ended at 10:00 p.m. each day (see Figure 1). Each day encompassed a full day of holistic curricula. We created a detailed schedule for all personnel to know exactly where girls should be and what they were doing to ensure accountability and safety of the girls. They were broken into smaller groups, the Lavender (rising 7th grade) and Periwinkle (rising 8th grade) groups, to keep the classroom ratios at a place for high level learning.

Figure 1. Sample daily schedule

7:00 am	**WAKE – UP / Rise and Get Ready for the Day!**	
7:15 – 7:20	TRANSITION: Walk to Dining Hall	
7:20 – 7:50 am	**Breakfast**	
Location:	Dining Hall	
7:50 – 8:00 am	TRANSITION: Walk to Science Labs	
8:00 – 9:30 am	**Science Labs**	
9:30 – 9:35 am	TRANSITION: Walk to Campus Center (CC) Building	
9:35 – 9:45	**BREAK / SNACKTIME**	
Location:	Campus Center Courtyard	
	LAVENDER GROUP	**PERIWINKLE GROUP**
9:50 – 11:05 am	Art/Movement	Financial Literacy
Location:	CC 233	CC 243
11:05 – 11:10 am	**BREAK / SWITCH CLASSES / USE THE RESTROOM**	
11:10 – 12:25 pm	Financial Literacy	Art/Movement
Location:	CC 243	CC 233
12:25- 12:30 pm	TRANSITION: Walk to Dining Hall	
12:30 – 1:25 pm	**Lunch**	
Location:	Dining Hall	
1:25 – 1:30 pm	TRANSITION: Walk to Campus Center	
1:30 pm – 3:00 pm	Research Fundamentals	Critical Consciousness
Location:	CC 243	CC 233
3:00 – 3:15 pm	**BREAK / SNACKTIME**	
Location	Campus Center Courtyard	
3:15 pm – 4:45 pm	Critical Consciousness	Research Fundamentals
Location:	CC 233	CC 243
4:45 – 5:00 pm	**BREAK / TRANSITION TO ACTIVITY ROOM**	
5:00 – 6:20 pm	**Black Girl Magic – chaperone led activity**	
Location:	Campus Center Activity Room	
6:20 – 6:25 pm	TRANSITION: Walk to Dining Hall	
6:25 – 7:25 pm	**Dinner**	
Location:	Dining Hall	
7:25 – 7:30 pm	TRANSITION: Walk to Yoga Room	
7:30 – 8:30 pm	**Evening Centering**	
Location:	Yoga Room	
8:30 – 10:00 pm	SHOWER / Go to Dorm Rooms/ Settle in for night	
10:00 pm	**LIGHTS OUT**	

The ***Visual/Movement Arts*** (see Figure 2) utilized a mixed-methods (e.g., painting and drawing, sculpture, murals, photography, dance, music, poetry acting) approach to center Black girls' experiences, amplified their voices, and realized their dreams. The girls were encouraged to dream boldly and utilize the fullness of their experiences to celebrate their past triumphs, express their current existence, and visualize their future aspirations. Their final projects demonstrated how they participated and engaged in the world thus far juxtaposed with how they intend to engage in the world moving forward.

Figure 2. Self portrait

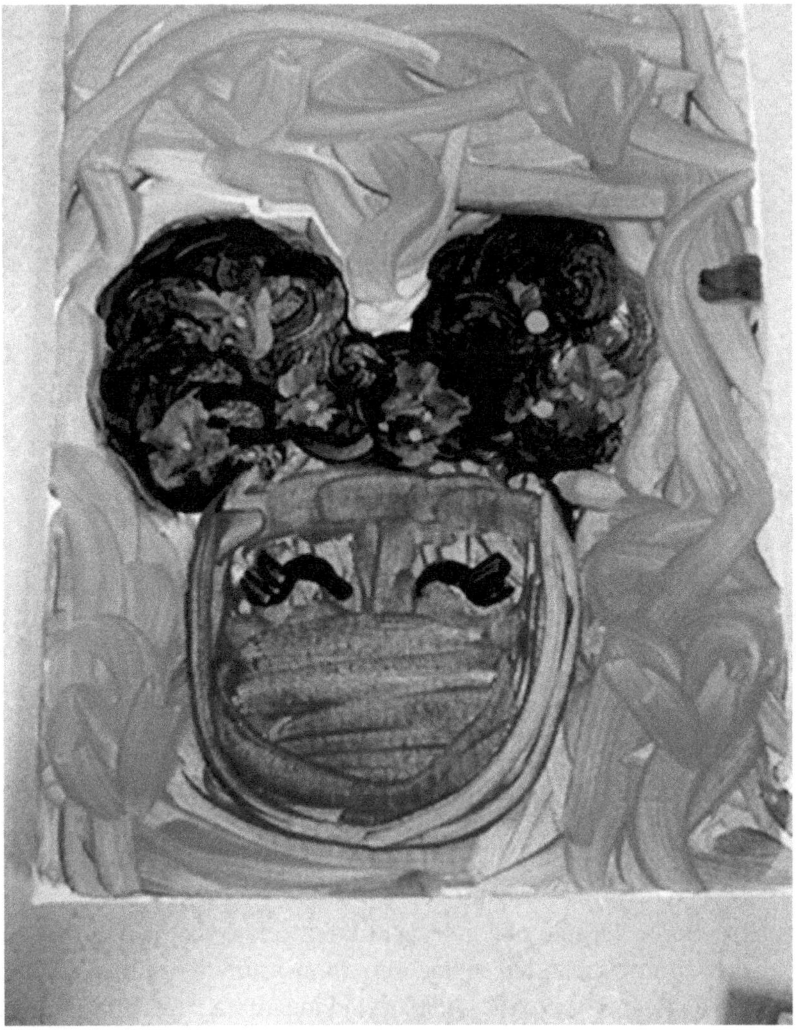

The ***Critical Consciousness & Activism*** (see Figure 3) course centered on Hooks' (1994) theory of Teaching to Transgress to equip the girls to work against discrimination and dismantle oppressive systems. The instructor of this class utilized Hooks' pedagogical approach and teaching techniques (e.g., generative themes, dialogue, conscientization) so the girls might gain expertise in critical thinking and praxis (action/reflection) to reflect upon their social reality and transform it through action.

This curriculum was designed to help the girls understand identities, intersectionality, activism, different themes of activism, and different roles of social activism (e.g., the citizen, rebel, change agent, reformer). Some of the questions explored in this course included:

(1) What does it mean to be critically conscious of who you are as a Black girl?
(2) What does it mean to be critically conscious of who you are as a Black girl in the world, specifically within the United States context?

Figure 3. Community work on Black girls' identities

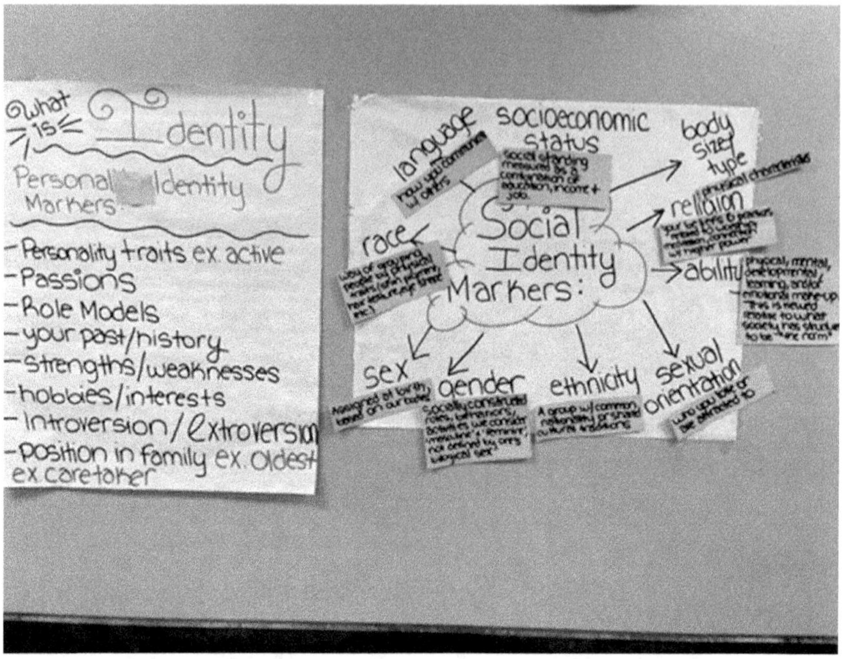

(3) How does your identity fit within the world (e.g., praxis)?
(4) How can you utilize your identity to engage in effective and inclusive activism?

This course also aimed to support the girls in becoming aware of navigating and exploring personal behaviors, self-efficacy, group efficacy, and stereotypes.

The ***Science*** courses applied a culturally relevant framework to the field of basic sciences. Lessons and experiments ranged from utilizing bioengineering to measure light propagation through different materials to utilizing analytical chemistry to separate and identify a peptide or liquid. Students applied their learned knowledge from these foundational science lessons to further understand themselves, and their identities, through a STEM lens. For example, students had a lesson on the properties of Black hair. This lesson included understanding the biologic and structural components (e.g., hair shaft shape, follicular shape) of hair, the genetic components that lead to different hair textures and patterns for people with African ancestry, different physical properties of hair (e.g., porosity, density, color), and chemical composition of hair care products.

Research Fundamentals was a course supporting the girls in understanding basic ideas of social science research such as developing problem statements, creating research questions, and reporting their findings. Questions explored in this course were "What is Research and Who is Research for? "How do people use research?" And "What role do our everyday experiences play in research?" These questions were designed to push back on hegemonic understandings of objectivity in traditional research conceptualizations.

Finally, ***Financial Literacy*** (see Figure 5) exposed the girls to a variety of financial skills, including personal financial management, budgeting, and investing.

Figure 4. Conceptualizing Black hair science experiment

Other curricula included Black Girl Magic activities (see Figure 6) created by the chaperones. Each night, after dinner, the girls engaged in activities that brought them joy, such as karaoke and dancing, making bracelets, and daily reflections and bonding. On the weekend, the girls also participated in a campus tour in the form of a scavenger hunt. They were led by Black undergraduate women who created clues for a major or career, and then the girls had to locate the building. For example, one clue stated: "I was made for people who wish to become entrepreneurs. I am sometimes referred to as the school of management. Which building am I? This type of campus tour produced much excitement for the girls. On the weekend, they also created vision boards, and watched "Queen Katwe," a film depicting the life of Phiona Mutesi, a girl living in the slums of Uganda. She learns to play chess and becomes a Woman Candidate Master after her victories at World Chess Olympiads. The girls did a pre-assessment of big ideas in the movie and engaged in a debrief and questions afterwards. They identified themes such as overcoming challenges, community support, and persistence as important for Black girls' lives.

Each day ended with what we called ***Evening Centering*** (see Figure 7). This curriculum involved a multi-sensory centering experience that emphasized a gentle and restorative approach to self-awareness. Centering began with 30 minutes of healing yoga and meditation (including audio that promotes stillness), followed by 10 minutes of journaling to a self-reflection prompt, and 20 minutes of sharing and listening. This curriculum promoted mental, physical, and emotional wellbeing and the girls left this class with a more in-tune sense of self, increased emotional intelligence, comfortability with vulnerability, and an immense love of Black girlhood and sisterhood.

Overall, the curriculum and activities of BGB promoted a living and learning environment of liberation, love, and freedom. The learning environment allows the girls to be children, to play, to laugh, to think hard, ask questions important to them, and most all, to elevate their voices and narratives. We reminded the girls daily of their worth and humanity.

Figure 5. Financial literacy lesson plan

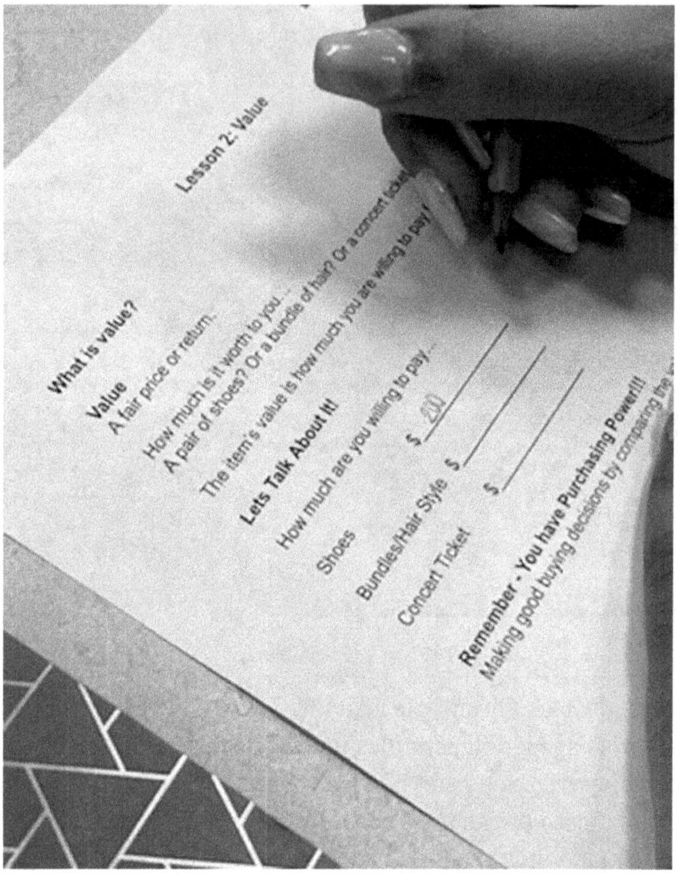

Black Girls Becoming Summer Research Institute Scholarship

Rationale

Beyond its programming, BGB advances new knowledge through social science research. Black girls and Black women are critically underrepresented in STEM. Research shows that only two percent of Black women have STEM careers (National Center for Science & Engineering Statistics, 2020). The purpose of the research arm of BGB is to assess an identity-centered curriculum as a method to increase Black girls' interest and participation in science, technology, engineering, art, and mathematics (STEAM). Assessing identity-centered curriculum's impact on Black girls' interest and participation in STEAM can provide timely information to educators on the importance of incorporating race, gender, and disciplinary belonging in STEAM curricula to provide more meaningful educational opportunities for Black girls. The study can provide actionable insights to educators, legal guardians, and the girls themselves as they navigate the concurrent developmental trajectories associated with race, girlhood, and STEAM identities. This study is a longitudinal study, and lasts for nine years of follow-up (e.g., after the girls graduate from high school and, if they choose, enroll in a college/university). Through this study, we investigate the girls' experiences in BGB, their parents' perspectives of their experience in BGB and any lasting changes, and if creating an identity-focused STEAM curriculum increases the girls' longitudinal interest and participation in STEAM. The girls and their families are compensated for their time. Here, we focus on results from the first wave of data collected in summer 2022. We asked and answered the following research questions:

(1) What do Black girl participants report as their perceptions of the BGB?
(2) What might their perceptions tell us about their own STEAM identities, the STEM workforce, and overall influence of a Black girlhood identity-centered curriculum?
(3) What aspects of the BGB appeared to influence Black girls' level of interest and participation in STEAM over time?
(4) What do parents and caregivers report as their perspectives of the BGB?

Figure 6. Black girl magic activities

Figure 7. Evening centering

Setting/Context

BGB took place at Vanderbilt University, an elite private research university offering undergraduate, graduate, and professional degrees. Vanderbilt University is located in the Southeast, a historical space in which violence against Black women and girls during slavery and beyond has not only led to their marginalization, silence, and oppression, but also their invisibility. Vanderbilt University, however, has invested substantial effort into addressing its racist and sexist past through symbolic gestures such as removing the term *Confederate* from a prominent building's name on campus out of respect for the racialized pain such a word inflicts. Structural and programmatic changes are also evident with the addition of a chief diversity officer to the leadership team, and the development of the Office for Equity, Diversity, and Inclusion. The numerical representation of Black students and faculty has also steadily increased. As of 2023, 10.4 percent of undergraduate students identify as Black or African American, and Vanderbilt University earned the accolade of having the third-highest percentage of Black students in the 2021 first-year cohort among high-ranking universities (Vanderbilt University, 2023c). With regards to race and gender diversity among faculty, a 2018 report revealed that 5.46 percent of all faculty university-wide identified as Black or African American (Vanderbilt University, 2018). Only 2.95 percent of all faculty (inclusive of tenured, tenure-track, and non-tenure stream) are Black women (Vanderbilt

University, 2018). This demographic snapshot of Black women's representation on campus provides contextual "data" with regards to the setting in which BGB was developed.

Participants

The participants in the study were the girls in the program and their parents and caregivers. We recruited by sending out a flier asking for volunteers to be a part of the study. For this inaugural year, we had 19 participants. All parents gave their own consent as well as their permission for their daughters to participate in the study. All 19 girls gave their assent to participate in the study. The girls represented 14 local middle schools in Nashville, including one private, one charter, and one Christian. There was also variation in socioeconomic and guardianship status (i.e., students raised by single parents, grandparents, and siblings).

Data Sources

Data sources included three surveys completed by the girls during the second week of the program, parent focus groups, and follow-up interviews with the girls after they completed BGB. Survey #1 is the *Is Science Me?* science identity survey created by The Caltech Pre College Science Initiative Research Group and explores perceptions of science and scientists, interest in STEM, desired college majors and careers, family and peer expectations related to science interest and activities, perceptions of science classes and teachers, and self-confidence in STEM abilities. Survey #2 evaluated the girls' experience of BGB using a mixture of a Likert scale and open-ended questions. Students shared feedback on workshops, ways in which BGB can be improved, concepts they took away from BGB, the elements that positively impacted them the most, and how BGB influenced their perceptions of STEAM. Survey #3 determined the girls' relationship and engagement with themselves, others, and the world using a Likert scale and open-ended questions. Students shared their perceptions of women in STEAM, Black women in STEAM, and if they feel as though they can fully bring their narratives and their identities into STEAM fields. Interview #1 included virtual parent focus groups of four parents/guardians discussing their views on BGB. Parents shared their perspectives of BGB, any observations that they noticed in their child after participating in BGB and gave any recommended feedback and constructive criticism of BGB. Interview #2 occurs annually for nine years and consists of one-on-one interviews with the girls' expressing their level of interest and participation in STEAM. We plan to inquire about academic course selection, participation in STEAM-related extracurriculars, career interests, interactions with teachers and peers, and learned practices from BBGP that they currently utilize and apply in the classroom, if applicable.

Data Procedures

All three surveys were collected during the second week of the BGB. The surveys were designed in REDCAP and links were provided to each girl to complete her responses. Survey one took approximately 30 minutes. Surveys two and three took about 20 minutes to complete. The parent focus groups were conducted the following Saturday after the end of the program. They were paired for a virtual one-hour focus group. Five focus groups were conducted.

Data Analysis

For all three surveys, descriptive statistics were performed to examine means and distribution percentages of various items. Parent focus groups were transcribed and coded to look for patterns and themes.

RESULTS

This section reports findings from the surveys and parent focus groups. The results show variation in the girls' responses to science identity, their perspectives of the program, and how they see themselves and others in STEM.

Survey #1: Black Girls Perspectives of Science Identities

This survey assessed the extent to which the girls identified as a scientist and what factors might influence a science identity. While nearly half (i.e47%) of girls strongly agreed with the statement "I think science is interesting," only 32% strongly agreed with the statement "I think I could be a good scientist one day." However, nearly three-quarters (73.7%) reported that a teacher made it exciting to learn science, and 90% of the girls strongly agreed or somewhat agreed that "it is important that my teacher likes me," emphasizing the potential impact that welcoming and inclusive learning environments can have on Black girls' beliefs in their own scientific ability.

Survey #2: Black Girl Perspectives of BBGP

The second survey was used to evaluate the short-term effectiveness of the program from the perspective of its participants. A few data points in the aggregate suggest that the program had a positive impact. First, the girls found participation both enjoyable and motivating: 94.7% of girls agreed that they enjoyed participating in the program. More specifically, in terms of how the program influenced their perceptions of future academic plans, 84% of girls agreed that participating in BGB "makes [them] more interested in going to college," and 58% agreed that BGB made them more interested in STEAM. Second, in terms of increasing self-confidence as a Black girl, survey results were also illustrative: 100% agreed that "participating in BBGP made [them] realize more ways in which being a Black girl is a good thing," and 90% agreed that they "feel more confident in their abilities after participating in BGB." Last, BGB's influence was particularly powerful in terms of Black girls learning from Black women instructors. For instance, 90% of girls agreed "having Black women teachers enhanced my ability to learn in my classes" and 100% agreed that they "feel safer having teachers that are Black women." Moreover, 79% of girls agreed that "I believe that teachers at my school would teach better if they saw how BGB teachers teach and interact with students."

Survey #3: How Black Girls See Themselves and Others in the STEM Workforce

This survey assessed the girls' relationship and engagement with themselves, others, and the world using a Likert scale and open-ended questions. Here, we highlight some of the most salient findings. One item that had unanimous strong agreement (100%) was the statement, "A Black girl can be anything she puts her mind to." Another statement, "Being a Black girl is a good thing" also had strong agreement (94.7%). These responses are encouraging, as it suggests that all 19 Black girl participants firmly believed that they are limitless in their abilities. However, despite the strong agreement with Black girls having high capabilities and seeing Black girlhood as positive, these sentiments did not translate to how Black girls are viewed in formal classroom contexts. When asked whether Black girls are taken seriously and respected in the classroom, participants were more ambivalent, with nearly half of participants reporting "neutral" responses on the Likert scales. When asked about their comfortability participating in class, ambivalence was again present. When asked whether they personally felt comfortable raising their hand in class or speaking/participating in class, 47% and 32%, respectively, were neutral in response to these statements.

Parent Focus Group: Which One is My Kid?

The focus groups aimed to evaluate the impact and effectiveness of BGB participation on adolescent Black girls' identity formation and STEM interest. As BGB's curriculum interrupts outdated and ineffective learning and classroom norms, parents evaluated the effect that inclusive and innovative pedagogy, particularly centering Black girl narratives and Black girlhood, had on their daughters. Parents had an overwhelmingly positive response to BBGP's ideology and curriculum, as many stated that it was the first time their child engaged in STEM in a majority-Black women environment. Overall, parents found BGB to be refreshingly disruptive for their daughters by dismantling teaching styles that uphold white supremacy and silence Black girls in the classroom,

DISCUSSION

These findings suggest that middle school girls in the Black Girls Becoming Summer Research Institute understand that Black girlhood is not celebrated in classroom and schooling spaces, and that their voices and what they bring to the learning process are not always welcomed. These findings could be explained by prior studies that have examined Black girls' schooling experiences, including Carter Andrews et al. (2019) robust study whereby they listened to 70 high school Black girls describe and make meaning of their schooling experiences as racialized and/or gendered. Their findings showed that Black girls faced toxic racialized and gendered experiences with adults and peers in their schools. For example, Black girls pointed out that teachers and other educators considered them unmotivated and therefore unworthy of high-quality learning experiences. The girls described instances where teachers appeared to be uninterested in providing rigorous or engaging curriculum and instruction or dismissing their questions and concerns. This might explain why many girls in the BGB program were not confident in their opinions about feeling respected and safe participating in math and science classes. BGB was an informal learning space where these negative experiences were dismantled. It aimed to meet the academic, social, and identity needs of middle school Black girls in ways that honored their full humanities and images of self. Black girls had opportunities to name, unpack and address aspects of oppression Black girls face within the school context. More educational policy work is needed to design and fund such spaces that offer liberation and healing for many Black girls.

CONCLUSION

Black Girls Becoming Summer Research Institute engaged social media platforms during the two-week institute, specifically, Instagram. It was our goal to go get attention with the girls creating TikTok videos as well as instructors making videos of what the girls were doing during the institute. We also tagged former first lady Michelle Obama as her book inspired the name of the institute. During that week, a high school Black girl entrepreneur, the founder of the Original Black Dolls Project and her mom reached out to Nicole Joseph because she was so impressed with the program. She wrote: "…I absolutely love this program and the work that you're doing in the community with girls that look just like me! I'm pretty sure that your participants are too old for dolls. But I'd love to donate something to each of your participants like perhaps a hardback copy of Michelle Obama's Becoming book?" This young lady and her mom purchased 19 *Becoming* books written by Michelle Obama and then we mailed these books to the girls as this occurred once the institute was over. Several people interacted with our Instagram account mentioning positive comments about the need for more programming such as this. That week, we gained over 100 followers. Parents of the girls send us videos of activities their daughters are doing related to STEAM, and we post them. We hope that BGB can one day become an influencer of Black girl excellence in STEAM. There was other media press that covered BGB, including Vanderbilt University's Provost tweeting about the program and Nicole Joseph's work translating into practice. There was also a reunion event during the fall after the summer program and the girls took another science workshop and continued to discuss critical consciousness around Black girlhood.

Figure 8. Vanderbilt University provost's tweet

CONCLUSION

This chapter aimed to unpack the Black Girls Becoming Summer Research Institute, including the theoretical foundations of the design of the curriculum, activities, and learning environment. We also argued that conducting research coupled with STEM community partnerships is imperative to add to the literature. We showed through the methods and activities that BGB is a community and university partnership focused on Black girlhood empowerment and agency. Because formal schools seldom provide middle school Black girls learning opportunities to shine and bring their full humanities to the fore, informal spaces appear to be locations where counter-narratives can be created. Schools are oppressive for Black girls; thus, to create STEAM learning environments that are liberating requires the active creation of critically conscious designers, curriculum developers, teachers, educators, and policy makers who form innovative ideas and visionary possibilities beyond the white gaze. There can be much promise when STEM outreach partnerships center these dispositions.

REFERENCES

Bowers, E. P., Li, Y., Kiely, M. K., Brittian, A., Lerner, J. V., & Lerner, R. M. (2010). The five Cs model of positive youth development: A longitudinal analysis of confirmatory factor structure and measurement invariance. *Journal of Youth and Adolescence*, *39*(7), 720–735. doi:10.100710964-010-9530-9 PMID:20397040

Brown, R. H. (2013). *Hear Our Truths: The Creative Potential of Black Girlhood*. University of Illinois Press. doi:10.5406/illinois/9780252037979.001.0001

Butler, T. T. (2018). Black girl cartography: Black girlhood and place-making in education research. *Review of Research in Education*, *42*(1), 28–45. doi:10.3102/0091732X18762114

Carter Andrews, D. J., Brown, T., Castro, E., & Id-Deen, E. (2019). The impossibility of being "perfect and white": Black girls' racialized and gendered schooling experiences. *American Educational Research Journal*, *56*(6), 2531–2572. doi:10.3102/0002831219849392

Collins, K. H. (2018). Confronting color-blind STEM talent development: Toward a contextual model for Black student STEM identity. *Journal of Advanced Academics*, *29*(2), 143–168. doi:10.1177/1932202X18757958

Crenshaw, K. (1990). Mapping the margins: Intersectionality, identity politics, and violence against women of color. *Stanford Law Review*, *43*(6), 1241–1299. doi:10.2307/1229039

Crooks, N., Sosina, W., Debra, A., & Donenberg, G. (2022). The Impact of COVID-19 Among Black Girls: A Social-Ecological Perspective. *Journal of Pediatric Psychology*, *47*(3), 270–278. doi:10.1093/jpepsy/jsab139 PMID:35182154

Davis, R. (2020). New Study Shows Black Women Are Among The Most Educated Group In The United States. *Essence*. https://www.essence.com/news/new-study-black-women-most-educated/

Epstein, R., Blake, J., & González, T. (2017). *Girlhood interrupted: the erasure of black girls' childhood*. SSRN., doi:10.2139srn.3000695

Gershenson, S., Hart, C. M., Hyman, J., Lindsay, C., & Papageorge, N. W. (2018). The long-run impacts of same-race teachers (No. w25254). National Bureau of Economic Research.

Gholson, M. L.Maisie L. Gholson. (2016). Clean corners and algebra: A critical examination of the constructed invisibility of black girls and women in mathematics. *The Journal of Negro Education*, *85*(3), 290–301. doi:10.7709/jnegroeducation.85.3.0290

Gholson, M. L., & Martin, D. B. (2019). Blackgirl face: Racialized and gendered performativity in mathematical contexts. *ZDM*, *51*(3), 391–404. doi:10.100711858-019-01051-x

Halliday, A. S. (2019). *The Black Girlhood Studies Collection*. Women's Press.

Harding, S. (1992). After the neutrality ideal: Science, politics, and" strong objectivity. *Social Research*, *59*(3), 567–587.

Hooks, B. (1994). *Teaching to Transgress*. Routledge.

King, N. S., & Pringle, R. M. (2019). Black girls speak STEM: Counterstories of informal and formal learning experiences. *Journal of Research in Science Teaching*, *56*(5), 539–569. doi:10.1002/tea.21513

Lane, T. B., & Id-Deen, L. (2023). Nurturing the capital within: A qualitative investigation of Black women and girls in STEM summer programs. *Urban Education*, *58*(6), 1298–1326. doi:10.1177/0042085920926225

Milner, H. R. (2021). *Start where you are, but don't stay there: Understanding diversity, opportunity gaps, and teaching in today's classrooms*. Harvard Education Press.

Morris, M. W. (2015). The relevance of sacred inquiry in the education of delinquent Black girls. *International Journal of Human Resources Development and Management*, *15*(2-4), 185–193. doi:10.1504/IJHRDM.2015.071156

Morris, M. W. (2016). *Pushout: The criminalization of Black girls in schools*. The New Press.

Morris, M. W. (2019). *Sing a rhythm, dance a blues: Education for the liberation of Black and Brown girls*. The New Press.

Onyeka-Crawford, A., Patrick, K., & Chaudhry, N. (2017). *Let her learn: Stopping school pushout for girls of color*. National Women's Law Center.

Ray, R. (2022). School as a Hostile Institution: How Black and Immigrant Girls of Color Experience the Classroom. *Gender & Society*, *36*(1), 88–111. doi:10.1177/08912432211057916

Thomas, A. (2021). Out of the box: Countering the silence of Black girls with memory-work. In D. Apugo, L. Mawhinney, & A. Mbilishaka (Eds.), *Strong Black Girls: Reclaiming Schools in Their Own Image* (pp. 9–19). Teachers College Press.

Vanderbilt University. (2018). *Report on the Faculty: Composition and Demographics of Provost Reporting Vanderbilt University Employed Faculty Members*. Vanderbilt University.. https://cdn.vanderbilt.edu/vu-wp0/wp-content/uploads/sites/241/2018/05/08164600/ReportOnTheFaculty2018.pdf

Vanderbilt University. (2023a). *Intersectional Study of Black Women and Girls in Society*. Vanderbilt University. https://www.vanderbilt.edu/ibwgs/

Vanderbilt University. (2023b). *Trans-Institutional Programs (TIPs).* Vanderbilt University. https://www.vanderbilt.edu/strategicplan/trans-institutional-programs/

Vanderbilt University. (2023c). *Vanderbilt at a Glance: Diversity and Inclusion.* Vanderbilt University. https://admissions.vanderbilt.edu/profile/

APPENDIX

Student Application Rubric

Student:________________________________ **Evaluator:**________________________________

Use the following scale of 1 – 10 to evaluate the first five statements.	
1 – ***very*** strongly disagree 2 – *strongly* disagree 3 – *moderately* disagree 4 – *slightly* disagree 5 – **completely neutral**	6 – *slightly* agree 7 – *moderately* agree 8 – agree 9 – *strongly* agree 10 – ***very*** strongly agree

	Rating:	**Notes:**
1. The student has a strong, healthy, and positive self-perception/sense of self.		
2. The student's learning style/awareness of their learning style is compatible with BBGP.		
3. The student has constructive, positive, and ambitious goals/outlook for their future.		
4. The student's attitude towards school/learning is conducive to the climate of BBGP.		
5. The student's greatest strength(s)/accomplishment and desired area(s) of improvement are in alignment with focus areas of BBGP programming		
Total	/50	

Rate the final five statements using a scale of 1 – 10		
(1) Poor	(5) Average	(10) Excellent

	Rating:	Notes:
6. Student's ability to work well with others		
7. Student's desire to learn/grow academically		
8. Potential value of BBGP to the student's growth/ development (personally and academically)		
9. Student's concept of Black girlhood and sisterhood		
10. Overall impression of the application		
Total	/50	

RAND TOTAL:	/100
INTERVIEW (yes or no):	

Chapter 8
"There Are a Lot of People Like Me…": The Impact of a STEM Program for Underrepresented Girls

Kelly L. Knight
George Mason University, USA

Padmanabhan Seshaiyer
George Mason University, USA

Danielle Craddock
Girls Inspired and Ready to Lead, LLC, USA

Claudette P. Davis
LaGuardia Community College, USA

ABSTRACT

There is a disparity between the number of men and women in STEM fields. The disparity is further exacerbated when considering the percentage of women from underrepresented backgrounds in the STEM workforce. However, early and persistent exposure to STEM through out-of-school programming can potentially increase the likelihood of a young girl, and particularly girls of color, pursuing a STEM degree and ending up in a STEM career. This chapter describes an out-of-school STEM program designed for middle school and high school girls. The design of the program is described as well as the critical role of community and corporate partnerships. Participants complete both pre- and post-program participation surveys which are used to gather data on demographics, STEM experiences, as well as their interests and motivation. The data presented provides evidence of the impact of the program. This chapter may also be a useful resource for others in the field who are designing similar programs which are specifically curated for girls.

DOI: 10.4018/978-1-6684-7771-7.ch008

INTRODUCTION

The underrepresentation of groups such as women and those from other marginalized identities in scienceculturally relevant pedagogy, technology, engineering, and mathematics (STEM) continues to be problematic. As of 2015, the National Science Board (NSB, 2018) reported that 28% of the STEM workforce identified as female, and 11% identified as either Black, Hispanic, Alaska Native, or American Indian. Different approaches over the years have been taken to improve these statistics and yet despite these efforts, these groups continue to be underrepresented in STEM fields (Tsui, 2007). One approach that may alleviate this issue is the development and effective implementation of out-of-school STEM programs and the subsequent engagement of historically underrepresented program participants.

Out-of-school programming includes programming that occurs outside of the school setting either before school, after school, on weekends, or during seasonal breaks such as winter or summer. In 2015, the National Research Council (NRC, 2015) published a report by the Committee on Successful Out-of-School STEM Learning. The purpose of this report was to establish criteria that could be used to identify and support productive out-of-school STEM programs. The Committee reported that a review of the literature reflects the ability of out-of-school STEM programs to (a) contribute to young people's interest in and understanding of STEM; (b) connect young people to caring adults who serve as role models; and (c) reduce the achievement gap between young people from low-income families and high-income families.

Additionally, The Afterschool Alliance (2008) reviewed over 30 afterschool programs and found that they (a) improved school attendance and engagement in learning; (b) improved test scores and grades; and (c) students considered to be at greatest risk often show the greatest gains (2008). While these and other reviews report positive outcomes for students participating in out-of-school programs, there is less research available on the impact of these programs on historically underrepresented students.

The NRC (2015) also described three criteria to develop out-of-school programs with positive outcomes. The programs must be engaging intellectually, academically, socially, and emotionally; responsive to the participants' interests, experiences, and cultural practices; and they must make connections to STEM learning in the various settings of a participant's day-to-day life including in and out of school as well as at home. While the NRC report provides a detailed overview of criteria for effective programs as well as recommendations and limitations for evaluation, there is still a need for literature that describes the work done with out-of-school STEM programs specifically for underrepresented students.

In this review, we examine studies of out-of-school programs which evaluate interest, attitudes, and motivation in STEM, but we also need to consider how that translates to pursuing a STEM career. This is an important differentiation as this is often a desired outcome of providing out-of-school STEM programs, to go beyond the first level of triggered situational interest and generate an interest that translates to a well-developed interest eventually leading to the pursuit of a STEM career and thereby increasing the representation of minorities in these fields (Hidi & Renninger, 2006). For example, the research by Kang et al. (2018) revealed consistent common predictors of students' identification with basic biology, applied biology, basic physical, and applied physical science careers. The students' perception of self was associated with higher identification with STEM careers in all four categories. Participation in science-related activities at home and important for this study, outside of school, predicted higher identification with STEM careers in all categories except for applied biology and basic biology careers, respectively.

Out-of-school STEM programs are also helpful in exposing students to careers that may not be discussed or explored in traditional school settings. For example, in a 2015 study of a university pipeline program for African American high school students, participants reported "discovering occupations they never knew existed, thus expanding their range of career choices within health careers" as well as "reconsidering specific occupations" they had originally chosen to pursue but had lost interest in (Wallace et al., 2015, p. 492).

With these ideas in mind, this chapter's co-authors joined to create Females of Color and those Underrepresented in STEM (F.O.C.U.S.) in 2014. F.O.C.U.S. began as a week-long camp that exposes students to a variety of disciplines within science, technology, engineering, and mathematics (STEM). The camp's original target population was middle school students, an age where studies have shown an interest in STEM begins to significantly decline, however, due to its increasing popularity additional programs have been created including the F.O.C.U.S. Academy for high school students as well as the Project Mason Outreach in STEM (M.O.S.T.) program for teachers.

In this chapter, the authors describe the various programs in more detail and present program participant data. For example, the demographic data is useful for better understanding the types of students that attend the program. The authors also sought to learn more about the participants' experiences with STEM before attending this program as well as how their interests changed, if at all, from the start of the program to the end. In addition to discussing some of the data collected, the chapter discusses programmatic development and the importance of leveraging community and corporate partnerships to develop a sustainable program. Before getting into these specifics, however, a literature review on the current research related to underrepresented students in out-of-school STEM programs is provided. While our program covers the different components of STEM, many of the programs described in the literature review focus on one or two areas within STEM (e.g. science only, or mathematics and science) but this research provides an important foundation for the work done with this program.

Literature Review

The literature review is framed using three themes including the unique needs of students in out-of-school programs; the importance of the quality of these programs; and program evaluation. Each of these themes has played a role in the development of the out-of-school STEM program discussed in this chapter.

Defining "Underrepresented"

Before delving into the research, the definition of an underrepresented student needs to be clarified. Researchers use the word 'underrepresented' in various ways which is one limitation of this work. A literature search for the term 'underrepresented' returns results for underserved students (e.g. from rural, immigrant, and low socioeconomic backgrounds), studies that evaluated students from various ethnic backgrounds (e.g. Black or African American, Indigenous or Native, Hispanic or Latino, etc.), first-generation students, as well as gender-specific (non-ethnic) studies (e.g. white girls and women). Also, while not observed in the search results, other groups of students may fall in this category as well such as individuals with disabilities and other 'unseen' differences (such as sexual orientation; Trytten et al., 2014). These categories also fall under the blanket terms of non-dominant, marginalized, or minoritized which could include any of the previously mentioned descriptions (individuals with disabilities, underrepresented minorities, underserved students, etc.). Also to consider, there is sometimes the issue of racially minoritized students being conflated with and/or assumed to be at-risk, urban, of low socioeconomic status, and underserved which although this may sometimes be true, is not always the case. Because of this, racially minoritized students who do not fall into those categories and the issues affecting them specifically are often left out of the research.

Overall, each of these groups may be considered underrepresented in STEM, however, the issues affecting each one are complex, separate, and unique. The intersectionality of these students relating to multiple identities also complicates the research (e.g. being an African American girl with low socio-economic status from an urban community). There is limited research on out-of-school STEM programs which focus on individual underrepresented groups, likely due to the intersectionality previously mentioned, which can make it difficult to evaluate the impact of these programs given the many social and cultural variables introduced when amalgamating these groups or considering the participants' multiple identities. Due to these limitations, this review includes programs with students from varying gender identities that identify as belonging to a historically underrepresented race and/or ethnicity in STEM (Black or African American, Hispanic, as well as indigenous or native groups).

Also important to understanding the term *underrepresented* is an examination of the systemic factors associated with the various STEM environments that continue to perpetuate issues that maintain the underrepresentation of marginalized individuals. For example, STEM identity, which is discussed at length in this review, is sometimes viewed through a deficit lens when considering the ways students can gain the skills needed to develop a STEM identity that will allow them to assimilate or integrate into the dominant STEM norms. In contrast, a critical perspective leverages the capital students bring to the STEM classroom and interrogates the ways social structures mediate identity formation (Larnell et al., 2016). In other words, how do we move from asking "What do STEM students need?" to "What do STEM students have?" Further, Mejia et al. (2018) suggest that participation in STEM be redefined by shifting the focus from being solely on STEM identity to a focus on how the environments of STEM students can be positively changed to be more inclusive.

The Unique Needs of Out-of-School STEM Students

One way to make STEM environments more inclusive is to specifically design out-of-school STEM programs with the unique needs of the populations of students being served in mind. STEM identity can provide some insight as to how to approach this process. A growing amount of research supports the idea that the underrepresentation of minoritized individuals in STEM may have less to do with achievement gaps and more to do with those students' self-perceptions and STEM identities or their ability to see themselves as a part of the STEM community (Hayden et al., 2011; Kang et al., 2018; Pinkard et al., 2017; Rahm & Moore, 2015; Scott & White, 2013). STEM identity is described as not only the content and procedural knowledge one has but also the recognition within the discipline by self and others as a contributing member (Carlone & Johnson, 2007; Ortiz et al., 2019). With the drastically low numbers of women and minoritized individuals in STEM, students are unable to 'see themselves' or others who look like them in STEM fields thereby directly impacting their STEM identity development.

Carlone and Johnson (2007) examine the development of STEM identity through competency in STEM, performance in STEM practices, and recognition as a STEM person, however, STEM identity is complex and is described differently in the literature. For example, Kim and Sinatra (2018) describe that STEM identity is based on the interactions one has with their various environments whereas Stewart (2022) proposes that it is a communication process. Regardless of how it is defined, exploring STEM identity through an intersectional lens becomes critically important as it is shaped by our various other social identities such as race and gender.

Directly addressing the dominant STEM stereotypes is one approach to developing STEM identity. Pinkard et al. (2017) did this by co-designing narratives with their participants in the *Digital Youth Divas* program while Scott and White (2013) used a culturally relevant pedagogical approach to challenge stereotypes and reshape the self-perceptions of their participants. Pinkard et al. (2017) described three challenges that programs must overcome in their effort to spark the interest of 'nondominant girls in STEM': (a) mitigate stereotypes and concerns about fitting within STEM; (b) provide high-quality STEM activities; and (c) develop ways to keep them engaged beyond the "narrow physical and temporal confines of the intervention" (p. 482). To address these challenges, they describe *Digital Youth Divas*, an out-of-school STEM program focusing on computational activities for 'non-dominant' middle school girls. Using a project-based curriculum, the participants acted as co-designers, along with mentors and professional authors, to create fictional narratives of young girls that mirrored the complexity of their identities as well as the project activities they were working on at the time. These stories and the diverse, relatable characters found within them allowed participants to visualize alternatives to the dominant stereotypes of those engaged in computing. Through session observations, online activity summaries, and semi-structured interviews, the researchers found that the narratives provided context and purpose for the girls to complete the STEM learning activities and in turn, improve their self-perceptions and ability to identify as someone working in STEM.

Using standpoint theory and a culturally relevant pedagogical approach, Scott and White (2013) sought to evaluate how *CompuGirls*, a multimedia enrichment program for girls from urban districts, affects the self-perception of the participants as future technologists and maturing women of color. Standpoint theory describes how those from marginalized backgrounds, such as girls and women of color, generate specialized knowledge based on their experiences and how perceptions differ from dominant or mainstream perspectives (Harding, 2004). Ladson-Billings (1995) defines culturally relevant pedagogy as a "...theoretical model that not only addresses student achievement but also helps students to accept and affirm their cultural identity while developing critical perspectives that challenge inequities that schools (and other institutions) perpetuate" (p. 469). In a mixed methods study of 41 African American and Latino girls (ages 13-18) over two years, the participants were given digital opportunities to solve problems affecting their communities and were encouraged to provide peer feedback. Through this curriculum, the data revealed that their self-perceptions and identities shifted in a way that reflected their connectedness to the digital community. Scott and White (2013) found that the more the girls challenged themselves and the stereotypes around their STEM abilities, the more motivated they were.

In other examples, both Rahm and Moore (2015) and Hayden et al. (2011) designed out-of-school STEM programs to uniquely fit the needs of their participants. Rahm and Moore's (2015) longitudinal study on first-generation college-bound students highlighted the transformative impact of a Math and Science Upward Bound program. This program provided exposure and resources for students to envision themselves as valuable contributors to STEM. Video, narrative, interview, and written reflection data were collected at three points: during their application process to the out-of-school STEM program, during their program participation (three years), and once mid-way through college. Four dimensions of identity in science were focused on: figured worlds, positionality, authoring of self, and making of new worlds. Two of the students identified as Caucasian (one girl, one boy). The other two students identified as Hispanic (both boys). These students were described as both underserved and underrepresented due to their status as first-generation students coming from underserved communities. The results indicated that before participating in the program, the students figuring *of science* (i.e. their epistemic understanding of science and who might pursue it) and authoring *of selves* in science was low. However, participating in the program provided the exposure and resources students needed to see their potential and envision themselves positioned as valuable contributors to STEM.

Hayden et al. (2011) were interested in evaluating whether student self-perceptions were improved following attendance at the *iQuest* summer camp for underserved middle school populations. To uniquely support their participants, *iQuest* program project teachers were provided professional development to support them in designing innovative curricula, but several other key features made the program unique such as the hands-on experiences, collaborative learning approach, and focus on daily reflection.

Overall, the literature reviewed here supports the concept that not only is the development of positive self-perceptions and identities in STEM fundamental for underrepresented students but, more importantly, that this is something that can be achieved through participation in out-of-school STEM programs. Pinkard et al. (2017) offered an explanation as to why this may be: an out-of-school STEM program provides "non-isolating, identity-rich learning spaces...to experience and participate in rich STEM activities" (p. 481). Additionally, there is evidence that culturally relevant programming plays an important role in the development of STEM identity (Rahm & Moore, 2015; Scott & White, 2013; Stevens et al., 2016).

Complicating this research, however, are the multidimensional aspects of the identities of underrepresented students. It is important to consider the intersectionality of students' social identities, such as race and gender, when exploring STEM identity development. Assessing the impact and value of an out-of-school STEM program for these students can become challenging when taking into consideration the interaction of the various social, cultural, and economic identifiers of each student. What is lacking from this research is an evaluation of historically underrepresented groups that are not first-generation, from low socioeconomic backgrounds, and are not from rural, urban, or otherwise underserved communities. It would be just as valuable to understand the development of the STEM identities of minoritized students who do not fit within those groups.

Another important consideration of the unique needs of out-of-school STEM participants is an understanding of how to spark and sustain interest in STEM using the effective engagement techniques that are sometimes more difficult to incorporate in the traditional classroom setting. Going back to the Hidi and Renninger (2006) model of interest, they describe that engagement emerges from the interaction between individuals and their environment. Pinkard et al. (2017) point out that this model does not account for the diversity of environments and how various sociocultural factors influence these interactions, something that is very important to consider when discussing underrepresented students.

Calabrese Barton et al. (2017) describe three forms of engagement observed through their afterschool, community-based maker space program *Making 4 Change* with primarily African American participants from low-income families: critical, connected, and collective engagement. In their study, the participants experienced 'interest-powered learning' due to their critical engagement in projects regarding issues that affected their daily lives. The youth also exhibited connected engagement in which they connected with others (peers and community members) as well as connected their making to broader social issues. And lastly, through collective engagement, the participants worked together on problems via group interactions and leveraged others' experiences and the struggles therein.

So how does one 'do effective engagement' in an out-of-school STEM program for underrepresented students? There needs to be more research in this area, however, some suggestions, both explicit and implicit, can be gleaned from the study on the *Digital Youth Divas* program: (a) create a supportive learning environment that affords participants the time, activities, peers, and expectations to explore unfamiliar potential areas of interest; (b) design learning environments that build off the participants' current interests as well as considers their broader learning interests; (c) explicitly acknowledge/debunk stereotypes that may prevent underrepresented minorities from participating in and identifying with STEM; (d) provide high quality and challenging STEM experiences that spark interest; and (e) provide ways for participants to take ownership and play an active role in their learning (Pinkard et al., 2017). Stevens et al. (2016) also describe the important role culturally relevant pedagogy plays in the increased engagement of underrepresented students.

Also, worth noting is that engagement with underrepresented students must extend beyond the out-of-school program setting if the goal is to translate their interest to their studies in school and eventually a STEM career. Interest in STEM alone will not sustain students from minoritized backgrounds through the difficulties they will encounter during their pursuit of a STEM degree. To make this connection between a K-12 out-of-school STEM program and a STEM career, engagement needs to persist throughout their academic years as well because despite being interested in a STEM field, the lack of financial resources and social support at the higher education level as well as their fear of failure may play more of a role in their decision to terminate their pursuit of a STEM career than a loss of interest (Bystydzienski et al., 2015). There is limited research on out-of-school STEM programs that have taken this approach of a successful STEM ecosystem for this population of students which supports students through higher education and beyond (Hargrave, 2015).

The Role of Teachers and Mentors

For effective engagement in an out-of-school STEM program to occur, the important role of teachers and mentors must be emphasized as it is critical. This is not surprising as a review of any foundational and/or prominent literature on STEM education generally acknowledges the impact, whether small or large, that teachers have on the learning experiences of their students (Lederman & Abell, 2014). Sanders and Rivers (1996) demonstrated the long-lasting effects of having a high-performing teacher versus a low-performing teacher. Fifth-grade achievement scores were compared between students who had consistently low-performing teachers beginning in second grade to those who had consistently high-performing teachers beginning in second grade. Per Sanders and Rivers (1996), the difference in the mean student percentiles in fifth grade between the two groups (low-performing vs. high-performing teachers) was between 52 and 54 points (percentile range of 0-100). This result is both shocking and disturbing. Of concern, specifically, is the disproportionate number of students from underrepresented backgrounds that are often assigned to inadequate teachers. Further, the low expectations those teachers often have of underrepresented students (Gilbert & Yerrick, 2001; Zuniga et al., 2005).

While acknowledging that it is not a complete solution, impactful teachers and mentors in an out-of-school STEM program can have as much of a lasting effect on the sustainability and achievement of underrepresented students in STEM as teachers in the traditional classroom. Teachers have been described as the "backbone" of these programs (Hargrave, 2015). Mentors generally fall into one of three categories in out-of-school programs: student mentors/counselors, teachers, and STEM professionals.

The important role of the student mentor in an out-of-school STEM program should be especially highlighted. Student mentors who are more experienced and knowledgeable than the current participants and have participated in the program previously are both the distributor and recipient of benefits. As a 'benefit distributor', these students provide a meaningful connection between the participants themselves and the instructors/administrators of the program. Participants can often connect with the student mentors in a way that they cannot connect with the instructors because they can *see* themselves in them. As a 'benefit recipient', student mentors are also able to not only give meaning to their purpose in pursuing STEM but they can obtain a deeper level of understanding of the material by becoming instructors themselves either through one-on-one assistance or by leading their own sessions. Stevens et al. (2016), in their study of a culturally relevant science program for Native American students, described that the ability to reteach others as mentors not only increased their self-efficacy but also increased their pride in their culture. These two elements of student mentorship, specifically as 'benefit recipients' may contribute to their sustainability in STEM. Further research should be done to investigate this further.

According to Clark and Sheridan (2010), effective student and teacher/professional mentors in out-of-school STEM programs should be able to facilitate learning through project exploration and implementation, collaborative learning, troubleshooting, reflection, and critique/feedback. The authors found this requires them to be both flexible and knowledgeable. This also implies that mentors should not only be knowledgeable of discipline content material but should also have some training in pedagogical techniques in STEM education. Additionally, having the participation of STEM professionals, especially from underrepresented groups, in these programs as mentors and instructors is important for students to be able to 'see themselves' in STEM.

The Importance of Quality Out-of-School STEM Programs

Studies evaluating the impact of out-of-school STEM programs also shed light on the importance of the quality of the program itself. Kang et al. (2018) conducted a longitudinal study on middle school girls from low-income communities participating in afterschool science programs. The authors established the theoretical STEM identities model to investigate the development of STEM identities. STEM identity was described as the ways in which individuals not only see themselves currently in STEM but also how they see their possible future selves as being in STEM. The model defines identity negotiators, or constructs, which shape individuals and, in their opinion, determine an individual's STEM identity. The identity negotiators described include personal and family backgrounds, participation in science-related activities, perceptions, and their possible future selves in STEM. Overall, the data from the "Is Science Me?" survey showed no significant racial/ethnic differences in the relations of STEM identities, however, there were some observations that were of interest to this review. For example, their study showed that of the girls surveyed, African American girls reported the lowest self-perception in relation to science. Additionally, the authors argue that the underrepresentation of women in STEM can be correlated to 'missing opportunities' to form their sense of self in science but that these opportunities can be provided through out-of-school programs.

Kang et al. (2018) also posit that *more* afterschool exposure is not necessarily the answer, but instead the *kind* of exposure should be the focus. More specifically, the exposure should be designed in such a way that it "leverages the assets of the girls, their families, and their communities" (p. 434). Scott and White (2013) also emphasized the importance of the quality of the exposure when evaluating *CompuGirls*. These authors describe that the richness and relevancy of the content in their program led to the participants' ability to improve their self-perceptions and visualize themselves as 'technological innovators.'

Evaluation of Out-of-School STEM Programs

Evaluating out-of-school STEM programs requires careful attention to research design. Studies by Clark and Sheridan (2010) and Leonard et al. (2016) highlight the need for appropriate instruments and pre-program data to accurately assess changes in attitude and interest. In their study of 44 middle and high school students (93% male, 80% African American) in a Saturday Game Design program, Clark and Sheridan (2010) measured interests and attitudes toward STEM content. At the end of the program, most students reported positive attitudes and a strong interest in technology, math, and science. Similarly, Leonard et al. (2016) reported interest was high in focus groups following participation in a geoscience program for ages 8-12, however, neither study had preliminary group data to compare to their results. By comparing their final ratings of interests and attitudes to pre-program data, these studies could have better elucidated what type of impact the program itself had on changes in attitude and interest towards STEM.

While using a pre-program evaluation as well as a post-program evaluation is beneficial for understanding the impact of a program on attitudes towards STEM and other aspects, it is also critical to use the appropriate instrument to adequately perform this evaluation. Researchers have historically discussed the difficulties of assessing out-of-school STEM programs (NRC, 2015). Choosing the appropriate instrument to use is one such difficulty. In a study done by Moreno et al. (2016), although observations supported the sustained enthusiasm and engagement of the participants throughout their *Think Like an Astronaut* after school program (158 students in grades 3-5, 68% Hispanic and 22% Black), the data showed no significant improvement in students' attitudes. They argue that this could be due to the instrument itself failing to capture aspects of the students' attitudes that are specific to the after school environment (Moreno et al., 2016). While this may be true, it may also be a valid argument that perhaps the participants began the program with positive attitudes like the Clark and Sheridan (2010) game design study, therefore, there was no change in attitude to detect.

To assess the *iQuest* summer program for 7th and 8th-grade students, Hayden et al. (2011) also used a pre- and post-survey format to assess student attitudes towards science and technology in a summer camp targeting 7th and 8th-grade classrooms. The sample group included both boys and girls from various races and ethnicities including African American, Native American, Asian, Filipino, Hispanic, and Pacific Islander. The Hispanic population made up most of the participants (100% in the first year and 81% in the second year of the program) and was described as an underserved population in this study. The Test of Science-Related Attitudes (TOSRA) survey, which evaluates student attitudes and interest toward science careers, showed differences in interest gains over time based on gender and grade level. Boys showed a significant increase as well as the 8th-grade students overall when compared to the 7th graders. It would be of interest to delve deeper into this research and this program to examine the cause of the lack of interest in the girl participants.

Hayden et al. (2011) also used the Information and Communication Technology Attitude (ICTA) survey in their study. The results indicated that while there were no significant differences in self-perception by gender, grade, or ethnicity, the aggregate scores for all students combined increased significantly. In other words, the entire sample group, including all genders and ethnicities, improved in their self-perceptions regarding their information and communication technology skills. This study is also an example of the terms 'underserved' and 'underrepresented' being used interchangeably, as described previously, as well as the difficulties that can arise with assessment interpretation when there are a multitude of variables to consider. When assessing groups with multiple identities, researchers should take into consideration the sociocultural factors that may be affecting their results.

Bystydzienski et al. (2015) demonstrated the importance of doing both pre- and post-evaluations of a program in their study of their *Female Recruits Explore Engineering* (FREE) and *Pathways* after school intervention. The participants included 131 diverse young women beginning in grade 10 and ending in grade 12. They noted that at the outset of the program, only 18% of the girls were considering engineering as a career whereas by the end of the 2nd year of the project, 57% of the participants were seriously considering a career in an engineering field. Without a pre-evaluation, it is difficult to gauge what results are due to participation in a program and what results are due to prior experiences.

Females of Color and Those Underrepresented in STEM (F.O.C.U.S.)

Community Partnership With GIRL, Inc.

Planning for the first F.O.C.U.S. (Females of Color and those Underrepresented in STEM) camp began in the fall of 2013 at George Mason University in Fairfax, Virginia in partnership with Girls Inspired and Ready to Lead, Incorporated (GIRL, Inc., n.d.). George Mason University is the largest public research university in the state of Virginia. Girls Inspired and Ready to Lead is a 501(c)(3) organization with the mission "To mentor and empower teen girls for future success through promoting career exploration, confidence building, and leadership development" (GIRL, Inc.,, n.d., para. 1). The partnership with GIRL, Inc. has been critical to the success of the program, particularly in recruiting participants, providing programmatic support, as well as providing financial support in the form of scholarships for participants.

GIRL was founded in 2010 and offered workshops, career conferences, and worksite visits to middle and high school girls to expose them to careers they did not know existed. When the George Mason University (GMU) STEM Accelerator Program was looking for community groups for outreach initiatives, the director at the time contacted the founder of GIRL to discuss how GMU could serve the community. The founder wanted to create a summer camp for her organization and so the director convened the co-authors of this chapter, referred to as the "F.O.C.U.S. Four", to create a plan. The founder was able to recruit girls who were already participating in GIRL programs to attend F.O.C.U.S. GIRL was able to even reach out to its partner community centers in low-income neighborhoods to ensure those girls also had the opportunity to attend the F.O.C.U.S. camp because the community center provided transportation which their families could not. In addition to recruiting, the founder was able to assist with planning and securing speakers through GIRL's corporate partners and provided on-site management the week of the camp to support logistics. Also, GIRL was able to fund students to attend the camp through grants and donations given to GIRL.

The partnership between GMU and GIRL demonstrates the importance of colleges working with small local nonprofits to contribute to the ecosystem of potential students in STEM who could matriculate at GMU. Partnering with large national nonprofits can often mean missing the students that are in the community where the university is located. The girls who attended the camp could continue engaging in STEM and career exploration throughout the year by attending other GIRL programs and they felt a sense of community. One participant said, "Not only did this program teach me content related to these rather technical fields, but it allowed me to meet female leaders of color within the STEM community who ultimately became my mentors. Attending GIRL's F.O.C.U.S. program helped me fall in love with learning as well as the researching process."

Corporate Sponsorship

In addition to partnering with the community, pursuing corporate sponsorships has been the lifeline of the F.O.C.U.S. program. Although the program is not incorporated into the university budget, by obtaining funding from external organizations the F.O.C.U.S. team has been able to drastically decrease the cost of the program from $400 for the week to as low as $100 for the week. Included in their registration in addition to the STEM programming are various supplies (such as backpacks, notebooks, pens, and pencils), camp t-shirts, and lunch every day. When combining registration fees with sponsorship funding, the F.O.C.U.S. team is also able to cover the cost of any workshop supplies needed by our presenters. Additionally, the team can cover various staffing costs such as counselors as well as GMU students, faculty, and external workshop presenters. Further, and perhaps most importantly, we provide many scholarships to our participants who communicate a need for financial support.

Corporate sponsorship began in 2014 when the STEM Accelerator program was awarded a grant in the amount of $20,000 from the Northern Virginia's Business Women's Giving Circle to expand the F.O.C.U.S. program. With those funds, the cost was reduced by almost 60% for the 2015 camp. The funds also allowed the program to grow significantly to 75 girls and 20 counselors.

With $20,000 in funding received from Battelle Corporation and $1000 from Virginia Bio, the F.O.C.U.S. camp grew to 93 girls and 22 undergraduate learning assistants in 2016. In 2017, the camp grew to 100 girls and 24 undergraduate learning assistants with $20,000 in funding from Battelle Corporation. With $15,000 in funding from Battelle Corporation again in 2018, the F.O.C.U.S. camp hosted 73 participants and 16 undergraduate assistants. Since that time, Battelle and Virginia Bio have continued to provide financial support to the F.O.C.U.S. program in various ways. Other organizations that have supported and/or financially sponsored this program over the years include State Farm, the National Science Foundation, the Quantum Materials Center, The Stafford Foundation, the New Wave Foundation, the Moore Family Advised Fund of Grand Rapids Community Foundation, R&R Star Holdings, LLC, and the George Mason University Disability Services.

BACKGROUND

F.O.C.U.S. Program Overview

After months of planning, the first program launched in July 2014 with 19 girls and four *learning assistants* who served as mentors/counselors for the camp participants. Learning assistants are undergraduate students in the George Mason University College of Science who are hired by faculty to engage in undergraduate instruction, course innovation, peer-to-peer tutoring, and mentoring (George Mason University College of Science, n.d.). The first camp was a huge success. The positive feedback received from the parents and the students was overwhelming. Given the success of the program, the authors applied for a $20,000 grant from the Business Women's Giving Circle (BWGC) to expand the program in 2015 and were awarded the grant. Over the years with the support of BWGC and other corporate sponsorships from organizations such as Battelle, the program has been held almost every summer since. As described in the previous section, corporate sponsorships have been integral to the development of the program.

F.O.C.U.S. is a weeklong program held in the summer (Monday-Friday). After registration, the attendees are broken down into four groups of 25. Students from the same or adjacent grades are generally grouped. Each of the four groups is assigned a lead counselor who is the main point of contact for that group. The lead counselor helps to guide the other counselors, maintains all emergency contact information, and is responsible for checking program participants in and out each day.

The curriculum for the F.O.C.U.S. program changes each year, however, the structure generally stays the same. There are four hands-on and interactive workshops held Monday through Thursday with each day focusing on a different letter of STEM: Monday is Science day, Tuesday is Technology day, Wednesday is Engineering day, and Thursday is Mathematics day (Figure 1). Workshops are generally about an hour and fifteen minutes and presenters are recruited in various ways. For example, some workshop presenters are recruited through the STEM faculty listservs. Solicitations for workshop presenters are also sometimes shared on social networks such as LinkedIn and oftentimes, workshop presenters are recruited through direct contact. Those recruited are from varied backgrounds such as students, faculty, and STEM professionals. Presenters are given the general guidelines that any 'lecture time' should be 15 minutes or less. Program coordinators emphasize to presenters that for most of their workshop, the students should be engaged in hands-on and/or interactive activities. Presenters apply to present their workshops online and their applications are reviewed before confirmation. When reviewing the workshop applications, various components are evaluated such as evidence of cultural relevance, feasibility, cost, age-appropriateness, as well as engagement and interactive components.

Figure 1. Example schedule for F.O.C.U.S.

Example F.O.C.U.S. Schedule

MONDAY	
Time	Session
8:30-9:00 am	Registration
9:00-9:20 am	Welcome
9:30-10 am	Pre-Assessment
	Team Building
10 am-10:30 am	Team Building
	Pre-Assessment
10:30 am-11:00 am	Team Building
	Team Building
11:00 am-11:30 am	Team Building
	Team Building
11:35-12:40 pm	LUNCH
12:45-1:55 pm	Spineless Wonders
	DNA Necklaces
2:05-3:15 pm	DNA Necklaces
	Spineless Wonders
3:15-3:30 pm	PICK UP
	PICK UP

	TUESDAY	WEDNESDAY	THURSDAY
Time	Session	Session	Session
8:30-8:55 am	Group A-DROP OFF	Group A-DROP OFF	Group A-DROP OFF
	Group B-DROP OFF	Group B-DROP OFF	Group B-DROP OFF
9-10:10 am	Forensic Science	Weather & Climate	MEGL
	Digital Storytelling	Polarized Light	Microbial Zoo
	Digital Storytelling	Weather & Climate	Microbial Zoo
	Forensic Science	Polarized Light	MEGL
11:35-12:40 pm	LUNCH	LUNCH	LUNCH
12:45-1:55 pm	Cybersecurity	3d Printing	2D Sprites
	Drones	Gravity is optional	Rocket Design
2:05-3:15 pm	Cybersecurity	3d Printing	2D Sprites
	Drones	Gravity is optional	Rocket Design
3:15-3:30 pm	Group A-PICKUP	Group A-PICKUP	Group A-PICKUP
	Group B-PICKUP	Group B-PICKUP	Group B-PICKUP

FRIDAY	
Time	Session
8:30-8:55 am	Group A-DROP OFF
	Group B-DROP OFF
9-10:10 am	Careers in STEM
	FOCUS Academy Panel & Poster Preparation
	FOCUS Academy Panel & Poster Preparation
	Careers in STEM
11:35-12:40 pm	LUNCH
12:45-2:00 pm	12:45-1:00 pm-Post-Assessment
	1:00-1:45 pm-Finish poster & practice talk
	12:45-1:00 pm-Finish poster & practice talk
	1:00-1:15 pm-Post Assessment
	1:15-1:45 pm-Finish poster & practice talk
2:05-2:35 pm	Poster Session
2:45 pm-3:00 pm	Closing Session
3:00-3:15 pm	Certificates
3:15-3:30 pm	Group photo

The four larger groups of 25 students rotate so that each group has an opportunity to participate in each workshop. Each of the larger groups is further broken down into smaller groups of five with each smaller group having their own dedicated counselor. This is a particularly important component of the F.O.C.U.S. program. Not only do the counselors help guide the program participants through their workshops but as described in the literature review, positive role models are important for STEM identity development but also can also have a positive impact on participants' attitudes towards STEM (Clark & Sheridan, 2010). Out-of-school STEM programs like F.O.C.U.S. which connect participants with mentors in higher education (such as counselors and faculty) can positively impact their persistence in the STEM pipeline (Hargrave, 2015).

This can be further explained using situated learning as our theoretical framework which views learning and identity development within communities of practice (Lave & Wenger, 1991; Roberts et. al, 2018; Sadler, 2009). Wenger (1998) describes a community of practice as an environment in which its members engage in authentic learning experiences with peers, near peers, and experts. This is precisely the environment in which F.O.C.U.S. operates. Participants engage with one another but also with the undergraduate students who serve as counselors as well as with faculty and STEM professionals who serve as experts leading the workshops. It is through these interactions that we believe their attitudes towards STEM are influenced and the development of their identities as individuals who study and work in STEM fields is nurtured.

On the last day of the program, we hold special sessions that focus on STEM integration, global STEM perspectives, STEM careers, entrepreneurship, and leadership. Participants are often given opportunities to meet successful women who currently work for leading organizations and/or own their own businesses. They also spend the last morning preparing for their poster presentations. The participants have an opportunity to collaborate with their groups and present one of the topics from the week in a poster session with invited faculty, students, and family. This activity helps reiterate content learned during the week but it also helps build their confidence and their communication skills.

Due to the COVID-19 pandemic, the 2020 F.O.C.U.S. program was canceled, however, in 2021, we wanted to bring the program back despite the ongoing pandemic so we decided to take F.O.C.U.S. virtual. Earlier that year in the spring of 2021, the Project M.O.S.T. (Mason Outreach in STEM for Teachers) program was held. This program will be described in more detail later in the chapter but the teachers that participated in the program developed virtual workshops that they would lead during the virtual F.O.C.U.S. program. Each group of teachers that was leading a workshop created a list of supplies to be ordered for the students which would be added to the STEM kits. A STEM kit which contained the supplies for each workshop as well as other items such as STEM books was prepared for each participant. Due to the size of the kits, the kits could not be mailed so participants picked up their kits the week before the program. Although we preferred to be in-person, our virtual programs allowed us to connect the participants with mentors all around the country such as the women from STEMNoire (n.d.), a group of Black women in STEM who are pursuing or are currently in STEM professions.

F.O.C.U.S. Academy

Three years after the first F.O.C.U.S. middle school program, we piloted the first official F.O.C.U.S. Academy residential program for high school students in 2017. The F.O.C.U.S. Academy was born out of the numerous requests from our F.O.C.U.S. middle school camp alumni who loved the program so much that they wanted to continue through high school. This program was designed to be more advanced and immersive in the college experience than the F.O.C.U.S. middle school programs. The STEM workshops are longer and more advanced, but many other enhancement sessions are included such as college and SAT prep, goal setting, fitness classes, a farewell party, and more. The full program launched in 2018 and has continued since that time.

Building on the mounting success of the past years, the F.O.C.U.S. leadership team has taken the time between sessions to learn and develop plans for expansion in several areas. The team continues to seek ways to create a STEM ecosystem which begins in middle school and continues through the professional years. The authors are also working to develop the F.O.C.U.S. Scholars program which will recruit underrepresented women studying STEM including targeted recruitment of Academy students. As described previously in the literature review, sustained engagement with students beyond the confines of the program is important. In addition to working as mentors for the F.O.C.U.S. mini-camps, middle school camp, and high school Academy, the F.O.C.U.S. Scholars will be assigned faculty mentors through the fall and spring semesters and will remain engaged in STEM mentorship and research activities throughout the year.

Project MOST: Mason Outreach in STEM for Teachers

Another exciting expansion includes Project MOST (Mason Outreach in STEM for Teachers) which will help establish a new generation of teacher leaders who will help disseminate best practices in integrated STEM education to improve student learning. The objective is to create a network of teachers in Virginia who will become ambassadors and help expand STEM learning and innovation across the state. The MOST network supported the professional development of middle school teachers selected from Title I schools in the Northern Virginia area that serve high English Language Learners (ELL) populations. Project MOST will help establish a new generation of teacher leaders who will help disseminate best practices in integrated STEM education to improve student learning. MOST is building best practices to grow successfully into a state-wide network in future years to model other successful state-based STEM networks, such as the Ohio STEM Learning Network (OSLN) and Tennessee STEM Innovation Network (TSIN) as well as the national network STEMx.

The inaugural Project MOST Institute was held in April 2021. It was a 3-day professional development institute targeting grade 6-12 teachers in Virginia. The institute, which was sponsored by Battelle, focused on culturally responsive teaching practices as well as effective mentorship and successful practices for engaging girls of color and those marginalized in STEM. During the institute, the selected teachers learned how to design STEM curricula and activities that included effective pedagogical and instructional approaches through learning materials, STEM-lab activities, and integrated lesson plans. Additionally, they learned how to create performance-based assessments that were equitable and non-biased. The teachers used what they learned during the institute to prepare a workshop for the virtual F.O.C.U.S. programs held in August 2021.

Using collaborative real-world problem-based learning, the teacher participants worked with university scientists and F.O.C.U.S. program leaders to acquire enhanced skills to engage students from underrepresented populations in exploring driving questions in STEM that require them to formulate problems, collect and analyze data, present and justify their reasoning, and critique others' ideas to evaluate their conclusions. The objective was to create a network of teachers in Virginia who will become ambassadors and help expand STEM learning and innovation across the state. The MOST project empowered these teacher participants with the latest scientific findings on the learning characteristics of underrepresented populations, awareness of cultural differences, understanding of students with multiple learning preferences, and the use of equitable and non-biased assessments. They learned how to recognize their own biases while being enlightened about local communities, and recognizing how talents appear in various cultures. They also learned how to become more effective mentors and proven techniques to engage females of color and those underrepresented in STEM in their schools. The project helped teachers to create STEM curricula and activities that will include effective pedagogical and instructional approaches through learning materials, STEM-lab activities to support learning by doing, and performance-based assessments that are both formative and summative.

In August 2021, the teachers had the opportunity to participate in presenting STEM activities at the F.O.C.U.S. virtual programs for middle and high school girls. The participants each built a STEM-focused lesson plan. The 2021 MOST pilot was successful and we hope to bring these teachers back to serve as mentors to a new cohort of teachers in 2023.

MATERIALS AND METHODS

Data Collection

There are both quantitative and qualitative aspects of the data collection for this program that we used to address three questions:

(1) How do the F.O.C.U.S. camp participants describe their race/ethnicity?
(2) What were the participants' thoughts of and experiences with STEM prior to attending the F.O.C.U.S. program?
(3) Are there changes in student interest, confidence, and career goals?

Before data was collected, approval from the Institutional Review Board was obtained. Each year before beginning the program, the participants are asked to complete an electronic questionnaire. All parents and/or guardians must sign and submit a parental consent form to participate in the research before the participant completes the questionnaire. The pre-program questionnaire provides a baseline that can be used to compare to the data obtained from the questionnaire administered on the last day of the program (Table 1). It includes Likert-scale, multiple choice, and open-ended questions to gather data on participant demographics, previous experiences with STEM (both formal and informal), as well as their interests and motivation to pursue STEM.

Table 1. Example questions from the 2019 pre-program questionnaire

Question	Response Options
Are you interested in a STEM career?	Yes, No, I'm not sure at this time
If you answered yes to the question above, what type of STEM career are you interested in?	Different career options provided and also the options to select other or "I know I'm interested in a STEM career but I'm not sure of the career right now
What does STEM mean to you?	Open-ended
What subjects do you enjoy most in school?	Multiple select
Do your parents encourage you to participate in STEM activities?	Yes, No, Sometimes
What do you expect to gain from this week's F.O.C.U.S. camp?	Open-ended
What are you looking forward to during this week's F.O.C.U.S. camp?	Open-ended
What type of STEM activities have you ever participated in?	Multiple select including other
Do your teachers encourage you to participate in STEM activities?	Yes, No, Sometimes
As you think about college, what would you like to major in?	Multiple select
The table below lists things you can do when you are working on school activities or assignments. Choose the most appropriate response to each of these things.	

continues on following page

Table 1. Continued

Question	Response Options
When I see a new math problem, I can use what I have learned to solve the problem.	
I can use what I know to design and build something mechanical that works.	
In lab activities, I can use what I have learned to design a solution to a problem.	
I can effectively lead a team to design and build a hands-on project.	
I know where I can find the information that I need to solve difficult problems.	Never, Sometimes, Very Often, Always
I can use what I have learned to teach myself to program a computer game.	
I can explain math or science to my friends to help them understand.	
I can get good grades in math.	
I can get good grades in science.	
Reading the following topics and check the box if you see any of these activities in your future:	
Using Math to design and build things	
Using technology and computers to make new designs	
Working with others to invent new products and solutions	
Designing or building new things: like an engineer might do with new ships, cars, tools, buildings & machines	I can see this in my future, Perhaps if I knew more, No way
Any kind of Science with technology and engineering while working with teams and problem solving	
Read each statement below and make a prediction about your friends in class today, not the entire class!	
Someday, in their career my friends will have to use MATH.	
Someday, in their career my friends will have to use TECHNOLOGY.	
Someday, in their career my friends will have to use SCIENCE.	Yes, Maybe, Never
Someday, in their career my friends will have to use RESEARCH to solve a problem.	

The post-program questionnaire reassesses their interest in STEM and seeks to obtain information about the various aspects of the program that they did and did not enjoy (Table 2). By having both quantitative and qualitative aspects to the questionnaire, we can gain a more in-depth understanding through triangulation of a variety of evidence. The questionnaire was designed and reviewed by experts in STEM and STEM education both in K-12 and higher education areas. Since the program began in 2014, while some questions have been added and/or removed, we have continued to use the pre-program and post-program questionnaire method for collecting data. A descriptive analysis was done to gain a broad understanding of the program. Future research will include more detailed and systematic analyses as well as additional methods of data collection such as focus groups and formalized observations.

Table 2. Example questions from the 2019 post-program questionnaire

Question	Response Options
After this week's camp, are you interested in a STEM career?	Yes, No, I'm not sure at this time
If you answered yes to the question above, what type of STEM career are you interested in?	Different career options provided and also the options to select other or "I know I'm interested in a STEM career but I'm not sure of the career right now
Has your definition of STEM changed after this week? If yes, how?	Open-ended
What subjects do you enjoy most in school?	Multiple select
Do your parents encourage you to participate in STEM activities?	Yes, No, Sometimes
How would you rate the F.O.C.U.S. camp?	Much better than I expected, I learned a lot and want to participate next year. Much better than expect, I learned a lot. Better than I expected. What I expected. Other.
What activities did you find most interesting during this week's F.O.C.U.S. camp?	Multiple select
Do your teachers encourage you to participate in STEM activities?	Yes, No, Sometimes
What would you change about the camp?	Open-ended
The F.O.C.U.S. camp:	
Helped me understand STEM better.	
Led me to a better understanding of my own career goals.	
Increased my interest in studying STEM in college.	
Made me think more about what I will do after graduating from high school.	
Made me decide to work harder in school.	
Made me decide to take different classes in school (including college) than I had planned to.	
Made me more confident in my ability to participate in STEM projects or activities.	
Not at all, Slightly, Moderately, A great deal	

Participants

A voluntary selection method is used to recruit the program participants who identify as rising 6th-8th graders for F.O.C.U.S. and rising 9th-12th graders for F.O.C.U.S. Academy through flyers, emails, and word of mouth. Partial to full funding is available for most participants who communicate a financial need. The program has a website and information for these programs can be found there as well as on the university summer program website. Middle school girls were originally selected as the target population for this program as previous studies describe that students with positive attitudes towards STEM at this age are more likely to persist in STEM (Maltese & Tai, 2011; NRC, 2012; Tai et al., 2006). The research presented in this chapter highlights data collected from the middle school program. Additional demographic data describing the race/ethnicity of the participants can be found in the results section.

RESULTS

The F.O.C.U.S. program outcomes are to: (a) provide students with the educational and social enrichment opportunity, (b) provide positive role models so they can feel confident in their knowledge and competencies in STEM, (c) increase their interest in STEM, (d) increase their confidence level in their STEM skills, and (e) to help them to better understand their career goals. With those outcomes in mind, we considered three questions for this study: (1) How do the F.O.C.U.S. camp participants describe their race/ethnicity? (2) What were participants' thoughts of and experiences with STEM prior to attending the F.O.C.U.S. program? and (3) Are there changes in student interest, confidence, and career goals?

Question 1: How do the F.O.C.U.S. Camp Participants Describe Their Race/Ethnicity?

Although the program's targeted population is girls from underrepresented backgrounds, participation in the program is not exclusive, therefore, any student who would like to participate in the program may do so. On average, most of our participants identify as being Black or African American (46% using data from 2016-2019 and 2022) as shown in Table 3. Approximately 18% identify as Asian or Pacific Islander, 12% as Hispanic or Latino, 11% as multiracial, and 3% as white. The remaining participants either preferred to not respond or identified as "other".

Table 3. Camp participant race and/or ethnicity

	2022 (%) (n=44)	2019 (%) (n=47)	2018 (%) (n=59)	2017 (%) (n=93)	2016 (%) (n=90)	Average (%)
Black	57.4	55.1	33.0	40.0	45.0	46.1
Hispanic	11.1	12.2	21.0	10.0	6.0	12.1
Asian and/or Pacific Islander	16.7	14.3	14.0	23.0	22.0	18.0
White	1.8	2.0	1.0	6.0	6.0	3.4
Multi- racial	11.1	12.2	12.0	13.0	8.0	11.3
Prefer not to respond	0.0	2.0	19.0	8.0	13.0	8.4
Other	1.8	2.0	0.0	0.0	0.0	0.8

Question 2: What Were Participants' Thoughts of and Experiences With STEM Prior to Attending the F.O.C.U.S. Program?

For this question, we considered the participants' responses to questions about their parents and teachers encouraging them to pursue STEM, the STEM activities they have participated in previously, and their definitions of STEM. Most of the students over the years indicate that both their parents and their teachers encourage them to pursue STEM although the participants report their parents to be much more encouraging than their teachers. For example, in 2019, when asked whether their parents encourage them to participate in STEM activities, 74.5% answered yes, 23.4% said sometimes, and only 2.1% said no. In contrast, when asked whether their teachers encouraged them to pursue STEM, only 29.8% answered yes, 51.1% said sometimes, and 19.1% said no.

We also asked participants the question: "What type of STEM activities have you ever participated in?" Participants had the option to multiple select from several options including STEM summer camps, science fair, robotics, video game challenges, afterschool STEM activities, or they could select 'other' and write in an option. In 2019, while 66% of the participants reported participating in at least one or two types of STEM activities, 11% of the participants said they had never participated in any STEM activity prior to F.O.C.U.S. The remaining 23% of participants indicated participating in 3 or more different types of activities.

The participants were also asked in the post-program questionnaire whether their definition of STEM changed since the beginning of the week. In the beginning of the week, most participants defined STEM by the stating the words that make up the acronym (science, technology, engineering, and mathematics), however, by the end of the week, participants often indicated that their definition did change and they began to include terms such as "problem-solving" and "creativity." One student wrote: "My definition has changed because I realized that stem isn't just science, technology, engineering, and mathematics, it's problem solving and working together to create a product." Another student wrote: "STEM is a way that people can solve problems in the world by being creative using science, technology, engineering, and math."

Question 3: Are There Changes in Student Interest, Confidence, and Career Goals?

Each year, students are asked about their interest in STEM in both the pre-program and post-program questionnaires. Overall, the percentage of students interested in STEM increases by the end of the program. Also, their uncertainty about whether they are interested in STEM decreases as shown in Table 4. This is an important finding because the goal is not to push every student into STEM but to provide them with positive exposure in a way that helps them to better understand their goals, whether their goals are to pursue STEM or not. For example, the students were asked whether they were interested in STEM. The response options were "yes", "no", or "I'm not sure at this time."

Table 4. Camp participants' interest in STEM

	2022 (%) (n=44)		2019 (%) (n=47)		2018 (%) (n=59)		2017 (%) (n=93)		2016 (%) (n=90)	
	Pre	Post	Pre	Post	Pre	Post	Pre	Post	Pre	Post
Yes	81.8	88.6	55.0	80.0	64.5	81.4	54.7	73.1	65.3	78.9
No	4.5	4.5	11.0	9.0	9.7	5.0	12.6	7.5	10.5	7.8
Unsure	13.6	6.8	34.0	11.0	25.8	13.6	32.6	19.4	24.2	13.3

In 2019, the percentage of students that answered "yes" was 55% on the first day of the program and it increased to 81.4% by the end of the week. Also, the percentage of students who were unsure of whether they were interested in STEM started at 34% in the pre-program questionnaire but decreased to 11% by the end of the week. Additionally, on average, 41% of the students attending the program between 2015 and 2018 reported that the F.O.C.U.S. program not only increased their confidence in participating in STEM activities "a great deal" but it also led to a better understanding of their career goals (Table 5).

Table 5. Impact on confidence in STEM participation and understanding of career goals

	2018 (%) (*n*=59)		2017 (%) (*n*=93)		2016 (%) (*n*=90)		2015 (%) (*n*=74)	
Participant's perceived amount of impact	**Confidence**	**Career Goals**	**Confidence**	**Career Goals**	**Confidence**	**Career Goals**	**Confidence**	**Career Goals**
Not at all	6.8	6.8	6.7	9.7	7.8	3.3	8.1	6.8
Slightly	20.3	18.6	28.9	25.8	24.4	18.9	23.0	14.9
Moderately	30.5	33.9	22.2	30.1	30.0	31.1	24.3	32.4
A great deal	42.4	40.7	42.2	34.4	37.8	46.7	44.6	45.9

Quality of the Program

As described in the literature review, maintaining the quality of an out-of-school program is important and has been the focus since its inception in 2014. Counselors are provided with training in advance of the program. The training covers basic program operations but also incorporates various scenarios involving cultural relevancy which are discussed as a group. It also covers guidance on how to ensure that participants feel included throughout the week. During the week, participants are provided with intentional opportunities to reflect, bond with their fellow participants, and have their voices heard during their group sessions at the beginning and end of the day.

Earlier in the chapter, we described expectations for our presenters. For example, we request 15 minutes or less of 'lecture time' with hands-on and/or interactive activities that are culturally relevant and age-appropriate. Each workshop is carefully reviewed and feedback for implementation is provided when needed to ensure the session will be high quality, engaging, and appropriately challenging. While there are no formal mechanisms in place yet to ensure adherence to these guidelines, we do ask our counselors and staff present in the workshops to informally provide their input during our end-of-day debriefings. We also ask the participants to indicate which sessions they attended were their favorite sessions on the post-assessment questionnaire. For future programs, we plan to incorporate more formal methods of collecting this data such as asking the participants to provide open-ended responses explaining why they selected specific workshops to be their favorite and having planned observations of the workshops.

Limitations

The selection method and research design used in this study limit the ability to generalize beyond this program. Additionally, the sampling method may lead to selection bias, however, using the results of the pre-program questionnaire the researchers can consider how potential rival hypotheses may have influenced the findings. A complete and systematic analysis of the qualitative data is still in progress and will be reported in future research. Despite these limitations, an initial review of the findings is still impactful. Limited generalizability is not a concern as the results are used to continue to improve upon this specific program and to provide information on the program's effectiveness. The findings also allow the researchers to identify the most and least influential components as well as inform how these components could be modified. Lastly, others who may be interested in designing an out-of-school STEM program that is effective for students from underrepresented programs could use this data as a resource.

DISCUSSION

This chapter described three themes including the unique needs of students in out-of-school programs; the importance of the quality of these programs; and program evaluation. The literature provides evidence that incorporating these themes could potentially increase the numbers of students from underrepresented populations in STEM, therefore, they were meaningfully incorporated into the planning and design of the Females of Color and those Underrepresented in STEM (F.O.C.U.S.) programs. The F.O.C.U.S. programs have been designed using culturally relevant pedagogy and are interwoven with mentorship, as well as high-quality, engaging, and challenging STEM activities using project- and or problem-based learning. Further, to effectively increase program participant sustainability and persistence in STEM, we continue to move towards developing a STEM ecosystem supporting both students and teachers beyond the scope of the program and through higher education. These ecosystems will immensely benefit from community partnerships and corporate sponsorships.

Over the years, there have been many lessons learned. The amount of preparation and logistical work needed to implement a program of this nature cannot be underestimated. Planning should begin at least one year in advance with registration for the program opening three to six months before it starts. A team who can manage the various aspects of the planning is critical. In addition to recruiting presenters for the STEM content, additional team members who can manage administrative tasks such as hiring, communications, training, registration, parking, catering, and supply orders will ensure a smoother process. Further, having a team of individuals dedicated to soliciting funds from sponsors as well as maintaining communication with them and preparing post-program reports will help ease the load of those involved with the logistical planning. It is also helpful to be in constant communication with individuals who are familiar with handling the risk management side of the planning process. Hosting minors in any program is a significant responsibility and those involved with planning and implementing the program should be prepared for all possible scenarios.

CONCLUSION

The data presented in this chapter demonstrates the impact of a STEM program for students from underrepresented populations. Participation in the F.O.C.U.S. program not only increased interest in STEM but also increased their confidence and helped participants to better understand their career goals. Lastly, it should not be taken lightly the importance of targeting this specific population as one student wrote, "What I liked and appreciated most about this camp is how there are a lot of people like me that I can relate [to] as [an] African American girl. I also got to learn about other fields in STEM and learn things I haven't had time to thoroughly think about."

Supplemental Materials

For more information on the F.O.C.U.S. program at GMU, please see the website at: https://www.focusonstem.org/focuscamp

REFERENCES

Bystydzienski, J. M., Eisenhart, M., & Bruning, M. (2015). High School Is Not Too Late: Developing Girls' Interest and Engagement in Engineering Careers. *The Career Development Quarterly*, *63*(1), 88–95. doi:10.1002/j.2161-0045.2015.00097.x

Calabrese Barton, A., Tan, E., & Greenberg, D. (2017). The Makerspace Movement: Sites of Possibilities for Equitable Opportunities to Engage Underrepresented Youth in STEM. *Teachers College Record*, *119*(6), 1–44. doi:10.1177/016146811711900608

Carlone, H. B., & Johnson, A. (2007). Understanding the science experiences of successful women of color: Science identity as an analytic lens. *Journal of Research in Science Teaching*, *44*(8), 1187–1218. doi:10.1002/tea.20237

Clark, K., & Sheridan, K. (2010). Game Design Through Mentoring and Collaboration. *Journal of Educational Multimedia and Hypermedia*, *19*(2), 5–22.

George Mason University College of Science. (n.d.). *Learning assistants*. GMU. https://science.gmu.edu/academics/special-study-options/learning-assistants

Gilbert, A., & Yerrick, R. (2001). Same school, separate worlds: A sociocultural study of identity, resistance, and negotiation in a rural, lower track science classroom. *Journal of Research in Science Teaching*, *38*(5), 574–598. doi:10.1002/tea.1019

Girls Inspired and Ready to Lead, Incorporated. (n.d.). *About GIRL*. GIRL. https://girlsinc.org/about-us/

Harding, S. G. (2004). *The feminist standpoint theory reader: Intellectual and political controversies*. Psychology Press.

Hargrave, C. (2015). Counter Space: Analysis of Educational Structures of an After-School Program that Fosters Black Academic Success Narratives. *The Journal of Negro Education*, *84*(3), 348–361. doi:10.7709/jnegroeducation.84.3.0348

Hayden, K., Ouyang, Y., Scinski, L., Olszewski, B., & Bielefeldt, T. (2011). Increasing student interest and attitudes in STEM: Professional development and activities to engage and inspire learners. *Contemporary Issues in Technology & Teacher Education*, *11*(1), 47–69.

Hidi, S., & Renninger, K. A. (2006). The four-phase model of interest development. *Educational Psychologist*, *41*(2), 111–127. doi:10.120715326985ep4102_4

Kang, H., Calabrese Barton, A., Tan, E., Simpkins, S., Rhee, H., & Turner, C. (2018). How do middle school girls of color develop STEM identities? Middle school girls' participation in science activities and identification with STEM careers. *Science Education*, *103*(2), 418–439. doi:10.1002ce.21492

Kim, A. Y., & Sinatra, G. M. (2018). Science identity development: An interactionist approach. *International Journal of STEM Education*, *5*(51), 1–6. doi:10.118640594-018-0149-9 PMID:30631740

Ladson-Billings, G. (1995). Toward a Theory of Culturally Relevant Pedagogy. *American Educational Research Journal*, *32*(3), 465–491. doi:10.3102/00028312032003465

Larnell, G. V., Bullock, E. C., & Jett, C. C. (2016). Rethinking teaching and learning mathematics for social justice from a critical race perspective. *Journal of Education*, *196*(1), 19–29. doi:10.1177/002205741619600104

Lederman, N. G., & Abell, S. K. (Eds.). (2014). *Handbook of Research on Science Education*. Routledge.

Leonard, J., Chamberlin, S., Johnson, J., & Verma, G. (2016). Social Justice, Place, and Equitable Science Education: Broadening Urban Students' Opportunities to Learn. *The Urban Review*, *48*(3), 355–379. doi:10.100711256-016-0358-9

Moreno, N., Tharp, B., Vogt, G., Newell, A., & Burnett, C. (2016). Preparing Students for Middle School Through After-School STEM Activities. *Journal of Science Education and Technology*, *25*(6), 889–898. doi:10.100710956-016-9643-3

National Research Council. (2015). *Identifying and supporting productive STEM programs in out-of-school settings*. National Academies Press.

National Science Board. (2018). *Science & engineering indicators*. National Science Foundation. https://nsf.gov/statistics/2018/nsb20181/report/sections/science-and-engineering-labor-force/women-and-minorities-in-the-s-e-workforce#minorities-in-the-s-e-workforce

Ortiz, N. A., Morton, T. R., Miles, M. L., & Roby, R. S. (2019). What about us? Exploring the challenges and sources of support influencing Black students' STEM identity development in postsecondary education. *The Journal of Negro Education*, *88*(3), 311–326. doi:10.7709/jnegroeducation.88.3.0311

Pinkard, N., Erete, S., Martin, C., & McKinney de Royston, M. (2017). Digital Youth Divas: Exploring Narrative-Driven Curriculum to Spark Middle School Girls' Interest in Computational Activities. *Journal of the Learning Sciences*, *26*(3), 477–516. doi:10.1080/10508406.2017.1307199

Rahm, J., & Moore, J. (2015). A Case Study of Long-Term Engagement and Identity-in-Practice: Insights into STEM Pathways of Four Underrepresented Youths. *Journal of Research in Science Teaching*, *53*(5), 768–801. doi:10.1002/tea.21268

Sanders, W., & Rivers, J. (1996). *Cumulative and residual effects of teachers of future student academic achievement*. University of Tennessee Value-Added Research and Assessment Center.

Scott, K. A., & White, M. A. (2013). CompuGirls' Standpoint: Culturally Responsive Computing and Its Effect on Girls of Color. *Urban Education*, *48*(5), 657–681. doi:10.1177/0042085913491219

STEMNoire. (n.d.). *Home*. STEMNoire. https://www.stemnoire.org/

Stevens, S., Andrade, R., & Page, M. (2016). Motivating Young Native American Students to Pursue STEM Learning Through a Culturally Relevant Science Program. *Journal of Science Education and Technology*, *25*(6), 947–960. doi:10.100710956-016-9629-1

Stewart, C. O. (2022). STEM Identities: A Communication Theory of Identity Approach. *Journal of Language and Social Psychology*, *41*(2), 148–170. doi:10.1177/0261927X211030674

The Afterschool Alliance. (2008). *Evaluations backgrounder: A summary of formal evaluations of the academic impact of afterschool programs*. ASA.

Trytten, D. A., Foor, C. E., Murphy, T. J., Shehab, R. L., Walden, S. E., & Pan, R. (2014). UNSEEN DIFFERENCES. *ASEE Prism, 24*(3), 0-0. https://www.proquest.com/scholarly-journals/unseen-differences/docview/1628648512/se-2

Tsui, L. (2007). Effective Strategies to Increase Diversity in STEM Fields: A Review of the Research Literature. *The Journal of Negro Education*, *76*(4), 555–581.

Wallace, E., Perry, J., Ferguson, R., & Jackson, D. (2015). The Careers in Health and Medical Professions Program (CHAMPS): An Impact Study of a University-Based STEM+H Outreach Program. *Journal of Science Education and Technology*, *24*(4), 484–495. doi:10.100710956-014-9536-2

Zuniga, K., Olson, J. K., & Winter, M. (2005). Science education for rural Latino/a students: Course placement and success in science. *Journal of Research in Science Teaching*, *42*(4), 376–402. doi:10.1002/tea.20064

Chapter 9
Enhancing Middle Grade Students' Interest and Learning in a Community-Engaged Summer STEM Competition

Rebecca L. Hite
https://orcid.org/0000-0002-6275-3804
Texas Tech University, USA

Jessica L. Spott
https://orcid.org/0000-0002-5184-8816
Texas Tech University, USA

Levi E. Johnson
https://orcid.org/0000-0003-2894-1358
Texas Tech University, USA

Kaylee Johnston
Lubbock Independent School District, USA

Lane J. Sobehrad
Lubbock Independent School District, USA

ABSTRACT

This chapter presents a curriculum collaboration between Texas Tech University (STEM CORE and the TTU College of Education) and Lubbock Independent School District (Lubbock ISD) for a week-long STEM challenge, focused on bringing STEM into their communities and reducing barriers to accessing STEM opportunities. The goal of the summer 2021 iteration of Lubbock ISD STEM Challenge was to engage students in a summer STEM experience designed to increase their interest in STEM-related education and career pathways while positively affecting their learning about STEM topics. Readers will discover ten graphic organizers; detailed teacher notes that provide step-by-step facilitation of the event; learning objectives aligned to national standards; a materials list and preparation checklist for activities; detailed answer keys to related student assessments; and five exit tickets. Significant improvement was found among students' use of 21st century skills, STEM learning, and STEM interest. This activity may be used for in-school or community-based (out-of-school) youth STEM enrichment.

DOI: 10.4018/978-1-6684-7771-7.ch009

INTRODUCTION

STEM Challenges provide a project-based setting for groups of students to develop open-ended questions, propose solutions to those questions, and test the effectiveness of each solution. The result of each project is a culminating product with an applicable real-world setting. Each iteration of STEM Challenge asks students to create a product that solves a different problem. In the summer of 2021, students were asked to design and build a cardboard boat that held all four members of the student group and kept their delicate cargo safe in transport across a pool in order to fulfill a supply order for their client. Their designated cargo was an egg, which could not break if they were going to fulfill the terms of their contract. Next, students had to develop driving questions that would help them successfully create their product and complete their task. At first glance, many students' initial driving questions may appear overly simplistic - "How are we going to ensure that our egg [cargo] is protected?" However, the parameters of the project required students to consider more complex factors. Protecting their cargo across a body of water necessarily included a boat that did not sink, could support their weight of four people, and a cargo container that could protect the egg from significant movement. This ideation was the starting point for a week-long, multi-stage design process as seen in Figure 1.

Figure 1. Students collaborate to determine the best build to keep cargo safe

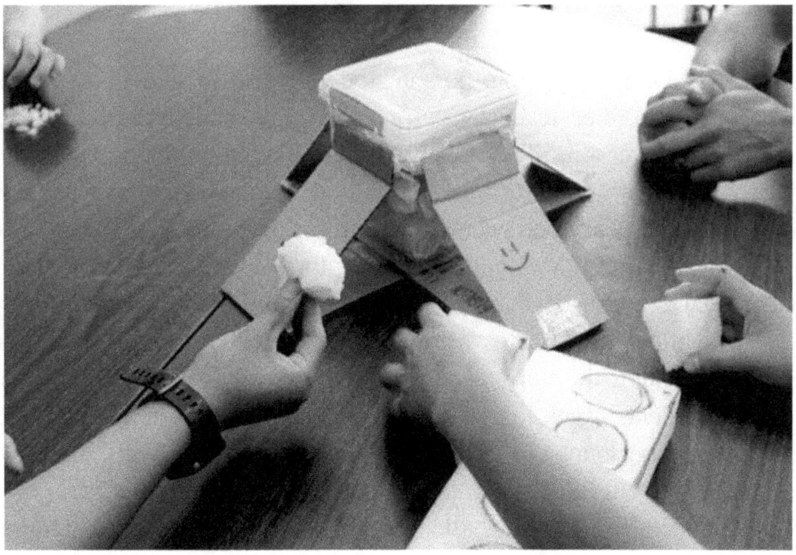

Inspired by the Massachusetts Institute of Technologies' *Cardboard Boat Regatta* in which undergraduate engineering students build and race a boat from minimal, limited materials (Trafton, 2007), the Lubbock ISD STEM Challenge carefully designed bi-daily, hands-on STEM experiences to guide middle-school students in (1) employing *scientific* concepts like buoyancy, propulsion, volume, weight, speed and surface area; (2) making appropriate and precise measurements with *technology and tools* including meter sticks, stopwatches, and video capture; (3) engaging in *engineering* principles such as design and prototyping; and (4) using *mathematical* concepts like ratios and proportions to meet the task of this year's STEM Challenge (as seen in Figure 2).

Figure 2. Student's graphic organizer for designing a cargo container (day three mini challenge)

Think before you do:

Tilt and Crush are important aspects when transporting fragile and/or dangerous cargo.
Imagine you are shipping a glass container of flammable liquid. You don't want the container to tip over (tilt) while you are moving it and pour out the contents in transit. You also don't want the container to break (crush) if someone stacks a box on top of it.

Your shipping company will be asked to transport two types of cargo. Today you will prepare to transport 2 raw eggs within a sealable plastic cargo container. Consider how you will address issues of crush and tilt when transporting this delicate cargo across the pool.

Develop your ideas by sketching out TWO designs below to protect the 2 eggs (cargo) in the container.

Table 3.1 *Ideas to reduce tilt and crush concerns of the cargo within your boat design.*

TILT	*CRUSH*

During the 2021 STEM Challenge, four-person student teams of students were recruited from ten Lubbock ISD middle schools. Project leaders specifically targeted students who were interested in STEM but did not have a history of success in STEM at school. These student teams collaboratively designed and constructed cardboard boats with support from a teacher facilitator and undergraduate mentor. Students' specific tasks were to transport a payload twenty-five yards, safely exchange the payload for an alternate (heavier) cargo, then transport the new cargo another twenty-five yards back to the starting position. The student team that completed the journey in the least amount of time and with the least amount of damage to their cargo won the competition. Successful teams worked collaboratively to incorporate concepts from the week's daily activities to develop and construct the best boat and payload support designs within the budget constraints. Students' week-long work in STEM culminated in a community event at which teachers, parents, university personnel, and others were able to view the boats, discuss design processes with students, and cheer on teams for the competitive boat races (see Figure 3).

The intent of the Lubbock ISD STEM Challenge is to provide middle grade students an opportunity to engage in out-of-school STEM activities that enhance their interest and learning in STEM during their summer break, when students in the predominantly economically disadvantaged school district have historically been less likely to engage in, or have access to, academic enrichment. All told, forty students participated in the 2021 STEM Challenge. Demographics of participating students was diverse, consisting of 17 male, 21 female, and two non-binary students. Approximately one-third identified as White, Hispanic, and African-American, respectively, with most (17) as rising eighth graders. Each group included four middle school students, a science/math teacher from the students' school and one undergraduate student mentor. To understand the impact of STEM Challenge on students' STEM interest and learning, students completed three assessment tools: (1) a STEM interest pre/post-test; (2) daily exit tickets that asked students to reflect on their learning; and (3) their performance at the competition that demonstrated their understanding of the learning that occurred during the week. Each activity included in this chapter on the Lubbock ISD STEM Challenge is rooted in research demonstrating the positive impacts of community-engaged, out-of-school STEM learning. In addition, teacher facilitators and undergraduate mentors were provided instructional materials intentionally scaffolded for STEM learning in middle grades. The positive results found from research on the Lubbock ISD STEM Challenge, along with the templates and tools appended, provides evidence-based strategies and a model for others to engage their stakeholders in providing novel access to STEM enrichment in their own communities.

Figure 3. The Community cheering on a student team on the final day of the STEM challenge

BACKGROUND

Common assumptions contend that exposure to STEM learning in both in and out-of-school settings increase the likelihood that a student will be interested in STEM learning in secondary and post-secondary settings and, subsequently, more likely to take steps to pursue a STEM career. Such assumptions are no doubt held, in part, due to the long-standing and widespread popularity of out-of-school STEM enrichment activities, projects, and programs, including STEM competitions like the Lubbock ISD STEM Challenge. However, despite these efforts, "scant research has been conducted on the impact of these kinds of STEM competitions" (Miller et al., 2018, p. 99).

STEM competitions, specifically, show positive results in generating STEM interest and self-selection into other STEM opportunities. Miller et al. (2018), for example, found that adolescent students (in middle grades) who participated in STEM competitions were more likely to express interest in STEM-related careers at the end of high school. Further, these authors found that those students' interests increased with additional opportunities to compete in STEM competitions, and by "both improving students' academic achievement and fostering their interest in STEM," students are more likely to pursue a STEM field in post-secondary education (p. 98). Tai et al. (2006) found that eighth graders who had STEM experiences both in- and out-of-school strongly correlated to future career choices in a STEM-related field. Hite and Spott (2022) found that out-of-school STEM experiences helped to enhance the community's understanding of girls' interests and futures in STEM fields. Jones and Hite (2021, 2022) found that science interest was a critical factor when youth consider a science-related career. Miller et al. (2018) argued that participation in out-of-school STEM opportunities improves students' hard skills (e.g., inquiry and science process) and soft skills (e.g., communication, collaboration) while increasing students' interests in STEM. Thus, a rigorous and appropriate STEM curriculum is needed for students' future successes in STEM courses (Kennedy & Odell, 2014). This includes out-of-school enrichment programs (Moreno et al., 2016) and STEM design challenges (Lesseig et al., 2016).

Other studies on STEM competitions suggest benefits for enhancing students' learning outcomes (Nugent et al., 2011, 2016), interest in STEM (Melchior et al., 2015), and pursuit of STEM careers (Sahin, 2013). The 2021 Lubbock ISD STEM Challenge leveraged motivating elements of STEM competitions described by other scholars, such as employing avenues for challenge-oriented teamwork (Purzer, 2011), promoting active engagement in STEM through hands-on activities (Christensen et al., 2015), and taking notes during collaboration for metacognitive practice (Chu et al., 2020). By incorporating best practices in STEM competitions, we anticipated that participating students would also benefit from enhanced knowledge of STEM content and improved STEM interest.

The Lubbock ISD STEM Challenge was developed by TTU's STEM Center for Outreach, Research & Education (STEM CORE) leadership as an early effort to establish authentic relationships with STEM-focused educators in Lubbock ISD. STEM CORE provided access to programmatic funding for staffing, materials, and supplies, while Lubbock ISD provided access to project spaces and recruited students and teachers. Collaboratively, they developed standards-aligned curriculum that aimed to support district efforts to improve math and science outcomes and student self-selection into secondary and post-secondary STEM pathways. The Lubbock ISD science curriculum coordinator proposed building a week-long "challenge" for middle school students patterned after the popular television show *Junk Yard Wars.* Teams from Lubbock ISD middle schools would come together for a gameshow-like competition that offered interested students a chance to showcase math, science, and engineering skills that they might not have the opportunity to engage in traditional K-12 classrooms. Since its formation, the STEM Challenge has grown in scale and sophistication by adapting and transitioning into a robust STEM curriculum, providing opportunities for research and evaluation, and becoming an annually celebrated community event. However, the original core value—placing students in out-of-school STEM opportunities to engage and motivate student growth in STEM knowledge and STEM interest—remains central to the Challenge.

During the curriculum design process, each activity was intentionally and explicitly aligned to the Next Generation Science Standards (NGSS Lead States 2013), the International Society for Technology in Education (ISTE, 2016) standards for students, and the Common Core State Standards (2010) for Mathematics, English Language Arts and Literacy. The specific standards currently in use by the Lubbock ISD STEM Challenge are summarized below in Table 1. The Teacher's Guide in Appendix K provides greater detail regarding the aligning set of standards for each of the five Mini Challenges and five Design Challenge activities. School districts, state education agencies, universities, and funding sources may align themselves to varying, yet comparable, standards, and the model provided by Lubbock ISD STEM Challenge is a useful tool for aligning a community-based STEM enrichment experience to industry best-practices.

Table 1. Standards alignment to STEM challenge

Next Generation Science Standards				Common Core State Standards		International Society for Technology in Education
Disciplinary Core Ideas (DCIs) for middle grades (MS) physical science (PS)	Science and Engineering Practices (SEPs)	Cross Cutting Concepts (CCCs)	Engineering, Technology, and Applications of Science (ETS)	Mathematics	English Language Arts	
Matter and its Interactions (MS-PS1-2), like Structure and Properties of Matter (PS1.A); Motion and Stability: Forces and Interactions (MS-PS2-2), like Forces and Motion (PS2.A); and Energy (MS-PS3-5) related to transfer of energy	Science Models, Laws, Mechanisms, and theories Explain Natural Phenomena; Developing and Using Models; Engaging in Argument from Evidence; Analyzing and Interpreting Data; and Obtaining, Evaluating, and Communicating Information	Analyzing and Interpreting Data, Energy and Matter, Structure and Function, Cause and Effect, Influence of Science, Engineering, and Technology on Society and the Natural World, and Patterns	Criteria-constraint problem solving (1.1), evaluating competing designs (1.2), analyzing data from experimentation (1.3), and developing models (1.4). The aligned DCIs in engineering are: Defining and Delimiting Engineering Problems (ETS1.A), Developing Possible Solutions (ETS1.B), and Optimizing the Design Solution (ETS1.C).	Alignment in mathematics was for mental computation and estimation strategies (MP.2.7.EE.3), utilizing ratios (56.RP.A.1) and proportional relationships (7.RP.A.2), solving multi-step real-life mathematical problems (MP.2.7EE2), and modeling with abstract reasoning (MP.2.7.SP).	Integrating technical information by comparing and contrasting information (RST.6-8.7 and RST.6-8.9), writing arguments in a disciplinary context (WHST.6-8.1), developing persuasive multimedia presentations (SL8.5).	Empowered Learner (1a, 1c), Innovative Designer (4c, 4d), Computational Thinker (5a, 5c), and Creative Communicator (6c, 6d).

Materials and Methods

For the 2021 Lubbock ISD STEM Challenge, students used graphic organizers twice daily to scaffold both morning (*Mini Challenge*) and afternoon (*Design Challenge*) activities for each day of the five-day STEM Challenge. Graphic organizers were framed around the operative verbs provided by Bloom's Taxonomy to, "describe and classify observable knowledge, skills, attitudes, behaviors and abilities...By creating learning objectives using measurable verbs, you indicate explicitly what the student must do in order to demonstrate learning" (Utica College, n.d., para 1). Use of Bloom's Taxonomy assisted project staff in ensuring that students' activities, over the span of each day, were focused on higher order thinking skills appropriate to their grade level and aligned to STEM standards. Particular to the Lubbock ISD STEM Challenge, it was important that each day of student activities were not only aligned to NGSS, CCSS, and ISTE standards that help researchers determine learning outcomes, but also aligned to the Texas Essential Knowledge and Skills (Texas Education Agency, n.d.) that are critical in determine state accountability scores. Standards to curriculum alignment each activity provides students with opportunities to garner STEM knowledge and situated STEM-based skills and 21st century skills. Appendices A-J provide templates and examples of these graphic organizers for readers to use in their own STEM Challenges. Notably, this research study was approved by the TTU Institutional Review Board for human subject research as IRB2019-569 "Exploring Participants' Learning Outcomes of TTU/LUBBOCK ISD STEM Challenge" under the category 1 exemption for research within an established educational setting.

Each day of the STEM Challenge was separated into morning and afternoon activity blocks. Morning activities were collectively called the "Mini Challenge", while afternoon activities were collectively called the "Design Challenge." The goal of the Mini Challenge was to iteratively engage students in *recalling, recognizing*, and *reflecting* on their daily experiences through intentionally designed scaffolds and protocols that guide students from ideation to culminating product. First, students were given an activity task with an end goal, then prompted to *recall* their own prior knowledge during the prescribed morning activity and prompted to connect this existing knowledge with their work from the previous day. Second, student groups explored and made predictions about scientific phenomena based on discrete criteria as a critical design process in *recognizing* connections between existing knowledge and acquired knowledge. The project team asked group mentors to emphasize "*thinking before doing*" at this stage to reinforce concepts already covered in earlier activities. Third, students were prompted to *reflect* on their learning experiences using prescribed observation and reporting tools that facilitated generative discussions about the purpose of the Mini Challenge. Lastly, students were asked to complete a collaborative task utilizing the skills and knowledge assembled during the previous stages of the Mini Challenge resulting in a culminating product that was entered into a brief competition amongst the ten teams. The highest scoring student product received an advantage in the STEM Challenge competition at the end of the week.

After the Mini Challenge, students receive lunch and have informal team building time before beginning the Design Challenge in the afternoon activity block. Pedagogically, the Design Challenge was implemented using the same overarching concepts of *recalling, recognizing,* and *reflecting*, as well as the curricular tools to facilitate student thinking, writing, and creation. Unlike the Mini Challenge, however, which asked students to complete a standalone project each day, the Design Challenge provided an intentionally designed framework that asked students to complete predetermined portions of their week-long STEM Challenge boat build. Thus, morning activities introduced 21st century skills and scaffolded metacognitive and STEM concept development while afternoon activities reinforced students' newly acquired learning and skills through authentic engagement. The Design Challenge fostered best practices in boat and cargo design by intentionally developing students' proficiency in relevant concepts and skills in complementary scenarios. Design activities provided low-stakes opportunities for the project team to guide student groups to product completion from (Day 1) testing building materials to (Day 2) "purchasing" building materials, (Days 3-4) building design prototypes, and (Day 5) competing in the STEM Challenge. Table 2 provides an overview of the week's activities and Figure 4 shows a completed boat design and build from the LISD STEM Challenge.

Because the Lubbock ISD STEM Challenge is a multi-day project that requires a wide variety of activities, it requires significant preparation and planning. To assist readers who may start their own STEM Challenge, the authors have provided a detailed preparation section under "Supplemental Materials" in Appendix K both Mini Challenge and Design Challenge activities. The "preparation" tables are organized into rows labeled with a specific learning objective, scaffold, or protocol most pertinent to a given activity (e.g., recognize, recall, and reflect, think before you do, observe & report, formulate & create). Their associated columns include information about needed materials, time needed to complete an activity, and additional considerations for STEM Challenge facilitators based on lessons learned from the Lubbock ISD STEM Challenge. Day 1 and Day 2 include additional extension activities at the bottom of the preparation tables to provide additional enrichment opportunities for differentiation.

Table 2. Overview of STEM challenge activities by day

Day	Mini Challenge	Design Challenge	Purpose
1	Students explore the challenge of shipping cargo and how different objects float in water.	Students explore concepts of weight, weight distribution, and buoyancy by exploring variables (e.g., weight, people, oars)	Observations and estimations made in the morning and afternoon explorations are important for boat design and building.
2	Students are given cardboard, their primary boat building material, to determine how it interacts with water and how weight influences its buoyancy.	Students establish their team norms by determining their roles in the project (i.e., construction engineer, cargo specialist, materials engineer, and communications engineer). Teams are provided the materials they receive in their *base kit* (i.e., five large boxes, five medium boxes, one roll of duct tape and four oars) and determine what additional materials they may need within their limited budget.	Once students have an understanding of the constraints of the design challenge, they may begin collaboratively developing their boat designs.
3	Students learn about the concepts of crush and tilt as it relates to securing cargo and ensuring its safe delivery.	Students begin a cargo design, to ferry their payload across the pool, into their refined boat design. Students test prototypes for crush and tilt and are able to refine their designs.	Student teams have enough knowledge and experience to begin their boat builds.
4	Students utilize engineering design skills to incorporate the refined cargo design into their boat builds. Teams develop presentations of three engineering concepts they have incorporated into their build.	Teams continue and complete their boat and cargo builds.	Students gain experience with product prototyping and evaluating product designs.
5	Students reflect on their teamwork and individual roles in developing the boat and cargo designs and builds. Students use their observation skills to review best practices of other groups in their boat and/or cargo builds.	Teams test their boat and cargo builds; teams successful in getting their boats across the pool with intact cargo are ranked on their performance on the Payment Scale Challenge Rubric.	Students participate in a competitive environment.

Figure 4. A boat designed and built by a school team for the STEM challenge

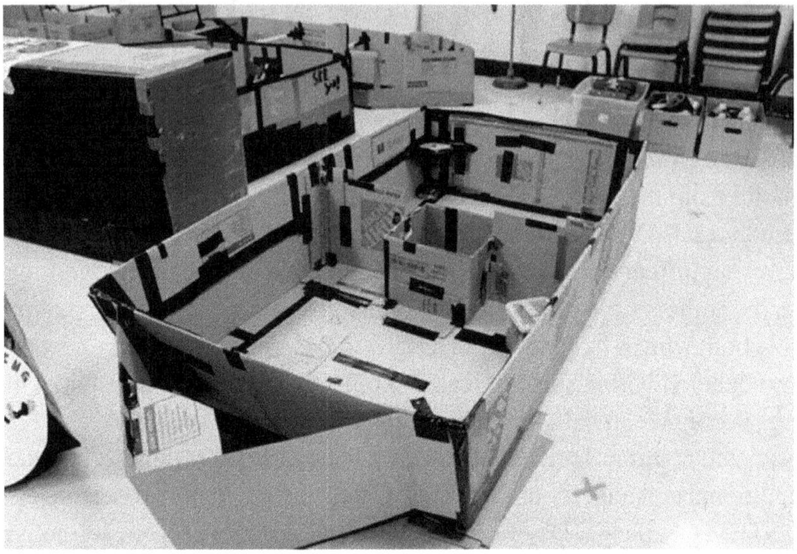

To be prepared for the STEM Challenge, we recommend developing notebooks containing all ten graphic organizers for students so they can record ongoing notes to refer to throughout the design, prototyping, and testing phases. Graphic organizers also allowed project staff to monitor students' record keeping more efficiently, including daily exit tickets. Colored pencils, paper, and other common school supplies are generally useful on each day for students to enhance their designs. Additionally, a cache of tools and power tools can provide additional support and assistance for measurements and estimation (e.g., clocks/stopwatches, rulers, meter sticks, calculators, screwdrivers, drills, circular saws). While not necessary for a high-quality STEM Challenge, students may wish to utilize their own smartphone or tablet technologies to capture photos and videos of work to enhance their design discussions, deliberations, and build hype for the final STEM Challenge race (e.g., Sketchup, Photoshop). The Lubbock ISD STEM Challenge is student-centered and student-driven, so project staff intentionally minimize the influence of the teacher mentor on student products. Undergraduate student mentors, in particular, serve as STEM learning mentors to mitigate teachers' influence on student work in the engineering design process (Hite et al., 2020). For STEM Challenge teams considering replicating the 2021 Lubbock ISD STEM Challenge referenced here, smaller pools will suffice in lieu of a large Olympic-sized pool needed for the cardboard boat race. However, lack of direct access to a twenty-five-to-fifty-yard pool is also an opportunity to work with another community partner who might have a larger pool, such as the YMCA, YWCA, a local school district or the city/township. Supplemental Materials in Appendix K include answer keys and scoring rubrics to assist the evaluation of students' STEM learning using this curriculum.

Evaluation of the 2021 Lubbock ISD STEM Challenge deployed a pre- and post-assessment using the Student Attitudes Towards STEM (S-STEM) Survey for Middle/High School (Friday Institute for Educational Innovation, 2012a). This instrument has been previously validated (Unfried et al., 2015) and widely used in similar STEM activities research studies (Han et al., 2021; Luo et al., 2019; Playton et al., 2021) and STEM Challenges (Hite et al., 2020). We used the S-STEM Survey to measure potential changes in participating students' confidence and efficacy in science and engineering (nine items) and 21st century learning skills (eleven items).

The project team separately added an additional battery of seven items to assess students' STEM interest in order to not impact the validity of the S-STEM Survey. These STEM interest questions related to students' affect towards STEM: (1) student interest in learning about STEM topics; (2) wanting to be known as a STEM person; (3) student interest in studying STEM in upper-level courses, both (4) high school and/or (5) college/university; and whether students wanted to do more STEM activities (6) at school and/or (7) in the community. Since the evaluation was focused on the construct level using nominal data (i.e., interests in engineering and technology, perceived growth in 21st century skills, and STEM interest), rather than at item levels (as ordinal data), we were able to utilize parametric t-tests for the analysis of aggregated data sourced from the Likert-style instruments.

Project staff also collected daily STEM interest and learning exit tickets to better understand how students used 21st century skills throughout the STEM Challenge process. Specifically, students ranked the use and utility of each of the five 21st century skills for each day of the STEM Challenge – collaboration, creativity, communication, and critical thinking. Once ranked, students answered questions on how they used their top reported 21st century skill during that day's Mini Challenge and Design Challenge activities. The Day Five exit ticket added additional end-of-project assessment items asking students what they liked most about the STEM Challenge and how they think the event could be improved. The open-ended nature of the exit ticket enhances student learning (Dixson & Worrell, 2016) by sharing what students learned during the day's activities (e.g., Green et al., 2019) while promoting student self-reflection and metacognition (Danley et al., 2016). Results of the exit tickets are particularly useful in adjusting curriculum and instruction for future implementation (Black & Wiliam, 2010).

RESULTS

The results of the S-STEM survey, STEM interest survey, exit tickets, and end-of project assessment provide evidence that helps validate existing studies that suggest participation in out-of-school STEM learning, specifically STEM competitions, increasing student affect towards STEM educational pathways and subsequent self-selection into STEM educational opportunities. With ten participating four-person teams, the S-STEM data capture included forty middle school students. Six students were not included in analysis because they did not complete the survey tool. The project team utilized the recommended S-STEM evaluation processes provided by the Friday Institute to determine whether student interest and affect in STEM was impacted by their participation in the Lubbock ISD STEM Challenge (see Friday Institute for Educational Innovation, 2012b). Upon conducting dependent sample t-tests, outputs indicated a significant improvement within students' interests in engineering and technology, $t(34) = 4.242$ with high confidence ($p = 0.003$). Similar results indicated significant improvement for students' self-perceived growth in 21st century skills, $t(34) = 7.792$ with high confidence ($p = 0.00001$), and a significant increase in STEM interest from before to after participation in the STEM Challenge, $t(39) = 4.007$ with high confidence ($p = 0.007$).

Exit tickets asked each student to rank 21st century skills according to their use and utility each day. Counts of student reported skills were summed to create Table 3. Overall, collaboration was consistently ranked as the skill used most often during STEM Challenge activities, followed by communication. Though these two skills are implicit in the structure of a team-based competition, it is important for the continuous improvement of the project to validate that students are engaging in STEM Challenge activities as intended. Creativity was ranked highest on Day 2 and Day 3, correlating to the specified tasks of those days that centered on boat design, developing a container to protect the ovate cargo, and discussing navigation strategies for the boat race.

The end-of-project assessment recorded sixty-nine responses detailing what students perceived as the best elements of the STEM Challenge. Categorization via two rounds of coding students' responses resulted in six response categories: (1) STEM Challenge activities; (2) Social-Emotional Learning; (3) New STEM knowledge ; (4) New 21st Century Skill Knowledge; (5) STEM Challenge Affect; (6) Experience on Texas Tech campus as shown in Table 4.

Table 3. Twenty-first century skill rankings reported on exit tickets by students each day of the STEM challenge

Reported Skill	Day 1 ($n = 28$)	Day 2 ($n = 29$)	Day 3 ($n = 26$)	Day 4 ($n = 31$)	Day 5 ($n = 27$)
Most	Collaboration	Collaboration	Collaboration	Collaboration	Collaboration
↕	Communication	Creativity and Critical Thinking (Tie)	Creativity	Communication	Communication
	Creativity		Communication	Creativity	Critical Thinking
Least	Critical Thinking	Communication	Critical Thinking	Critical Thinking	Creativity

Table 4. End-of-project assessment categorical response rates

What were the best elements of the Lubbock ISD STEM Challenge? (n=69)		
Category	n	**% of total**
STEM Challenge activities • Boat building (n = 14) • Boat race (n = 9) • Mini-challenge (n = 3)	26	37%
Social-Emotional Learning • Bonding w/ Friends (n = 7) • Teamwork/team spirit (n = 6) • Meeting students from other schools (n = 5)	18	27%
New STEM Knowledge	10	15%
STEM Challenge Affect	9	13%
New 21st Century Skills Knowledge	3	4%
Experience at TTU	3	4%
How could the Lubbock ISD STEM Challenge be improved? (n = 44)		
Category	n	**% of total**
Improved instruction	10	23%
More variety in Mini Challenges	9	20%
Working with students from other schools	9	20%
Unlimited Resources	5	11%
Miscellaneous	5	11%
No suggested changes	6	15%
*Mean response was 1.7 responses per student; median and mode was 2 per student.		

Many student responses (37%) relayed positive experiences with activities associated with the boat design and building processes, particularly those that have clear alignment with reported 21st Century Skill use. At the same time, 27% of coded responses related to the relationship skills competence area of Social-Emotional Learning (SEL), which describes the formation of positive relationships, working in teams, and communicating effectively (Collaborative for Academic, Social and Emotional Learning [CASEL], 2020). Such outcomes help demonstrate the impact of the Lubbock ISD STEM Challenge, as existing studies suggest that well-structured out-of-school time program are able to positively impact positive social behaviors such as teamwork, bonding with their friends, meeting new peers (Durlak et al., 2010). In addition, students reported how they liked that they gained new knowledge in STEM (15%) and 21st century skills (4%), developed positive affect to the Lubbock ISD STEM Challenge (13%), and enjoyed spending time at Texas Tech University (4%).

The end-of-project assessment also recorded forty-four coded elements, about one from each student, detailing what students thought could be improved for the next iteration of the Lubbock ISD STEM Challenge. Researchers open coded these responses in two rounds of review, which resulted in six response categories: (1) Improving Instruction; (2) More variety in Mini Challenges; (3) Working with students from other schools; (4) Unlimited resources; (5) Miscellaneous; and (6) No suggested changes. The most common response from students was that activities often lacked enough explicit instruction (23%). This is not unexpected, as project-based learning necessarily includes open-ended questions that cannot be overly prescriptive in order to facilitate effective student engagement and learning (Sobehrad & Sobehrad, 2022, p. 30-31). Nine of the ten middle schools included in the Lubbock ISD STEM Challenge employ a traditional middle school academic model, meaning students have likely had very little exposure to *formal* project-based learning experiences like the Lubbock ISD STEM Challenge (the remaining middle school is a Project-Based Learning STEAM Magnet school). Encouragingly, however, students also strongly recommended that the next iteration of the Lubbock ISD STEM challenge should increase the rigor, frequency, and variety of Mini challenges (20%). Students also wanted more opportunity to work with their peers from other campuses (20%), indicating that the non-content related learning student experienced during the Lubbock ISD STEM Challenge were perceived as most beneficial to the individual student and, as above, it was the collaborative, team-driven nature of the project that drove positive affect.

Other responses (22%), such as requesting unlimited resources with no constraints or having better snacks, are not possible within the project's budget and time constraints, though the recommendation to expanding the Lubbock ISD STEM Challenges do more schools does raise the possibility of restricting the event to include multiple teams from each campus or adding Grade 4-5 division.

DISCUSSION

Eight of the ten student groups who competed in the Lubbock ISD STEM Challenge successfully transported their cargo the length of the pool and back. Reflecting on their successes after the race, students cited having good balance due to equal weight distribution, a sturdy hull, and good group coordination with a fixed seating pattern. Student groups who were not as successful had cited needing more tape, needing a rudder for steering, and more rowing practice. Figure 5 shows a student group whose clearly defined roles and coordination led them to deliver the cargo without boat failure in a competitive amount of time.

Figure 5. A student team exhibiting teamwork during the STEM challenge race

Results suggest students utilized 21st century skills, primarily collaboration and communication. These results mirror soft skill development reported in concurrent studies of robotics competitions (Eguchi, 2014, 2016; Usart et al., 2019). Due to the nature of the competition, effective communication and collaboration was paramount for meeting success throughout the STEM Challenge. Similarly, Kandlhofer and Steinbauer (2016) found teamwork a significant construct when studying European students who participated in STEM activities using robots. Hoffman et al. (2021) have been able to empirically link social competencies and belonging to students' math and science efficacy and interest. Their research affirms and extends the salient findings of this studied intervention. That is, the STEM Challenge fosters belonging among participating students and thereby further enhances and possibly reinforces their interest and efficacy in STEM.

Moreover, significant gains in students' STEM interests as an outcome of the STEM Challenge are particularly encouraging. This STEM Challenge first originated from a university-district partnership focused on engaging talented students who had not excelled in the traditional STEM classroom (Hite et al., 2020). Therefore, improved STEM interest found among sampled participants suggests this intervention is an impactful opportunity in the context of time-bound (constrained), community-engaged, competition-based interventions. Meaning, students' active involvement in the event improved their STEM interest and learning, which may help participating students in improving engagement in their STEM-related classes in the subsequent school year.

The STEM Challenge partnership between TTU STEM CORE and Lubbock ISD continues to evolve in designing new challenges including designing gliders and hovercraft (Hite et al., 2020). The partnership credits the ongoing commitments of TTU and Lubbock ISD to STEM education and communication of student outcomes as this event enters its ninth STEM Challenge in June 2023, featuring push-pedal recycled derby cars. While this journey of almost a decade has not been seamless, the dedication of both STEM CORE and Lubbock ISD has made the support and the program stronger with time.

CONCLUSION

The STEM Challenge provided an opportunity for students to engage in summertime STEM enrichment and for the school and university communities to come together to support STEM in the City of Lubbock, Texas. The morning Mini Challenges provided opportunities for students to interact with STEM constructs that groups could incorporate into their afternoon design tasks. Students utilized skills of the 21st century, especially collaboration and communication, to take their designs and build tangible objects to race against other schools. Scaffolded by graphic organizers, students were able to capture, record and share their thinking throughout the five-day event. The teacher's guide provided vital insight to the facilitation of activities to ensure students had completed all pre-race activities such that each team was fully prepared on race day. We have updated the teacher's guide (Supplemental Material, Appendix K) with recommendations from participating educators and best practices from the pilot experience to improve fidelity of implementation. The race itself was attended by people across the Lubbock-area STEM ecosystem to cheer on students and celebrate their work from the week. Through our research and evaluation, we found that this competitive and community-engaged STEM Challenge curriculum enhanced middle grades students' interests and learning in STEM. We hope that by sharing this curriculum that other communities can establish their own in-school or community based (out-of-school) STEM Challenges to bring together community members to share the excitement of learning STEM and nurture their middle grade students' interests in STEM.

Supplemental Materials

All supplemental materials can be found at https://www.depts.ttu.edu/stem/challenge/activityguide.php

REFERENCES

Black, P., & Wiliam, D. (2010). Inside the black box: Raising standards through classroom assessment. *Phi Delta Kappan*, *92*(1), 81–90. doi:10.1177/003172171009200119

Chu, L., Ting, Y. L., & Tai, Y. (2020, July). Building STEM capability in a robotic arm educational competition. In *International Conference on Human-Computer Interaction* (pp. 408-421). Springer. 10.1007/978-3-030-50506-6_28

Collaborative for Academic, Social, and Emotional Learning. (2020). *CASEL's SEL Framework*. CASEL. https://casel.org/wp-content/uploads/2020/12/CASEL-SEL-Framework-11.2020.pdf

Danley, A., McCoy, A., & Weed, R. (2006). Exit Tickets Open the Door to University Learning. *InSight: A Journal of Scholarly Teaching, 11*, 48-58.

Dixson, D. D., & Worrell, F. C. (2016). Formative and summative assessment in the classroom. *Theory into Practice*, *55*(2), 153–159. doi:10.1080/00405841.2016.1148989

Durlak, J. A., Weissberg, R. P., & Pachan, M. (2010). A meta-analysis of after-school program that seek to promote personal and social skills in children and adolescents. *American Journal of Community Psychology*, *45*(3-4), 294–309. doi:10.100710464-010-9300-6 PMID:20300825

Eguchi, A. (2014). Learning experience through RoboCupJunior: Promoting STEM education and 21st century skills with robotics competition. In *Society for Information Technology & Teacher Education International Conference*, (pp. 87-93). Association for the Advancement of Computing in Education (AACE).

Eguchi, A. (2016). RoboCupJunior for promoting STEM education, 21st century skills, and technological advancement through robotics competition. *Robotics and Autonomous Systems*, *75*, 692–699. doi:10.1016/j.robot.2015.05.013

Friday Institute for Educational Innovation. (2012a). *Student Attitudes toward STEM Survey-Middle and High School Students*. Author.

Friday Institute for Educational Innovation. (2012b). *Student Attitudes Toward STEM (S-STEM) Survey: Tips for Using Your Data*. Author.

Green, K., Roller, C., & Cubeta, M. (2019). A plethora of fungi: Teaching a middle school unit on fungi. *Science Activities*, *56*(2), 52–62. doi:10.1080/00368121.2019.1682961

Han, J., Kelley, T., & Knowles, J. G. (2021). Factors influencing student STEM learning: Self-efficacy and outcome expectancy, 21st century skills, and career awareness. *Journal for STEM Education Research*, *4*(2), 117–137. doi:10.100741979-021-00053-3

Hite, R., Spott, J., Johnson, L., & Sobehrad, L. (2020). STEM Challenge: Two Years of Community-Engaged Engineering. *Journal of Research In Innovative Teaching & Learning*, *13*(1), 57–82. doi:10.1108/JRIT-12-2019-0080

Hite, R. L., & Spott, J. L. (2022). Improving Parents' and Teachers' Perceptions of Girls' STEM Activities and Interests before and after an Informal STEM Intervention. *Journal of STEM Outreach*, *5*(1), 1–13. doi:10.15695/jstem/v5i1.01

Hoffman, A. J., McGuire, L., Rutland, A., Hartstone-Rose, A., Irvin, M. J., Winterbottom, M., Balkwill, F., Fields, G. E., & Mulvey, K. L. (2021). The relations and role of social competencies and belonging with math and science interest and efficacy for adolescents in informal STEM programs. *Journal of Youth and Adolescence*, *50*(2), 314–323. doi:10.100710964-020-01302-1 PMID:32804294

International Society for Technology in Education. (2016). *ISTE National Educational Technology Standards for Students*. Author.

Jones, K. L., & Hite, R. (2021). A Global Comparison of the Circumscription and Compromise Theory of Career Development in Science Career Aspirations. *School Science and Mathematics*, *121*(7), 381–394. doi:10.1111sm.12492

Jones, K. L., & Hite, R. (2022). Exploring When, Why and How Students Circumscribe and Compromise Science Careers. *The Career Development Quarterly*. doi:10.1002/cdq.12311

Kandlhofer, M., & Steinbauer, G. (2016). Evaluating the impact of educational robotics on pupils' technical-and social-skills and science related attitudes. *Robotics and Autonomous Systems*, *75*, 679–685. doi:10.1016/j.robot.2015.09.007

Kennedy, T. J., & Odell, M. R. (2014). Engaging students in STEM education. *Science Education International, 25*(3), 246–258.

Lesseig, K., Nelson, T. H., Slavit, D., & Seidel, R. A. (2016). Supporting middle school teachers' implementation of STEM design challenges. *School Science and Mathematics, 116*(4), 177–188. doi:10.1111sm.12172

Luo, W., Wei, H. R., Ritzhaupt, A. D., Huggins-Manley, A. C., & Gardner-McCune, C. (2019). Using the S-STEM survey to evaluate a middle school robotics learning environment: Validity evidence in a different context. *Journal of Science Education and Technology, 28*(4), 429–443. doi:10.100710956-019-09773-z

Melchior, A., Burack, C., Gutbezahl, J., Hoover, M., & Marcus, J. (2015). *FIRST longitudinal study: Summary of preliminary findings–year 2*. The Center for Youth and Communities at Brandeis University.

Miller, K., Sonnert, G., & Sadler, P. (2018). The influence of students' participation in STEM competitions on their interest in STEM careers. *International Journal of Science Education. Part B, 8*(2), 95–114.

Moreno, N. P., Tharp, B. Z., Vogt, G., Newell, A. D., & Burnett, C. A. (2016). Preparing students for middle school through after-school STEM activities. *Journal of Science Education and Technology, 25*(6), 889–897. doi:10.100710956-016-9643-3

National Governors Association Center for Best Practices, Council of Chief State School Officers. (2010). Common Core State Standards (English Language Arts Standards and Mathematics Standards for Middle School. National Governors Association Center for Best Practices, Council of Chief State School Officers.

Nugent, G., Barker, B., Grandgenett, N., & Welch, G. (2016). Robotics camps, clubs, and competitions: Results from a US robotics project. *Robotics and Autonomous Systems, 75*, 686–691. doi:10.1016/j.robot.2015.07.011

Nugent, G., Barker, B., White, A., & Grandgenett, N. (2011, June). The impact of robotics competitions on youth STEM learning, attitudes and 21st century workplace skills. In EdMedia+ Innovate Learning (pp. 3614-3619). Association for the Advancement of Computing in Education (AACE).

Playton, S., Hite, R., & Klonowski, P. (2021). A case study of a researcher-practitioner partnership in teaching STEM+C to rural elementary students. *Electronic Journal for Research in Science & Mathematics Education, 25*(2), 30–46.

Purzer, Ş. (2011). The relationship between team discourse, self-efficacy, and individual achievement: A sequential mixed-methods study. *Journal of Engineering Education, 100*(4), 655–679. doi:10.1002/j.2168-9830.2011.tb00031.x

Sahin, A. (2013). STEM clubs and science fair competitions: Effects on post-secondary matriculation. *Journal of STEM Education: Innovations and Research, 14*(1), 5–11.

Sobehrad, L., & Sobehrad, S. (2022). *Medieval History in the Modern Classroom: Using Project-Based Learning to Engage Today's Learners*. Arc Humanities Press.

Tai, R. H., & Liu, Q. (2006). Planning early for careers in science. *Science, 312*(5777), 1143–1144. doi:10.1126cience.1128690 PMID:16728620

Texas Education Agency. (n.d.). *Texas Essential Knowledge and Skills*. TEA. https://tea.texas.gov/academics/curriculum-standards/teks/texas-essential-knowledge-and-skills

Trafton, A. (2007, October 24). *Cardboard Boaters sink or skim*. Massachusetts Institute of Technology. https://news.mit.edu/2007/regatta-tt1024

Unfried, A., Faber, M., Stanhope, D. S., & Wiebe, E. (2015). The development and validation of a measure of student attitudes toward science, technology, engineering, and math (S-STEM). *Journal of Psychoeducational Assessment, 33*(7), 622–639. doi:10.1177/0734282915571160

Usart, M., Schina, D., Esteve-Gonzalez, V., & Gisbert, M. (2019, May). Are 21st Century Skills Evaluated in Robotics Competitions? The Case of First LEGO League Competition. In *the Proceedings of the 11th International Conference on Computer Supported Education* (pp. 445-452).

Utica College. (n.d.) *Bloom's Taxonomy of Measurable Verbs*. Utica College. https://www.utica.edu/academic/Assessment/new/Blooms%20Taxonomy%20-%20Best.pdf

Chapter 10
Zoology Pathway Program:
Partnering for Student Success

Claire Lannoye-Hall
Texas Tech University, USA

ABSTRACT

There is an urgent need for more individuals to pursue science, technology, engineering, and math (STEM) careers, especially those who identify as members of historically excluded populations. This process needs to start well before a job application is filled out. Young people need to be interested in the field, be aware of opportunities, have the required knowledge and skills, feel they belong in the field, and have the support of mentors to help them succeed. This learning must extend beyond the typical K-12 classroom setting and include resources, such as community partners and immersive learning experiences, to make a measurable difference. A partnership between an informal science learning organization and a charter school that serves students primarily from historically excluded populations was formed to support the students' awareness and readiness to pursue STEM-based careers.

INTRODUCTION

For years, the U.S. federal government and Science, Technology, Engineering, and Math (STEM) industries have advocated diversifying the STEM workforce (Benish, 2018; National Academies of Sciences, Engineering, and Medicine, 2019). They are transparent in their call to action; there is an urgent need for more STEM-ready candidates to be employed in exponentially increasing vacancies as the field grows. It is critical to ensure that the next generation of STEM professionals extends beyond just ensuring that there are people in offices working on solving global issues. For a field that has historically excluded every gender, race, ethnicity, and other demographic besides White men, it matters who is in those jobs (Benish, 2018).

To prepare the next generation for STEM careers, young people need opportunities to engage in immersive learning experiences that build awareness and interest in STEM topics (Maltese & Tai, 2011). This process cannot wait until an individual graduates from high school and chooses a college major;

DOI: 10.4018/978-1-6684-7771-7.ch010

it needs to start during the formative years of elementary and middle school, then be built upon during high school to maintain and increase interest (Tai et al., 2006). Schools and teachers who partner with experts in their communities, such as informal science organizations, can provide programming that enhances students' engagement, skills, and conceptual knowledge of STEM content through immersive and relevant learning experiences that extend beyond the classroom (Bevan et al., 2010).

It can be very challenging to extend learning beyond the classroom. It defies the structure of the traditional school day, the logistics of transporting students are complicated and costly, and finding partners with sufficient capacity to provide programming that complements and enhances classroom curricula are few and far between. Despite all these challenges, a charter high school that serves historically excluded populations in a large city in the Midwest engaged with a zoo in a nearby suburb to provide their students with a three-year, immersive learning experience to prepare them for future careers in zoology or other STEM fields.

BACKGROUND

Building Interest in STEM

Students' interest in topics and access to opportunities guide what they engage with during their K-12 careers and beyond (Maltese & Tai, 2011). Most research on STEM career pathways focuses on the K-12 experience, specifically the middle and high school years (Jones et al., 2021; Maltese & Tai, 2011; Sadler et al., 2012; Schwartz et al., 2009). Students' learning within the traditional classroom is a significant factor in building their awareness and the content knowledge and skills they need to pursue STEM careers in the future (Than & Kyaw, 2021). Unfortunately, many teachers have low self-efficacy in teaching STEM content, resulting in limited time focused on it in the classroom (Kelley et al., 2020). This creates a deficit in foundational understanding of STEM concepts and limits awareness and interest in the field.

Maltese and Tai (2011) explored additional contributing factors to individuals pursuing STEM careers, including academic achievement, STEM coursework, and the overall learning experience. The authors' study indicated that most youth have favorable views of science in general. However, those views do not typically relate to their science experiences in the classroom. The challenge is that many teachers lack the expertise or self-efficacy to move past traditional lectures or textbook-reliant lessons into something more in-depth and meaningful. Studies indicate that classrooms with hands-on activities, collaborative learning opportunities, and relevant science content can improve student attitudes about science (Myers & Fouts, 1992; Piburn & Baker, 1993). In contrast, traditional classrooms can actually be detrimental to continued motivation in science.

Hands-on and high-quality classroom learning opportunities are especially important for historically excluded groups, with evidence that they have increased students' self-efficacy in science content while decreasing learning gaps between their peers who identify with the majority culture (Ballen et al., 2017). Unfortunately, many historically excluded individuals live in communities with low socioeconomic status, which often pairs with a lack of access to highly qualified teachers (Akiba et al., 2007). These multiple factors are known as intersectionality when social aspects such as gender, class, and race overlap and amplify disadvantages. This known disparity underscores the potential and importance of out-of-school learning experiences in building STEM interest for historically excluded populations.

Out-of-school time experiences are just as critical as classroom learning, if not more important, in the quest to diversify the STEM field (Ottinger, 2022). Across the United States, on average, students spend 6.7 hours a day and 179 days a year in school (National Center for Education Statistics, 2008). That means less than 30% of students' learning time is in the classroom, and most of their time is spent at home, at play, and in their communities, in informal learning settings. The entire educational experience is critical to consider when exploring opportunities to expand and support young people's interests and future pursuits. Combining formal and informal learning experiences could increase the number of young people pursuing STEM careers through innovative, persistent, and collaborative work.

Visits to cultural attractions, including museums, science centers, and zoos, provide engaging and unique learning environments. Through immersive learning experiences, hands-on programming, discrepant events, and the power of learning amongst social groups, cultural attractions have the potential to inspire, teach, and transform those who visit them. Informal science learning institutions, such as zoos, aquariums, science centers, museums, and other entities, provide opportunities for youth to build STEM awareness and interest during visits. Exhibits, programs, and interactions with institution staff can all prompt a sense of wonder and curiosity that prompts further discovery and understanding. By partnering with those in the field or institutions who have expertise, teachers can build their own STEM self-efficacy while simultaneously providing opportunities for their students to practice science skills.

Maintaining Interest in STEM

Tai's research on STEM interest indicates that learners come with prior experiences, attitudes, and motivations for engaging with content with varying levels of enthusiasm; Tai et al. (2006) found that as many as half of the eighth-grade students surveyed in the National Education Longitudinal Study of 1988 (National Center for Education Statistics, 2006) who shared their career intentions while in the eighth grade went on to earn a baccalaureate degree from a four-year program (Tai et al., 2006). Similarly, Sadler et al. (2012) found that the best indicator of post-secondary STEM career interest was students' interest as they started high school, meaning most students who start high school with the intention of pursuing a STEM career graduate with that same intention.

The high school years are essential in maintaining students' interest in STEM. This period is when most students refine their interests to be more specific and consider future careers and post-secondary plans in earnest (Sadler et al., 2012). Studies indicate that building interest before high school is critical (Tai et al., 2006), and maintaining or expanding interest during the teenage years is vital in pursuing STEM careers (Sadler et al., 2012). Unfortunately, most school science curricula follow an outdated approach that focuses on breadth instead of depth, prohibiting students from identifying or following their interests. Schools still use an early 20th-century approach when science instruction focused on sharing all the current scientific knowledge with students. Schwartz et al. (2009) suggest this strategy was feasible as scientific knowledge in the 1900s was quite limited compared to now. More than a century later, most curricula still focus on covering as much content as possible versus diving deeper into a single topic, limiting the opportunity to build interest and engage learners. These authors found that students who studied at least one science topic or concept in depth during high school performed better academically in post-secondary education. During the four years of high school, an extensive range of content can be covered while still allowing youth to choose and explore specific topics more in-depth.

While many schools may not have shifted how they teach science content, youth who participate in extracurricular STEM experiences during out-of-school time often have opportunities to explore content more in-depth. Additional research on how out-of-school science learning contributes to students who persist in pursuing science could identify ways to keep adolescents engaged and motivated (Fortus & Vedder-Weiss, 2012). Stake and Mares' (2004) study about a summer science program showed that students benefited from the engagement with content in a learner-driven setting. Notably, the impact of their intervention's effect on science achievement was more significant for students categorized as 'low achieving' and was most evident after the students returned to school with peers who had not had the program. Their findings reinforce the importance of pairing in summer and out-of-school learning, especially for groups who do not have equitable access to high-quality learning experiences.

Community Partnerships

Many schools collaborate with community partners, like informal science learning organizations, to bridge gaps in the learning experiences and provide opportunities for historically excluded populations. Partnerships between informal science organizations and schools can provide students with opportunities to build more robust science identities, maintain student interest in science, and have more options for learner direction (Bevan et al., 2010; Center for Advancement of Informal Science Education, 2021).

A publication by Bevan et al. (2010) highlights examples of such partnerships, all of which focus on creating more robust learning experiences for young people. One featured program used a green space near Washington High School in Chicago, Illinois, as the focal point for a partnership between the school and the Field Museum of Natural History (Bevan et al., 2010). The Calumet Environmental Education Program supported teachers in building their knowledge of local ecosystems as they explored the natural area near the school. They studied the plant life, a waterway, and the recreational areas that provided the basis for a thriving ecosystem while sharply contrasting the industrial pollution in the rest of the region from steel mills and landfills. The teachers who participated in the program increased their knowledge of ecology, which directly correlated to an increase in local environmental concepts in their curriculum. As a result, their students also demonstrated increased awareness and understanding of local ecology. The partnership consisted of ongoing summer workshops to support teacher development and provided students with field trips and stewardship opportunities. The program was resource-intensive but demonstrated a measurable change in teacher practice and student outcomes.

A partnership between the Monterey Bay Aquarium and three high schools in Watsonville, California, created the Watsonville Area Teens Conserving Habitats (WATCH) program (Bevan et al., 2010). The program started with a three-week summer workshop where students visited local habitats to conduct restoration work while studying the ecosystems. The partnership helped students develop scientific process skills like data collection, collaboration, and observation, and partnered teens with subject-matter experts who served in a mentorship capacity. Culminating projects allowed the teens to choose an issue relevant to them and plan how to address it. While the long-term effect of the partnership was not documented, the program provided a science-rich experience that primed students to develop identities as scientists.

Partnerships like these are likely more abundant than schools and organizations may realize. Initiatives like the CAISE helped bring attention to these partnerships and feature them as a model for schools and informal science organizations while increasing awareness of the importance of everyday science. Since 2007, CAISE has amassed more than 8,000 resources and made them available for free to enhance STEM learning. CAISE's commitment to improving the STEM learning experience for students and teachers was funded by the National Science Foundation (NSF) since its inception, underscoring its importance to the field. A recent NSF award has transitioned the work from CAISE to the in-development Informal STEM Learning Equity Resource Center (ISL-ERC; TERC, 2023).

Developing a Partnership

A large city in the Midwest is home to a charter school network that is comprised of ten schools, three of which are high schools, and serves more than 4,600 students. Each high school focuses on a general course path for students. The high school that was part of this study hosts a population of about 500 students, of which 97% identify as Black, and more than 75% are classified as low-income by the state. The school developed multiple math and science career pathways to support their high school students in identifying their interests and preparing them for future careers while connecting them with experts in each field. The students still have core science courses, with the pathways allowing opportunities for them to do more focused studies and research projects. The pathways are selected by the student or assigned by the school and can remain consistent from a student's sophomore to senior year, or the student can elect to shift pathways. Many of the pathways rely on partnerships with community organizations to enhance classroom offerings and provide consistent access to content area experts.

In early 2019, the charter school administration approached a local zoo about being a venue for a teacher appreciation event they were hosting. The zoo is large, over 100 acres, with more than 150 full-time staff, and is located in the suburbs just north of the city. The zoo is a popular cultural attraction with several award-winning habitats and is home to more than 2,000 animals of more than 200 species. It hosts more than one million guests in a typical year. While the teacher appreciation event ended up being hosted elsewhere that spring, the conversation about how the charter school and zoo could continue working together continued.

The Zoology Pathway Program evolved from these initial discussions. Funded by a generous grant from a family foundation, the school was able to outfit a classroom with science equipment and oversized, iconic imagery of animals from the local area and some who live at the zoo. The careful planning of the classroom space reflected the program's goal of providing an immersive learning experience. In addition to a well-equipped classroom, the program was designed to include regular out-of-the-classroom learning experiences at the zoo, a local aquarium, and greenspaces and waterways within walking distance of the school.

Enhanced Classroom Learning

The importance of building awareness and growing students' interest in STEM content was a driving factor in the structure of the pathway program. The program's first year included the high school hiring a teacher dedicated to planning and teaching the day-to-day content. The curriculum was modeled after zoology programs at local universities, providing students with rigorous, in-depth learning experiences to prepare them for the demands of college coursework. Hands-on learning experiences are the norm in the classroom experience, not the exception.

During the classroom portion of the program, students participate in regular hands-on learning experiences. One notable example is the Salmon in the Classroom program (Michigan Department of Natural Resources, 2023). The program offers opportunities for tens of thousands of students to learn about salmon as they are cared for, from eggs to young smolts, by the students and teachers. Teachers receive professional development and many other resources to be set up for success. Once the salmon have been head-started in the classroom, they are released into local waterways to support the control of invasive species, like alewife. The ecological story about the salmon in the Great Lakes offers opportunities to learn about invasive species, complex freshwater ecosystems, and how human decisions and actions impact natural systems. A large freshwater tank in the classroom is home to dozens of small salmon, with the students providing regular care for the fish. In addition to raising salmon, students have cared for other small animals in the classroom and aquatic and terrestrial plants. Caring for living things reinforces the importance of individuals' actions, with immediate cause-and-effect consequences if proper care is not provided and provides infinitely more contextual understanding than reading about the processes from text and other resources.

Learning Outside of the Classroom

The robust learning experiences in the classroom are supplemented by regular out-of-classroom experiences designed to shift learning into a real-world context, providing relevance to the content being covered. In addition to visiting the zoo, students visited a local aquarium that was recently renovated and reopened. They also visited an extensive riverfront park system with gardens and restored natural areas. This dynamic combination of learning environments has the potential to maintain and even increase students' interest in STEM during their critical high school years.

Regular outings for high school students take a considerable commitment on the part of the school. Finding days where students can miss multiple classes to leave the classroom is challenging, so the high school grouped their class schedules to make it more feasible for all their pathway projects to do in-depth studies and out-of-the-classroom experiences. The charter school also received funding from a generous family foundation committed to expanding learning opportunities for historically excluded populations to help cover the cost of bus transportation.

Zoo and high school staff coordinated programming through emails and meetings, but students in the first year of the partnership experienced episodic programming as the partnership evolved. Visits to the zoo were often self-guided, meaning students walked the zoo on their own. Without a consistent structure or reflective piece, there was little opportunity for impactful meaning-making, limiting the partnership's potential. By spring 2020, the program faced a new challenge: COVID-19. The pandemic halted programming for the remainder of the first year and only led to minimal virtual programs throughout the following school year. With drastic staffing cuts in the education department at the zoo and shifts in leadership at the charter school, planning conversations were not restarted until the end of the school year.

With the first cohort of students starting their senior year, there was an opportunity to rethink the partnership's potential. Several zoo educators were rehired in the months following the height of the pandemic, and additional new staff was hired to the department. Two zoo educators were assigned to the program for consistency and to provide opportunities to build mentorship-like relationships with the students. They worked collaboratively with the high school teacher to plan field trips to the zoo that complemented what was being covered in the classroom curriculum. All visits to the zoo transitioned to include a program led by zoo staff.

Wildlife Forensics, a program facilitated by zoo education staff, allowed students to analyze scenes staged to represent animals who died due to human impact. Disarticulated replicated skeletons were set out in natural settings in the zoo, amongst the dirt and other natural debris, with clues like plastic remnants tucked within a rib cage, ammunition, or broken bones carefully curated to tell a story. Students used string to grid the scenes, then drew what evidence they found in every grid square before removing and reassembling the skeleton to determine the species and hypothesize about the cause of the death. The program led to robust discussions about what happened to the animals and how humans were involved, prompting many questions to inspire future research questions.

Students also participated in a multi-part hybrid lesson about the ecological system of Isle Royale National Park. Isle Royale is an island ecosystem in Lake Superior, between the coastline of Canada and Michigan's upper peninsula, and home to the longest-running predator-prey study in the world. The study has provided a wealth of information about the dynamic relationships between animals and plants. Students used actual data from the study to analyze recent ecological decisions made by National Park Service staff to make recommendations about the next steps the Park Service should take in maintaining as healthy an ecosystem as possible. See Supplemental Materials, Appendix A for additional information about the program. After the session, the students visited the wolf habitat at the zoo. They watched the resident wolves interact with an enrichment item in their habitat while the students spoke with a zookeeper about the care and well-being of the wolves.

Throughout the school year, students focused on research questions they formed based on their personal interests. They collected data and learned about their topics through various resources, including content area experts. The zoo education staff arranged meetings between the students and the zoo's life science staff, grouping students according to their projects and the staff's expertise. Students asked the staff questions about their research project and learned more about the zoo staff's career paths and responsibilities. The zoo education staff also provided detailed feedback to each student on their research projects, suggesting ways to craft investigable questions and how to collect evidence for their respective studies.

The school year concluded with the senior students presenting their research at the zoo. To add excitement and novelty to the presentations, the zoo staff displayed the presentations on a large, 6' spherical display system typically used to demonstrate global phenomena through video data sets such as ocean currents and global weather patterns. Science On a Sphere (SOS) is a unique display system and was an innovative way to share their research with their peers and the zoo staff. While the school year was overall more successful with the coordination and frequency of programs, there were still opportunities to enhance and expand the program.

MATERIALS AND METHODS

Studying the Partnership

A community-engaged research project was designed to address the challenges and opportunities of the pathway program collaboration. Community-engaged research provides a framework for how to conduct research to ensure the results are beneficial for all involved, especially the community the research is intended to support (Boyer, 1996; Fitzgerald et al., 2016; Strand et al., 2003). This model challenges more traditional methods where researchers often design and implement research projects that exclude the input or needs of individuals who are the focus of the study. Using a community-engaged scholarship approach allows for more meaningful collaboration and outcomes, as the perspectives, priorities, and values of everyone involved are honored and incorporated.

The research project started by identifying all invested parties in the pathway program. The teachers, students, students' parents, the science coordinator at the school, and the zoo education staff were all identified. To structure the study, these groups were further combined into three main groups: teachers, zoo staff, and the science coordinator; the students; and the students' parents. From these three groups, the teachers, zoo staff and science curriculum coordinator were engaged in deciding what to focus on for the research project. After some discussion, the group decided to take a step back from the existing partnership to reevaluate the strengths and opportunities of the program and look for potential next steps. A collaborative research question was formed to guide the project: *Building on the current partnership, what other learning experiences would stakeholders find beneficial to enhance the Zoology Pathway Program?*

Three separate surveys were co-created in Qualtrics by the researcher and the charter school science curriculum coordinator to gather information from each invested party: the high school teacher and zoo staff, the students in the program, and the students' parents or guardians. According to the Texas Tech University's Internal Review Board that approved the research project, parent or guardian permission was needed for students under 18 to complete the survey. This process is critical to protect individuals during the research process. However, it can be limiting when collecting data and proved challenging in this study.

Procedure

The zoo staff and high school teachers were emailed directly with an information sheet and a request to complete the survey. One of the high school teachers in the program forwarded a contact list of parents and guardians to the researcher. The list was used to email an information sheet about the project and a direct link to the survey to all the parents and guardians. Parents were also asked to provide permission for their students to participate in a separate survey. No parents replied and provided permission, so only students 18 or older were allowed to complete the survey online while visiting the zoo during a scheduled learning experience. Of the 55 students in the pathway program, only six were able to participate in the survey because of this limitation. Survey response rates varied significantly between each stakeholder group: zoo and charter school staff had a 75% response rate (*n*=3); students had an 11% response rate (*n*=6); parents had a 2% response rate (*n*=1). See Table 1 for more detailed information about the demographics of the individuals who responded.

Table 1. Pathway study survey response demographics

Group	Response Rate	First Year in Program	Second, Third, Fourth Year in Program	Race/Ethnicity	Gender
Teacher/Staff	75%	2	1	1 White / 2 Black	two women/one man
Students	11%	3	3	6 Black	two girls/four boys
Parents*	2%	0	1	not reported	not reported

*Parents and guardians were calculated at one adult per student as information about the students' family composition was unavailable.

The staff survey included 11 open-ended questions about the planning and facilitation of the program and their perceptions of the students' experience. The parent survey had eight open-ended questions focused on their child's experience in the program and the parents' aspirations for their children. The student survey had nine open-ended questions about their experience in the program and two demographic questions to document gender and how long they have participated. The survey data were analyzed for common themes within and across the three groups.

Following the survey, the researcher met with the high school and zoo staff and shared the information. See Supplemental Resources, Appendix B for additional result information. The discussion prompted the high school staff to allot classroom time for in-person student interviews. The researcher directly corresponded with parents to request their student's permission to participate in the interview. Nine parents provided permission for their students to be interviewed. Due to time constraints at the end of the school year, students were interviewed in small groups about their experience in the pathway project. The interviews were conducted at the high school during their zoology class period. The same open-ended questions from the student survey were used in the interview, and follow-up questions were asked as needed for clarification and to elicit more details during the interviews. The interviews were recorded, then transcribed and coded by the researcher. An independent reviewer verified the coding after removing all identifying information from the data.

RESULTS

Overall, the survey and interview responses were positive for the pathway project across all three stakeholder groups. Due to the low overall number of responses, the survey and interview results were grouped into general themes across the three study groups.

The multiple visits to the zoo throughout the school year were noted as a positive outcome of the partnerships across all of the groups. Students noted how much they enjoyed learning with and from their peers in and out of the classroom setting. One student responded to a survey prompt about memorable learning experiences: "The out of school experience was something new and different." Students and staff mentioned the hands-on activities they participated in at the zoo, highlighting the Wildlife Forensics program and how its activities tied into a cohesive storyline, "[t]he memorable learning experience in general is… going hunting for the bones and putting the animal together."

The teachers and students appreciated the opportunities to meet with content area experts at the zoo, with the teachers hoping for more regular interactions in future years. A student shared in the survey that "[t]he most memorable learning experience of the program was going to the penguin room and sphere room. It was memorable because I've never experienced that before." Finding ways to coordinate long-term mentorship relationships between staff at the zoo outside of the education department is challenging due to their demanding schedules. Still, it is a worthwhile endeavor that will continue to be explored. There were consistent mentions of how the program increased the high school staff and students' understanding of the purpose and mission of modern zoos and how they provide and ensure excellent care and well-being for animals.

The survey and interview feedback also noted opportunities for future growth. The teachers and zoo staff identified the need for additional STEM skill development for the students. Despite the students reporting high self-efficacy in STEM, the staff noted gaps in conceptual understanding and identified common misconceptions that persisted. The program starts during the students' sophomore year, so additional STEM skill development could be targeted to the freshman science courses. Including more skill development within the pathway program is also a point for consideration.

Opportunities for long-term projects from school year to school year were noted as a desire for upcoming years. This could include research projects that stem from one another or programming themes that build purposefully from year to year. Even though the programs at the zoo were more purposeful than in previous years, during the meeting to share the study results, staff identified additional ways to complement classroom content through ongoing communication between the teams.

The interviews specifically revealed the students' appreciation for their teachers. They described the high school teachers as "...more engaging..." and their classes as "...more about nature..." and "... actually hands-on..." compared to their other high school courses. This model of shifting from a traditional classroom learning framework to one that has the potential to reach more students, supported by previously mentioned studies, is worth further exploration.

DISCUSSION

The interview and survey data were compiled and crafted into a presentation to be shared with the charter school and zoo staff. The teams met to outline key next steps to continue improving the program. Both groups agreed to increase the frequency of communication by having in-depth meetings at critical points of the partnership to ensure that programming was dynamic and responsive to the students' needs and interests. The program was slightly altered to provide students with additional time to practice the research process from start to finish, including forming investigable questions, collecting data, analyzing data, and disseminating the results. Staff at both institutions were committed to finding opportunities for the students to submit their work for publication in appropriate venues.

To provide students in the pathway program additional opportunities to engage in immersive learning opportunities, all of them were invited to join the teen volunteer program. A subset was invited to participate in a four-year internship at the zoo. The programs include regular opportunities during out-of-school time to participate in workshops about communication skills, STEM content, and in-depth exploration of zoo mission-related topics like conservation, sustainability, and animal welfare. In addition, the teens have opportunities to share their knowledge with zoo guests through on-grounds programming, reinforcing their understanding and practicing their communication skills.

The partnership between the zoo and the charter school continues to evolve and grow to meet the dynamic needs of each cohort of students who enter the program. Basing the partnership on current literature and agreed-upon best practices while leaving room to try new ideas, both in and out of the classroom setting, has the potential to make a significant difference in building and maintaining interest in STEM content for the students involved in the program. Regardless if the students decide to pursue a career in zoology at the end of the pathway program, they will have been immersed in STEM-rich learning experiences that have the potential to help them pursue college degrees, future careers, and overall prepare them to be more informed decision-makers in their future endeavors.

The approved IRB that provided the framework for the study was left open after the initial research period was complete. Provided all parties agree to continue investigating the project, subsequent years of survey and interview data will be collected and documented to look for change over time and provide robust feedback to all stakeholders on the program. Eventually, tracking the students' college degrees and career pursuits could also be documented for long-term results and future publications.

CONCLUSION

Partnerships between informal science organizations and schools, like the pathway program, can provide students with opportunities to build more robust science identities, maintain student interest in science, and have more options for learner direction. This is especially important for historically excluded audiences, who often benefit more than their peers by participating in robust science learning experiences. Additionally, non-traditional classroom approaches that include ample hands-on activities are more likely to maintain student motivation in pursuing STEM content and improve student attitudes about science.

The partnership also benefited the zoo in unexpected ways. Many out-of-school time programs, like those provided at a zoo, do not have sufficient interaction time to make a meaningful shift in people's preconceived ideas about zoos. Sharing the inner workings of the zoo and how much time and energy goes into providing excellent animal care is a large part of the shift in perspective. In addition, the program highlighted how many animals who live at the zoo are rescued from the wild due to non-releasable injuries or from private owners who do not have the skills or resources to provide proper care. Much of this information is surprising to program participants and made a lasting impression. All of this furthers the mission of the zoo in a positive direction.

Making a measurable difference in diversifying the STEM workforce is not simple. The intersectionality of issues that include race, gender, socioeconomic status, and access to a high-quality education must be addressed simultaneously. It will take collaborations between K-12 classrooms and out-of-school learning opportunities to make a meaningful change in the number of highly qualified students for future STEM careers and to meet the ever-growing need for a diverse and inclusive STEM workforce.

REFERENCES

Akiba, M., LeTendre, G. K., & Scribner, J. P. (2007). Teacher quality, opportunity gap, and national achievement in 46 countries. *Educational Researcher, 36*(7), 369–387. doi:10.3102/0013189X07308739

Ballen, C. J., Wieman, C., Salehi, S., Searle, J. B., & Zamudio, K. R. (2017). Enhancing diversity in undergraduate science: Self-efficacy drives performance gains with active learning. *CBE—Life Sciences Education, 16*(4). doi:10.1187/cbe.16-12-0344

Benish, S. (2018). Meeting STEM workforce demands by diversifying STEM. *Journal of Science Policy and Governance, 13*(1), 0-0. https://www.sciencepolicyjournal.org/uploads/5/4/3/4/5434385/benish.pdf

Bevan, B., Dillon, J., Hein, G. E., Macdonald, M., Michalchik, V., Miller, D., Root, D., Rudder, L., Xanthoudaki, M., & Yoon, S. (2010). *Making science matter: Collaborations between informal science education organizations and schools*. Center for Advancement of Informal Science Education.

Boyer, E. L. (1996). The scholarship of engagement. *Bulletin - American Academy of Arts and Sciences. American Academy of Arts and Sciences, 49*(7), 18–33. https://www.jstor.org/stable/3824459. doi:10.2307/3824459

Fitzgerald, H. E., Bruns, K., Sonka, S. T., Furco, A., & Swanson, L. (2016). The centrality of engagement in higher education: Reflections and future directions. *Journal of Higher Education Outreach & Engagement, 20*(1), 245–253. https://files.eric.ed.gov/fulltext/EJ1097202.pdf

Fortus, D., & Vedder-Weiss, D. (2012). Measuring students' continuing motivation for science learning. *Journal of Research in Science Teaching*, *4*(51), 497–522. doi:10.1002/tea.21136

Jones, M. G., Chesnutt, K., Ennes, M., Mulvey, K. L., & Cayton, E. (2021). Understanding science career aspirations: Factors predicting future science task value. *Journal of Research in Science Teaching*, *58*(7), 937–955. doi:10.1002/tea.21687

Kelley, T. R., Knowles J. G., Holland, J. D., & Han, J. (2020). Increasing high school teachers self-efficacy for integrated STEM instruction through a collaborative community of practice. *International Journal of STEM Education, 7*(14). doi:10.1186/s40594-020-00211-w

Maltese, A., & Tai, R. (2011). Pipeline persistence: Examining the association of educational experiences with earned degrees in STEM among US students. *Science Education*, *95*(5), 877–907. doi:10.1002ce.20441

Michigan Department of Natural Resources. (2023). *Salmon in the classroom*. MDNR. https://www.michigan.gov/dnr/education/educators/sic#:~:text=The%20Michigan%20DNR%20Salmon%20in,great%20natural%20resources%20each%20year

Myers, R. E. III, & Fouts, J. T. (1992). A cluster analysis of high school science classroom environments and attitude toward science. *Journal of Research in Science Teaching*, *29*(9), 929–937. doi:10.1002/tea.3660290904

National Academies of Sciences, Engineering, and Medicine. (2019). *Minority serving institutions: America's underutilized resource for strengthening the STEM workforce*. The National Academies Press. doi:10.17226/25257

National Center for Education Statistics. (2006). *National Education Longitudinal Study of 1988*. NCES. doi:10.3886/ICPSR09389.v1

National Center for Education Statistics. (2008). *Average number of hours in the school day and average number of days in the school year for public schools, by state: 2007-08*. NCES. https://nces.ed.gov/surveys/sass/tables/sass0708_035_s1s.asp

Ottinger, R. (2022). *Why education investors shouldn't ignore afterschool STEM*. Forbes Business Council. https://www.forbes.com/sites/forbesbusinesscouncil/2022/03/01/why-education-investors-shouldnt-ignore-afterschool-stem/?sh=2cd2377aafa6

Piburn, M. D., & Baker, D. R. (1993). If I were the teacher... qualitative study of attitude toward science. *Science Education*, *77*(4), 393–406. doi:10.1002ce.3730770404

Sadler, P. M., Sonnert, G., Hazari, Z., & Tai, R. (2012). Stability and volatility of STEM career interest in high school: A gender study. *Science Education*, *96*(3), 411–427. doi:10.1002ce.21007

Schwartz, M. S., Sadler, P. M., Sonnert, G., & Tai, R. H. (2009). Depth versus breadth: How content coverage in high school science courses relates to later success in college science coursework. *Science Education*, *93*(5), 798–826. doi:10.1002ce.20328

Stake, J. E., & Mares, K. R. (2005). Evaluating the impact of science-enrichment programs on adolescents' science motivation and confidence: The splashdown effect. *Journal of Research in Science Teaching*, *42*(4), 359–375. doi:10.1002/tea.20052

Strand, K., Marullo, S., Cutforth, N., Stoecker, R., & Donohue, P. (2003). Principles of best practice for community-based research. *Michigan Journal of Community Service Learning*, *9*(3), 5–15. http://hdl.handle.net/2027/spo.3239521.0009.301

Tai, R. H., Liu, C. Q., Maltese, A. V., & Fan, X. (2006). Planning early for careers in science. *Science*, *312*(5777), 1143–1144. doi:10.1126cience.1128690 PMID:16728620

TERC. (2023). *TERC selected to lead new informal STEM learning equity resource center* (ISL-ERC). TERC. https://www.terc.edu/news/terc-selected-to-lead-new-informal-stem-learning-equity-resource-center-isl-erc/

Than, W. W., & Kyaw, W. H. (2021). School environment, student engagement and educational aspiration of high school students. *Academia Letters*, *1712*, 1–12. doi:10.20935/AL1712

APPENDIX A: CONSERVATION PROJECT: ISLE ROYALE WOLVES SYLLABUS

Figure 1.

Isle Royale Wolves: Case Study

A course syllabus and guide for teachers and facilitators working with middle school learners.

This six-module course explores the longest running predator and prey relationship research project in the world. The on-demand learning experience addresses science, literacy, and math standards.

Figure 2.

Conservation Project: Isle Royale Wolves

Course Syllabus and Guide

Course Introduction: This case study is designed to provide middle school learners an in-depth experience as they explore the complex and dynamic relationship between predators and prey on Isle Royale, an island in Lake Superior. The relationship between the moose and wolves on the island is the longest continuously running predator/prey research project in the world, providing a rare look into these complex systems.

Instructors:
On-demand content available: URL
Content creation and site design curated by:
Claire Lannoye-Hall, Director of Education
Emily DeRosier, PK-6 Teacher
Contact: education@dzs.org, 248-541-5717 ext. 3800

Skills
The course provides opportunities for learners to practice the following skills:

- Reading and interpreting information from multiple sources
- Reading and interpreting data in graphs and tables
- Reading and navigating mapping software
- Constructing an argument based on evidence

Target Audience
Middle school learners

Course Requirements:

- Access to high-speed internet
- Electronic device able to access the internet, play audio and video, display text, and type text.

Course Description:
Two units, three modules in each. Optional unit to extend learning. Complete course will take approximately three and a half hours, over six learning sessions.

Learners will examine predator and prey relationships, consider humans' role in changing ecosystems, and explore island ecosystem dynamics through an interrupted case study built on real data and circumstances from the longest continuous study of predators and prey in the world. Learners will take on the role of a wildlife biologist, exploring the ecosystem of Isle Royale by analyzing pertinent data sets, examining artifacts and evidence, and reading historical and current accounts. This compilation of information will allow learners to make an assessment of the health of the island ecosystem and weigh the consequences of continuing to reintroduce wolves to the island, applying their knowledge to this ongoing project.

Figure 3.

Conservation Project: Isle Royale Wolves

Course Syllabus and Guide

Course Objectives:

- Learners will develop awareness of the complex cause/effect relationships in ecosystems and the consequences when they become unbalanced.
- Learners will develop awareness of how human actions impact ecosystems.
- Learners will interpret and analyze information to make a recommendation on continuing to reintroduce wolves to the island.
- Learners will develop awareness of conservation work and how their personal choices and actions make a difference.

Course Outline/Schedule

UNIT	MODULE	TOPIC	
Case Study	Explore	Watch wolf introduction video Read introduction section Review content about Isle Royale and analyze climate data Respond to prompts	45 minute session
	Assess	Review the wolf and moose population study chart and information Review the wolf moose population interactive model Respond to prompts	30 minute session
	Field Work	Watch backpacking video Complete aerial survey activity - record data Trail camera activity - record data. Complete pugmark activity - record data. Complete the track mystery activity. Complete moose bones activity - record data.	45 minute session
Action Plan	Wolf Reintroductions	Review information on wolf reintroduction plans Review information on wolf reintroduction progress Respond to prompts	30 minute session
	Environmental Decision Making	Review information on the ethics of reintroductions Review information on structured decision making	30 minute session
	Assessment	Write (or audio/video record) and submit assessment for next steps.	30 minute session
Extend	Compare Assessments	Discuss individual or small group assessments as a full group, if applicable.	Optional, session Length depends on group size
	Ask an Expert Scheduled, synchronous meeting	Students prepare questions to interact with an expert, a DZS staff member who has contributed to or is knowledgeable about the wolf conservation work.	Optional, 30 minute session

Figure 4.

Conservation Project: Isle Royale Wolves

Course Syllabus and Guide

Suggested Evaluation

Learners should respond to prompts, record notes, fill out data sheets, and thoughtfully consider the information they are reviewing. Information could be stored in a physical or virtual science journal for facilitator review. Final written assessments may be submitted to education@dzs.org for a written response within five working days.

Celebrating and Saving Wildlife

Gray wolves and humans have a long and complicated relationship. All animals are important and have a role in the ecosystem, we need to treat all of them with respect and consider their needs to find ways to peacefully coexist. The Detroit Zoological Society works with several other organizations to continue monitoring wolf and moose populations on Isle Royale, and to advocate for the protection of wolves in their native habitats.

Take Action - Extension Suggestions for Learners

1. Start a hiking club and visit local, state and national parks to see amazing, native wildlife firsthand.
2. Help identify and describe mammal species found in images as part of the Michigan ZoomIn citizen science project.
3. Wolves are currently listed as endangered species, helping to protect them in the Midwest. Stay current on developments with their listing status and reach out to your representatives to support keeping them protected.

Figure 5.

Conservation Project: Isle Royale Wolves

Course Syllabus and Guide

Next Generation Science Standards

MS-LS1-4 From Molecules to Organisms: Structures and Processes

Use argument based on empirical evidence and scientific reasoning to support an explanation for how characteristic animal behaviors and specialized plant structures affect the probability of successful reproduction of animals and plants respectively.

Wolves are pack animals, their social dynamics are key to their survival. The pack works together to take down large prey animals, such as moose. When wolves don't have a sufficient pack size, they can only hunt smaller prey animals, which requires hunting more often, therefore expending more energy. The course explores wolf social structures, reproduction and dietary needs.

MS-LS2-1 Ecosystems: Interactions, Energy, and Dynamics

Analyze and interpret data to provide evidence for the effects of resource availability on organisms and populations of organisms in an ecosystem.

This course includes real data, collected over the last five decades, to support the learners' understanding of the complex relationships between predators, prey, and producers, in an island ecosystem.

MS-LS2-2 Ecosystems: Interactions, Energy, and Dynamics

Construct an explanation that predicts patterns of interactions among organisms across multiple ecosystems.

The data collected by the wolf and moose project on Isle Royale shows a direct interaction between the two species through tables, charts and graphs. The data also shows impacts on producers, such as balsam fir trees, when the moose populations significantly increase.

MS-LS2-4 Ecosystems: Interactions, Energy, and Dynamics

Construct an argument supported by empirical evidence that changes to physical or biological components of an ecosystem affect populations.

Wolves first came to Isle Royale by an ice bridge that formed between Canada and the island. Ice bridges have continued to form, occasionally, which leads to the possibility of wolves coming and going from the island. As the Earth's climate continues to change, ice bridges will become less frequent, potentially eliminating the possibility naturally occurring genetic diversity among the wolf population.

MS-LS2-5 Ecosystems: Interactions, Energy, and Dynamics

Evaluate competing design solutions for maintaining biodiversity and ecosystem services.

The course asks learners to evaluate letting nature take its course, and leaving the island ecosystem to evolve without direct human interaction, or to continue introducing wolves to the island ecosystem by transporting them there from other locations. Learners will evaluate the benefits and concerns for both as part of their coursework.

English Language Arts Standards

CCSS.ELA-LITERACY.W.6.1

Write arguments to support claims with clear reasons and relevant evidence.

The recommendation portion (Chapter 6) of the course provides learners an opportunity to analyze and interpret all the information and data they collect throughout the course, then form an argument on whether or not wolves should continue to be introduced to Isle Royale. The data sheets and reflection prompts help learners organize and support their recommendation.

Figure 6.

Conservation Project: Isle Royale Wolves

Course Syllabus and Guide

CCSS.ELA-LITERACY.SL.6.1.A
Come to discussions prepared, having read or studied required material; explicitly draw on that preparation by referring to evidence on the topic, text, or issue to probe and reflect on ideas under discussion.
This course is designed to be completed individually, or in small groups. Learners will need to read material and review information in order to be an active participant in their learning process and to engage with their peers. Reflection prompts can be used to help guide group discussions, or to prompt individual meaning making.

CCSS.ELA-LITERACY.SL.6.1.B
Follow rules for collegial discussions, set specific goals and deadlines, and define individual roles as needed.
This can serve as a self-paced or facilitator-paced learning experience. Setting goals and deadlines is an important skill for online learning experiences.

Math Standards
CCSS.MATH.CONTENT.6.SP.A.2
Understand that a set of data collected to answer a statistical question has a distribution which can be described by its center, spread, and overall shape.
Data collected and shared by the Isle Royale research team is shared through charts and graphs. This data is interpreted through narrative descriptions, to help learners make sense of it.

Additional Readings/Resources
Detroit Zoological Society Blog: https://detroitzooblog.org/?s=isle+royale
How Wolves Change Rivers: https://www.youtube.com/watch?v=ysa5OBhXz-Q
Yellowstone Food Web:
https://drive.google.com/file/d/1ZGMuQ3fQ9Jni0iBw09IdQyJBjmskwNjy/view?usp=sharing
Wolf Quest Game: https://www.endangeredwolfcenter.org/educational-resources/wolf-quest-for-kids/

Thanks everyone!

APPENDIX B: SELECT STUDENT SURVEY QUESTIONS AND RESPONSES

Question	Response Groupings	# of times mentioned in survey	# of times mentioned in interviews
What is the main reason you chose to apply for the Zoology Pathway program?	Pathway was assigned by the school	1	2
	Seemed like a fun/unique experience	2	
	Opportunity to learn about animals	2	2
What experiences would you like to see more of in the Zoology Pathway program?	Hands-on activities	3	2
What experiences do you think could be eliminated from the Zoology Pathway program?	Final presentation		2
What is one of your most memorable learning experiences in general? (This could be at any age, through a school-related or out-of-school time experience.)	Trips to the zoo (social learning experiences)	4	5
	Trips to aquarium/boat/other outings		2
	Dissections in class		4

Chapter 11

Mutually Beneficial School–Industry Partnerships in STEM Education: An Aspirational Case Study

Anthony Michael Perry
https://orcid.org/0000-0002-6497-8345
Massachusetts Institute of Technology, USA

ABSTRACT

The case of Mid-America STEM High School showcases a partnership between one high school and a local STEM business. The partnership leverages the assets of both organizations to transform the quality of the high school STEM learning experience. As a result, learners regularly experience authentic learning opportunities which increases their odds of persisting in STEM-related postsecondary pathways. Cross-sector partnerships are often cited as a strategy to improve STEM education. The case demonstrates how one community overcame common collaboration barriers and suggests design principles that may be useful when developing new partnerships. The rich description of Mid-America STEM High School's partnership is derived from stakeholder interviews. Success can be attributed to intentional trust-building, a willingness to accommodate partner organizations' needs, and symbolic investments from organizational leaders. Other communities beginning new partnerships can learn from this case to build their own mutually beneficial school-STEM industry partnerships.

"[The national school transformation network] calls us freakishly connected, because everybody here is so collaborative and wants to work together and help youth." – District Career Academy Supervisor

DOI: 10.4018/978-1-6684-7771-7.ch011

INTRODUCTION

National education leaders look to Mid-America STEM High School[1] as an aspirational Science, Technology, Engineering, and Mathematics (STEM) education model. The school leverages strong community partnerships (especially with one large, local STEM business) to create authentic and engaging STEM learning experiences for all students. On any given day, youth might be learning alongside STEM professionals visiting the school, supervised by long-term mentors on business-related projects at professional worksites, or engaging with community stakeholders to root out important issues to be addressed through problem-based learning (PBL). Intentional adult collaborations are necessary to design and implement these experiences. Volunteer STEM professionals and school-based educators co-design PBL units for classroom use. In other cases, STEM professionals host workshops on professional skills like project management or design thinking which educators integrate into instruction.

The depth of school-STEM industry coordination at Mid-America affords all students at the school access to engaging STEM learning experiences. These changes to teaching and learning are improving student-level outcomes. Graduates from Mid-America STEM High School are outperforming state and district peers on several college and career readiness indicators. Previous national policy efforts have referenced cross-sector partnerships as an essential strategy to improve STEM education across the country (Committee on STEM Education, 2018). Yet, there is little evidence to suggest how these partnerships can be structured and sustained in a way that benefits youth.

This chapter argues that *mutually beneficial* school-STEM industry partnerships are critical to support student engagement and prepare youth to inherit a STEM-rich world. Mutually beneficial is introduced as a term to describe partnerships that (1) prioritize activities that affect the student learning experience during the regular school day and (2) require equitable participation from STEM professionals and school-based personnel to design, manage, and implement youth learning experiences (Radinsky et al., 2001). The prior literature suggests why such partnerships are difficult, but not how to overcome them. Investigating the case of one exemplary partnership provides the opportunity to fully describe one system in action and suggest design principles useful in other communities who wish to build their own mutually beneficial partnerships and transform the quality of STEM education in their schools.

BACKGROUND

"America's national innovation base depends more than ever on a strong, cross-sector collaboration around common STEM education interests and goals" (Committee on STEM Education, 2018, p. 1).

Developing Next-Generation STEM Talent

STEM education in American high schools is critically important for all youth. Regardless of postsecondary education requirements, the relative share of STEM occupations is growing (National Science Board, 2021). Non-STEM jobs, too, increasingly rely on STEM skills like computing and mathematics (Grinis, 2019). Outside of occupational aspirations, an increasingly technological world demands an increased level of STEM literacy for all adults (National Research Council, 2012). Despite the importance of STEM education, high school students report math and science are among their least engaging classes (Shernoff et al., 2003). Disengaged students are less likely to achieve and persist in STEM-related college and career pathways. Efforts designed to engage high school students in STEM are necessary to prepare young people for their STEM-rich futures.

Student engagement-related issues are amplified in schools that primarily serve youth underrepresented in STEM pathways by race and ethnicity. Black and Hispanic youth continue to be underrepresented among STEM bachelor degree holders and in the STEM workforce compared to the general population (National Center for Science and Engineering Statistics, 2023). Rather than frame this issue as a lack of matter of individual achievement to be remediated, Ladson-Billings (2013) suggests that a persistent *opportunity gap* is the root cause to be addressed. While no school or population is a monolith, there is evidence to suggest that Black and Hispanic students in urban schools have less access to high-quality and engaging STEM curricula compared to schools with primarily White students (Margolis, 2008; Montoya et al., 2018; Oakes, 1990). STEM industry partnerships with schools serving minoritized students are one possible approach to ensure more students have the opportunity to engage with meaningful and rigorous STEM learning experiences as part of the school day.

The Potential of School-STEM Industry Partnerships

STEM employers depend on the next generation to be STEM-ready and STEM-literate. These entities are stakeholders in the success of the K-12 STEM education system and have the incentive to work with school-based partners to tackle persistent issues like disengagement in secondary STEM education. Schools and STEM industry partners can create more engaging STEM learning experiences and improve student outcomes together than either organization could alone. The most recent federal STEM strategic plan makes the same argument suggesting that cross-sector partnerships with the K-12 system are vital to improving the quality of student learning experiences (Committee on STEM Education, 2018). Despite these claims, impactful school-STEM industry partnerships remain rare.

While STEM businesses and professionals have a long history of engaging with the K-12 education system in a variety of ways (Webber et al., 2020), not every partnership is equally impactful on the quality of the student learning experience. For example, while we should celebrate businesses donating funds to schools or hosting their own summer camps, these efforts fail to integrate the complementary assets of high schools and STEM businesses, reducing the potential impact. High school faculty and staff have expertise in curriculum, instruction, and positive youth development. STEM organizations have domain expertise and access to authentic STEM learning and working environments. Mutually beneficial partnerships apply these strengths with a shared commitment to supporting students during the school day. School-STEM industry partnerships can take many forms. Investigating the rare, but real, school-STEM industry partnerships which improve the student learning experience is important to support the broader adoption of similar practices. This study is driven by two research questions that address this need:

1) What systems enable Mid-America STEM High School to create and sustain a mutually beneficial school-STEM industry partnership?
2) How can the lessons learned from Mid-America STEM High School's successful partnership assist other communities in creating their own partnerships?

Literature Review

Collaborating with STEM industry partners boosts student engagement and achievement, as shown by the research summarized in the following paragraphs. These studies, however, do not provide a thorough description of how the partnerships emerged or are sustained. Where the evidence does exist, more is known about the organizational barriers to creating partnerships than the underlying structure of successful implementations. As a result, there are few examples and little evidence that communities who wish to build new partnerships can draw from. This study fills a critical need by contributing a rich description and analysis of one successful mutually beneficial school-STEM industry partnership.

Students Benefit From Mutually Beneficial School-STEM Industry Partnerships

Why should communities divert from the status quo to develop and sustain complex, cross-sector partnerships? The main benefit of a partnership is that students and teachers interact with local STEM professionals as part of a *community of practice* (CoP; Kelley & Knowles, 2016). This conceptual framework presents a model of integrated STEM education where students benefit from the STEM skills, knowledge, and experience embedded in their community. The CoP includes youth, their teachers, STEM professionals, community members, and university faculty. Authentic, PBL STEM learning experiences emerge from these partnerships. Emerging evidence from CoP research suggests that participating in integrated learning experiences during the school day improves student STEM content knowledge compared to traditional STEM learning environments (Kelley et al., 2022).

Most studies of school-STEM industry partnerships investigate the impact of smaller-scale partnerships rather than systems-level changes. Still, these partnerships are associated with improving student outcomes. For example, STEM industry partners ensure authentic contexts, skills, and tools are a part of PBL during the school day (Bowen & Peterson, 2018; Tan et al., 2019). Given recent causal studies demonstrating the positive effect of PBL on science academic achievement (Schneider et al., 2022), including industry partners in the design and implementation of PBL units should have an even greater benefit. STEM industry professionals can also have an indirect, positive impact through their sustained interactions with youth as caring adults. For example, work-based learning (WBL) experiences are often discussed in terms of occupational preparation and skills development (Darche et al., 2009). However, youth participating in a WBL program which requires one full-day internship each week suggest that relationships with caring adults, not skill-building, were the most important programmatic attribute (Bempechat et al., 2014). Shorter interactions between youth and STEM industry professionals can be also impactful. Youth view professionals as inspirational and expansive role models (Zhang et al., 2023) causing them to consider new STEM-related college and career pathways after graduation.

Figure 1 summarizes a conceptual framework that links the common learning activities arising from mutually beneficial school-STEM industry partnerships and persistence in STEM-related college and career pathways. When youth engage in these types of STEM learning experiences, engagement is heightened (Ewell et al., 2022). Engaged students are more interested, spend more time on tasks, and are more likely to pursue similar activities in the future (Sinatra et al., 2015). Situated expectancy-value theory can be applied to link the internal states associated with student engagement to performance-based outcomes (Eccles & Wigfield, 2020). The theory suggests that learning experiences (such as those generated from mutually beneficial school-STEM industry partnerships) create positive affective experiences with STEM learning that increase the likelihood of future engagement with similar activities. Previous studies on school-STEM industry partnerships have shown that this model can enhance students' interest in STEM-related careers (Smit et al., 2019). Mutually beneficial school-STEM industry partnerships are an effective strategy to improve critical student outcomes because they create more engaging STEM learning opportunities.

Figure 1. Conceptual Model Explaining How Students Benefit from Mutually Beneficial Partnerships

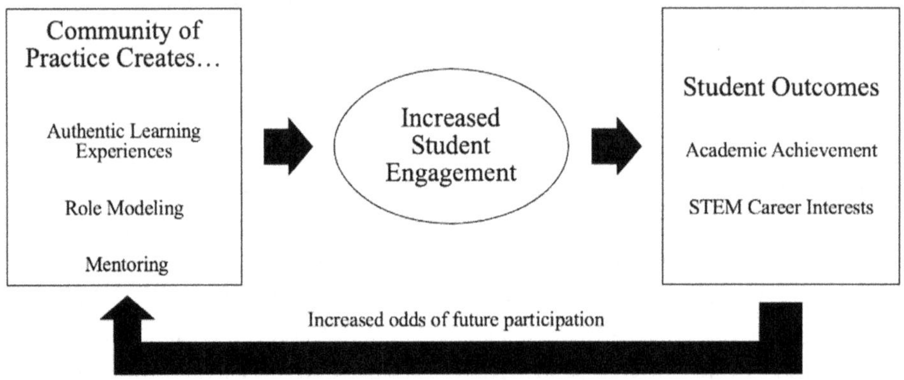

Reported Barriers to School-Industry Partnership

Despite the empirical evidence and conceptual framework suggesting mutually beneficial school-STEM partnerships are a net benefit to students, they remain rare. One study created a taxonomy to classify the role of school- and industry-based partners in 72 documented cases (Strobel & Sun, 2020). In their study, 80% of implementations lacked equal participation from school and industry partners. Researchers surveyed STEM industry leaders and education professionals in Europe to identify barriers to collaboration (Rios et al., 2014). They identified four common obstacles reported by both school and industry representatives: (1) *structural* (e.g., lack of resources), (2) *motivational* (e.g., real and perceived misalignment of partnership goals), (3) *procedural* (e.g., partner communication logistics), and (4) *cultural* (e.g., lack of mutual understanding regarding partner processes and constraints). A national initiative piloting school-STEM industry partnerships in Australia identified similar barriers as limitations to success (The Australian Industry Group, 2017). The prior research identifies criteria to evaluate design principles in school-industry partnerships but does not address how to tackle the associated challenges.

Partnership barriers are often caused by the traditional practices of American high schools. School principals are primarily trained as instructional leaders, not organizational leaders (Eacott, 2008) which limits their capacity to establish and maintain STEM industry partnerships. Lacking organization-level support, some external partnerships are managed by enterprising classroom teachers. In the current education system, many classroom teachers do not have the time or support needed to nurture and sustain meaningful external partnerships (Epstein, 2018). Teacher-centric partnerships, even when they create meaningful changes to the student learning experience, can be problematic in the long term. When one person has such a central role in operations, any changes to their roles and responsibilities can disrupt the work entirely. Rather than depending on the efforts of extraordinary actors, mutually beneficial school-STEM industry partnerships are resilient to changes in staffing.

Some prior case studies of existing school-STEM industry partnerships have been conducted. Watters and Diezmann (2013) describe the organizational conditions needed to establish mutually beneficial partnerships in four Australian schools. They find inter-organizational trust and a willingness from industry partners to accommodate the school's goals were critical to catalyzing strong partnerships. In the case of one career academy in the United States, the school-industry partnership arose out of a district's operational crisis and willingness to try a new approach to learning rather than a coherent vision amongst stakeholders (Hernandez-Gantes et al., 2019). Subsequently, the school struggled to sustain meaningful partnerships when district-level support waned. These case studies, however, lack detail documenting how the enabling organizational traits were enacted within the partnership.

The prior literature suggests that mutually beneficial school-STEM industry partnerships are a promising strategy to improve the quality of the high school STEM learning experience. However, such partnerships are rare and difficult to implement. The case study in this chapter provides a critical link between the common barriers to implementation, high-level organizational traits described in other settings, and ground-level information about the systems needed to operate and sustain mutually beneficial school-STEM industry partnerships which transform the student learning experience.

Systems Analysis in Education

Systems analysis is a useful framework to investigate complex educational issues like creating and sustaining mutually beneficial school-STEM industry partnerships. Systems are comprised of elements and relationships between those elements (Fuller & Kim, 2022; Meadows & Wright, 2008). Rather than studying events in isolation, systems analysis links organizational attributes to key outcomes. For example, education in the United States is often considered a complex system that interacts with other dynamic systems like cultures, economies, and politics (Cuban, 2013; Ghaffarzadegan et al., 2017). Systems analysis is applied here to link the configuration of adults in a cross-sector partnership with transformations in opportunities for engaging STEM learning experiences.

System analysis has received considerable attention to investigate other educational partnerships. One functional element, Intermediary Organizations (IOs; Mitra, 2009), is often central in these studies. IOs are standalone entities whose sole purpose is to mediate between two or more organizations. In education, IOs are typically youth-serving nonprofit organizations with the time and capacity to convene community stakeholders around a common vision for youth and improve the student learning experience (Mitra, 2018). Intermediary Organizations address communication and operational concerns that often plague cross-sector partnerships. In the context of school-STEM industry partnerships, IOs are commonly observed as system elements coordinating activities between schools and businesses (Strobel & Sun, 2020). This study aims to explore how IOs, and other design elements, can create mutually beneficial partnerships between schools and STEM industries by connecting system design with student-level outcomes.

MATERIALS AND METHODS

The prior literature establishes that mutually beneficial school-STEM industry partnerships can produce engaging learning experiences to increase the odds of youth persisting in STEM after high school. However, cross-sector partnerships are rare and difficult to implement, making Mid-American STEM

High School unusual. Single case study designs of extreme examples are meant to "reveal insights about normal processes" (Yin, 2014, p. 52). The present study selects one *existence proof* of a mutually beneficial school-STEM Industry partnership. The in-depth description generates new insights about the systems which may enable or inhibit school-STEM industry partnerships in other American high schools.

Case Selection, Data, and Analysis

Mid-America STEM High School was selected as a focal case because the student learning experience has been transformed by a long-lasting, mutually beneficial partnership with a major industry partner. In preliminary conversations with a national education leader who lives in the community, evidence of a mutually beneficial partnership was evident. They described specific activities which required staff engineers and school-based educators to collaborate on the design of student learning experiences that occur during the regular school day. Local news stories documenting the partnership corroborate this account. In one published story, an industry employee noted that they emphasize relationship-building, rather than gift-giving, in their approach to volunteerism. A district staff member noted in the same article that, as a STEM school, including the perspective of professional engineers in curricula design is essential to their success. Based on the evidence, it appears that the successful system is dedicated to a partnership that benefits both parties and enhances the learning experience of students during school hours.

The primary data come from hour-long semi-structured interviews with key actors involved with the partnership. The interview protocol is provided as part of the chapter Appendix. Study participants included adults who work for different institutions in the community and are involved in the school-STEM industry partnership. In total, the study includes interviews with five adults who work for the school district, four who work for the major STEM industry partner, and four who work for other community organizations. The district's STEM coordinator and the company's corporate social responsibility director participated in two interviews each.

The first interview was conducted with the STEM coordinator employed by the school district. Each participant was asked to suggest additional interviewees who may have information about the partnership creating a snowball sample (Parking et al., 2020). Data saturation (Francis et al., 2010) was reached when no new system elements, relationships, or study participants arose during the final interview. Study participant recruitment ceased at this time. Additional evidence from publicly available documents including news reports, school websites, and sociodemographic data was used to improve the trustworthiness and credibility of the analysis. All subjects consented to the interview before participation by the approved human subjects research protocol.

Data analysis occurred in two stages. First, the data were deductively coded to create a system diagram (Figure 2); a visual depiction of the system elements and connections. To ensure accuracy, the first interview participant was asked to review and confirm the information provided. This artifact represents the operations underlying the mutually beneficial school-STEM industry partnership (RQ1). Next, inductive coding of the system diagram and interview transcripts was used to identify themes that infer the functions of relationships and elements in this system. This analysis explains the reasoning behind the effective system and provides design principles that other communities can use when forming partnerships. (RQ2).

Relying on a single case study comes with limitations. Mid-America STEM High School is in a large, urban district with a major STEM business partner. Suburban and rural communities may have other constraints not present here. There may be other, unobserved attributes of the community that positively contribute to the success of this partnership. While the data collection process was designed to support data triangulation and improve the reliability of the partnership accounts, no direct observation data was available. The results presented here are meant to provide overall guidance for creating partnerships between schools and STEM industries, rather than specific instructions to follow. Think of it as a North Star to guide communities wishing to build their own school-STEM industry partnerships.

RESULTS

A Telling Case: Mid-America STEM High School

The school is in a mid-sized community near the Great Lakes, which is racially diverse and has around 200,000 residents. The median yearly household income in the community is approximately $40,000. Since the early twentieth century, the city has been a scientific innovation hub and manufacturing center. One STEM company keeps its global headquarters, and approximately 3,000 associates, in town. The partnership between the company and Mid-America STEM High School has been formalized, with the company now recognized as a significant named partner. This partnership was recognized at the state level for being an excellent STEM partnership in spring 2023.

Mid-America STEM is a public high school with 350 students. Students from across the city gain admission through a lottery with no academic requirements. Approximately 50% are underrepresented in STEM by race and ethnicity, two-thirds of the students identify as male, and 98% qualify for free or reduced-price meals. These demographics (other than gender) are representative of the school district's population.

District officials associate this partnership with success in three critical student learning outcomes measured by the state's school accountability plan. In the past five years:

1) The five-year high school graduation rate has continuously increased to over 95%; approximately 15% better than the district average.
2) The number of learners earning industry-recognized credentials has increased by more than 20%.
3) Nearly half of the 2021 graduates earned a job readiness stamp on their high school diploma the first year it was offered. This honor recognizes students who demonstrated proficiency in cross-cutting skills such as teamwork, problem-solving, creativity, and leadership. These skills must be verified by at least two community-based adult mentors.

The following sections describe the system in detail from the point of view of those most active in the partnership: youth (Learners), school-based educators (Learning Coaches), and STEM industry professionals. Throughout the narrative, moments of cross-sector collaboration are highlighted to demonstrate how the partnership affects the student learning experience. The case description emphasizes how Mid-America STEM High School creates engaging learning experiences for all students through the collaborative efforts of diverse stakeholders.

The Youth Learning Experience

"You never know what's going to turn the light on for a kid. And when that happens, it's a thing of joy."
– School District Administrator

The entire youth experience is backward designed for success in a senior capstone project. Students in the class identify a community problem they want to address in a year-long, open-ended project. The capstone provides an opportunity for students to apply disciplinary learning from prior coursework and launch students into their future as capable, caring, STEM-capable adults. An educator originally trained in English facilitates the capstone course though she has a master's degree in STEM curriculum & instruction. Without formal training in a STEM content area, she utilizes a deep network to connect students with community-based mentors related to problems her students are interested in solving. As course facilitator, she describes her role as a project manager providing just-in-time support to meet the needs of individual students as they work to solve real-world problems.

Before facilitating the capstone course for the first time, she attended an all-staff workshop on human-centered design hosted by the STEM industry partner. The workshop sparked an *aha!* moment: these principles, used by STEM professionals, can structure a process allowing students to, "do their own PBL" in an authentic way. In the second year of teaching the course, she formally added "community collaboration days" by inviting community members to attend interim presentations from learners on their progress-to-date and provide feedback. This change was primarily intended to reengage seniors in the second semester. However, after seeing the positive responses from adults and learners from these interactions, she gradually included more opportunities to interact with adults throughout the capstone process.

The full structure of the process-oriented capstone course is described in the Appendix. Briefly, juniors are given a summer assignment to identify local problems they might select to solve as their capstone. Once school begins, they begin researching and understanding the root causes of one problem they are interested in. The work involves working with community members who have practical experience and knowledge in the subjects that the learner is interested in. Spending months understanding the problems allows students to become "real problem solvers" rather than looking for "quick fix solutions" as is more typical in short-term school projects. In the second semester, students partner with technical experts to create real, working solutions which are publicly presented to the community at multiple stages of development including a final, spring event.

Preparing youth for capstone success begins when they first join the school community. Students are referred to as *learners* reflecting the active role they are expected to play in their education and within the school community. As freshmen, learners have a series of exploratory experiences with different STEM disciplines including visits from professionals and off-site career exploration activities. Collaboration, communication, and critical thinking skills are emphasized throughout the curriculum alongside typical academic standards. Many interview participants suggest these experiences lay the foundation for a strong culture of learning. All courses are designed using a PBL framework the whole school has adopted which dictates units must solve problems in partnership with community stakeholders. Over time, the units become more complex, open-ended, and interdisciplinary. A sample project is described in the supplemental files.

By their second year, learners select one of three career concentrations: biotechnology, environmental engineering, or information technology. Each concentration operates as a traditional career academy (e.g., Orr et al., 2003) meaning students experience an integrated curriculum linking concentration themes in core courses. Additionally, they interact with STEM professionals in the concentration as part of PBL and industry visits. Most students select a pathway associated with their future career interests. The instructional coherence provides learners with a sense of connection in the curriculum and a peer community within their concentration. One educator noted, "Our pathways are ... solution-oriented rather than problem...our students pick a social problem [for the capstone] but then end up using their pathway skills at the solution phase of the project."

Learners become capable and confident problem-solvers during four years at Mid-America STEM High School. They develop a sense of how STEM skills can be applied to address problems they are deeply interested in. Through partnerships, learners participate in a community that spans the four walls of their school and begin to see how they, too, can incorporate STEM as part of their personal and professional lives after high school.

Learning Coaches

"In the real world, you're going to collaborate with people, even if it's your project, you need other brains, you need other people to look at what you're doing and give you perspective." –STEM Capstone Learning Coach

Classroom teachers are referred to as *Learning Coaches* at Mid-America STEM High School. The language used is purposely designed to indicate the important role that the person is expected to have in enhancing the learning process. Instruction is primarily enacted through PBL with community partners. Some learning coaches find the school's version of PBL intimidating when they are first hired. As one study participant described, allowing learners to interact with adults in the community can be nerve-wracking. Many have no experience with PBL as a teacher or students. New coaches learn through experience that following the school's PBL approach is beneficial for learners. Learners have a sense of urgency and joy in their work knowing that it is real and treated seriously by adults who are not their parents or teachers. The community supports learners in each phase of the design process from problem framing to building solutions.

Facilitating open-ended problem-based learning experiences requires different instructional resources and different ways of working than typically occurs in high schools. Learning coaches are supported by STEM professionals from the major corporate partner. Formally, coaches attend a training led by engineers on professional topics like human-centered design and project management. They work with professionals to co-plan industry site visits or youth mentoring days at the school. Over time, coaches develop enduring partnerships with individual STEM professionals. They call on each other for ad hoc support to connect with the knowledge, skills, or other supports that emerge in real-time from learners' needs. Coaches can apply for small grants to create innovative experiences for students through their industry partner. The school curriculum becomes increasingly youth-driven over time to prepare learners for success on the open-ended, interest-driven, capstone project. While the major industry partner offers many resources to the learning coaches, there are times when a learner's needs dictate other partners are brought in to support learning. Coaches cultivate and draw from their networks to complement existing organizational partnerships. These additional experts are typically brought in for a one-time, short-term interaction with youth. However, coaches are taught to grow new connections into deeper, more committed school partners. Mid-America STEM High School learning coaches' primary role is to blur the line between the school and the broader community. As a result, learners have access to the resources they need to deeply engage with STEM practices and address personally meaningful problems.

STEM Professionals

"I call myself the switchboard operator" –STEM professional with major industry partner

The school district is committed to supporting school-industry partnerships at every high school in the district. They partnered with a national school re-design network to create a formal process that facilitates these arrangements. Through partnership agreements, local organizations are recognized for their work supporting local schools. Time and talent, rather than financial contributions, are the system's currency. One district-level administrator noted, "...we are accepting donations, we don't say no, but that does not make you a partner." A leader at the STEM business partner had a similar sentiment: "If you throw more money, you're gonna feel better. I believe in that effort that interface, the one-on-one contact, that's the difference maker." While a long-time school district employee acknowledges that the partnerships began with significant financial investments from community philanthropists, the money plays a less prominent role in the steady state.

The district's partnership process allows for multiple levels of commitment. The STEM business in focus for this case study has signed on as a named school partner, which represents the highest level of commitment. The three-year agreement requires them to provide at least 2000 hours of support to the school each year. The time is spent serving on a formal advisory board with the school, planning instruction with learning coaches, and direct youth mentoring. The school's website lists 21 other community and industry partners. These partners commit significantly fewer resources (two engagements over the school year). They play key roles in supporting specific initiatives but the interview data suggests they are not drivers of systemic change like the major STEM business partner. As a result, this section focuses only on the experiences of STEM professionals within the large, named business partner.

The partnership efforts at the business are organized through the corporate social responsibility (CSR) director and volunteer STEM education council. The CSR director developed a trusting relationship with the district's STEM coordinator over several years. They act as a bridge between organizations to ensure that strategic collaborations support the needs of organizations. For example, the STEM coordinator identifies stories of student success that can be celebrated in the media by the business partner and the CSR director ensures that internal processes are streamlined to quickly match requests for the time and talent of industry professionals to support learners' needs.

The CSR director also helped create a volunteer STEM education council within the company which strengthened the cross-sector partnership. The business had a long tradition of dedicated associates volunteering in local schools or hosting youth-serving STEM programs during out-of-school time. Rather than continue a series of ad hoc activities supporting STEM education, a manufacturing director suggested that STEM volunteerism could be centralized to better support employees and their partners. The STEM council vets and distributes requests from school-based personnel to ensure that industry professionals' time and talents are efficiently allocated. The STEM council can play matchmaker because they know the technical skills, personal passions, and capacity individual employees have as volunteers. The council also ensures learning coaches have one point of contact to fulfill their requests rather than needing to manage many personal relationships. The council is critical to managing the volume of interactions and the number of actors across institutions who create and implement STEM learning experiences.

Figure 2. System supporting STEM industry partnerships with Mid-America STEM HS

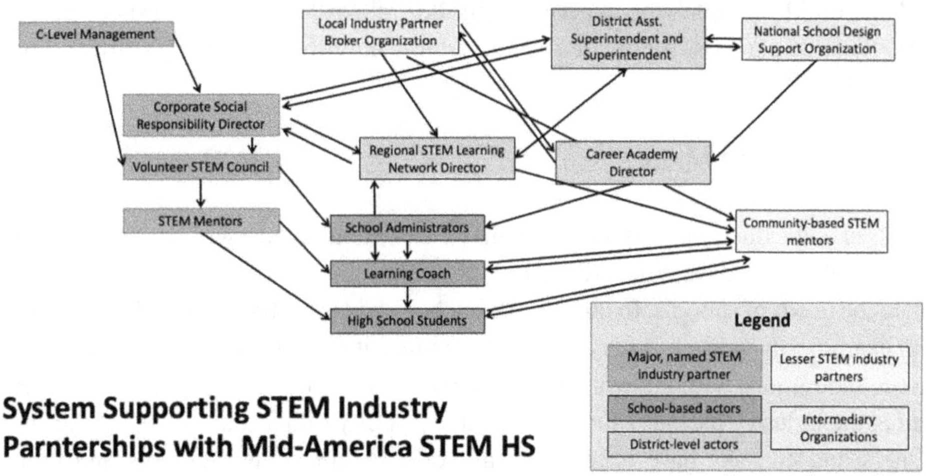

Designing a Successful, Sustainable School-STEM Industry Partnership

The previous section described the complex arrangement of activities between stakeholders in this mutually beneficial school-STEM industry partnership (learners, learning coaches, and local STEM professionals). The resources available in Mid-America STEM High School's community are critical to the unique system which has been built. The section analyzes the system to infer the design intent of this system (RQ2). Although this plan is connected to STEM resources in the area, the guiding principles could potentially be useful in different communities as well. Coding the interview data suggests three major activities between partners are critical to this case's success: trust-building, meeting partners in the middle, and institutional credibility. Discussion of these themes in relation to the prior literature is integrated within each sub-section.

Trust-Building Subsystems

Trust, and the ways actors intentionally developed trust between partners, was a recurring theme in the data. Interview participants from each organization described specific strategies they used to develop trust with their cross-sector counterparts. The school created opportunities for short interactions between community-based adults and learners including career talks, site visits, and public project showcases. The learning coaches see adults treating learners as colleagues rather than kids, which builds trust. STEM professionals are often impressed by the ability, interests, and passions of youths whom they don't often see as part of their day-to-day activities. They trust learners to be worthy colleagues and treat them as such. High-level officials use these brief interactions to celebrate their partners and highlight positive student stories resulting from the partnership to acknowledge their partners.

Trust is an essential resource in any cross-sector collaboration (Becker & Smith, 2017). In this case, trust continuously grows in a positive feedback loop. Over time, community members, STEM professionals, and learning coaches seek opportunities to deepen their collaboration and involvement with STEM learning experiences. Building trust through collaboration and involvement leads to continuous improvement of the STEM learning experience. The barriers to school-industry partnerships are often documented and told by people who experience hurdles rather than being in a successful system (Rios et al., 2014). Trust is, perhaps, a prerequisite to systems building before any of the common barriers can be addressed.

Meeting Partners in the Middle

The school-based and industry professionals in this case demonstrated a willingness to adapt to the needs of their partners. Watters and Deitzman (2013) described how local industry partners made adaptations to accommodate the needs of schools. In this case, the school *and* industry partners both made considerable changes. For example, the school adopted a block schedule with Fridays as flexible times to accommodate site visits or collaborations without losing instructional hours. School-wide professional development is often hosted at the industry partner's innovation center facility rather than in the school. These changes ensure learners and coaches have access to real-world learning environments and problem-solving expertise from working STEM professionals. Early in the partnership, the business partner "loaned" an engineer to the school for an entire year to work with learning coaches across the school to embed authentic STEM learning practices across the curriculum. The examples in this paragraph demonstrate ways in which the school and industry entities made structural changes to accommodate their partners' needs. Both parties demonstrate an understanding of each other's needs, strengths, and constraints which can overcome cultural barriers to collaboration (Rios et al., 2014).

Institutional Signals of Credibility

Figure 2 displays a systems map that encompasses important individuals such as C-level officials, a district superintendent, a local industry partner broker, and national school design support networks. While these entities do not serve a day-to-day function in the current partnership, they play an important role in key moments. The STEM industry partner CEO and district superintendent publicly celebrate the activities and student outcomes resulting from the partnership. Learning coaches and STEM professionals ascribe these actions as motivating and rewarding. The official third-party industry broker organization listed in the system does not play the IO role suggested in prior research (Darche et al., 2009; Strobel & Sun, 2020). Instead, several study participants suggested that this organization is powerful in the community and even symbolic participation is meaningful to align stakeholder interests in the partnership.

These groups also provide funding for essential infrastructure. District officials and STEM professionals interpret investment as another way to demonstrate support. Study participants stated that money is not the most important contribution from the system to support student learning, but strategic investment is critical. For example, allowing the volunteer STEM council to organize and award small grants on behalf of the company is interpreted by staff engineers as recognition that supporting local STEM education is critical to the company's mission. By paying a membership fee to participate in the national school design network, district officials communicate to staff how important mutually beneficial school-STEM industry partnerships are to their vision of positive youth development. Now that the partnership is mature, the district relies less on the technical support the network offers. Instead, it is seen as a national leader and hosts multiple visits each month from communities across the world who are interested in doing similar work. Being seen as a national leader in school-STEM industry partnerships motivates school-based and industry professionals to continue improving and honing their work in service of youth in their community.

DISCUSSION

The case of Mid-America STEM High School demonstrates how the systems one community built to support a mutually beneficial school-STEM industry partnership transformed the student learning experience and improved learner engagement. The system elements and connections described are a function of the local environmental conditions and are not meant to be generalizable. Rather, they provoke a new sense of what is possible in the context of mutually beneficial school-industry partnerships. While such work is notably challenging, the resulting partnerships can provide meaningful, authentic, and rigorous STEM learning experiences that prepare students to persist in STEM-related college and career pathways.

Future research can expand on the groundwork laid in this chapter. The study participants did not associate some design elements of the system found in the prior literature with the partnership's success. For example, Mid-America STEM High School is small (just over 300 learners) compared to a typical urban high school but study participants did not discuss school size. Others have suggested that keeping school size small is a necessary design element of career academies because it facilitates instructional coherence and supports a sense of belonging within the learning environments (Hernandez-Gantes et al., 2019). The study participants recognized they have a unique geographic setting (an urban community with a multinational STEM business headquartered in town) but did not see it as a unique advantage. Future research should investigate partnerships in contexts with different STEM assets. Small or mid-sized businesses, for example, might not have the capacity to streamline operations in the way this organization did. Finally, it is worth considering the role technology may have in facilitating mutually beneficial school-STEM industry partnerships. Through the difficulties of the COVID-19 pandemic, businesses and schools learned to leverage synchronous and asynchronous technologies that facilitate work and learning. To what extent is a sense of physical place needed to share a commitment to youth in the way observed in this case? It may not be necessary for business partners and STEM volunteers to come from the school's community.

CONCLUSION

For years, Mid-America STEM High School has been collaborating with an industry partner. However, it is unclear, from this study, how early stakeholders were convinced to invest their time and talent into building such a complex system. The system diagram in Figure 2 illustrates the complexity of the work and the significant investment required. The intention behind choosing a mature partnership was to showcase an example for communities that aim to develop similar partnerships. To that end, two recommended activities could be implemented with minimal time, funding, or structural changes that could benefit both schools and STEM industries. Creating short student-adult interactions and supporting personnel who act as institutional bridges can begin the process of creating mutually beneficial school-STEM industry partnerships.

Recommendation #1: Start With Short Students-Adult Interactions

Instructional processes and community partnerships exist in entrenched systems which can be resistant to change. To establish productive partnerships between schools and the STEM industry entities, begin by providing chances for students to engage with STEM professionals in their local community. Host a showcase event and invite the community to interact with students. Invite STEM professionals to present to students on a topic being studied and workforce relevant. On their own, these interactions are impactful for students. Over time, short engagements can spark organic growth and deepen the partnership commitment without any structural or financial investment.

Recommendation #2: Support Key Personnel to Manage Partnerships

Managing relationships is time intensive. Without institutional support, passionate employees trying to create mutually beneficial partnerships can quickly become overwhelmed and will resort to one-sided efforts rather than seeking to leverage the complementary assets of schools and STEM businesses. If a school or industry partner can make a financial commitment, allocate some staff time to building partnerships. For example, the P-TECH school model suggests dedicating a STEM business employee to support coordinating activities (Litow & Kelley, 2021). While not present in the Mid-America STEM High School case, prior literature suggests that local non-profits could also be institutional brokers and build cross-sector partnerships (Darche et al., 2009; Strobel & Sun, 2020). These organizations often have personnel who specialize in bringing diverse stakeholders together around a common goal and can take an active role to facilitate school-STEM industry collaborations when schools and STEM businesses are not ready to make investments.

Supplemental Materials

This research was supported by the Open Education Resources Foundation, Sarasota, Florida, a part of a multi-person project, "Equitable, Inclusive, Effective STEM Learning for All, Via Open Learning, Available to High Schools, Urban, Rural and Suburban." I would like to acknowledge the study participants who shared their time and expertise during each interview and our project team at MIT, especially PI Richard Larson, for their valuable feedback on the design of this project.

REFERENCES

Becker, J., & Smith, D. B. (2017). The need for cross-sector collaboration. *Stanford Social Innovation Review*, *16*, C2–C3. doi:10.48558/8473-JD42

Bempechat, J., Kenny, M., Blustein, D., & Seltzer, J. (2014). Fostering positive youth development through work-based learning: The Cristo Rey model. *Teachers College Record, 116*(13), 232–252. doi:10.1177/016146811411601310

Bowen, B., & Peterson, B. (2018). Exploring authenticity through an engineering-based context in a project-based learning mathematics activity. *Journal of Pre-College Engineering Education Research (J-PEER), 9*(1), 0-0. doi:10.7771/2157-9288.1073

Committee on STEM Education. (2018). *Charting a course for success: America's strategy for STEM education.* National Science and Technology Council. https://trumpwhitehouse.archives.gov/wp-content/uploads/2018/12/STEM-Education-Strategic-Plan-2018.pdf

Cuban, L. (2013). *Unlocking the Black Box of the Classroom.* Harvard Education Press.

Darche, S., Nayar, N., & Bracco, K. R. (2009). *Work-Based Learning in California.* James Irvine Foundation. https://www2.wested.org/www-static/online_pubs/workbasedlearning.pdf

Eacott, S. (2008). Strategy in educational leadership: In search of unity. *Journal of Educational Administration, 46*(3), 353–375. doi:10.1108/09578230810869284

Eccles, J. S., & Wigfield, A. (2020). From expectancy-value theory to situated expectancy-value theory: A developmental, social cognitive, and sociocultural perspective on motivation. *Contemporary Educational Psychology, 61,* 0-0. doi:10.1016/j.cedpsych.2020.101859

Epstein, J. L. (2018). School, family, and community partnerships in teachers' professional work. *Journal of Education for Teaching, 44*(3), 397–406. doi:10.1080/02607476.2018.1465669

Ewell, E., Haglid, H., Truszkowski, E., Walicki, C., De Meulder, P., De Meulder, M., Stephens, T., Zhang, X., Trama, C., Hamilton, A., Mo, D., Wohner, J., Furrey, R., Labowsky, M., Perry, A., Hu, S., & Labowsky, H. L. (2022). Invention as a complement to high school chemistry. *Journal of Chemical Education, 99*(5), 2012–2019. doi:10.1021/acs.jchemed.2c00135

Francis, J. J., Johnston, M., Robertson, C., Glidewell, L., Entwistle, V., Eccles, M. P., & Grimshaw, J. M. (2010). What is an adequate sample size? Operationalising data saturation for theory-based interview studies. *Psychology & Health, 25*(10), 1229–1245. doi:10.1080/08870440903194015 PMID:20204937

Fuller, B., & Kim, H. (2022). *Systems thinking to transform schools: Identifying levers that lift educational quality.* Brookings Institution. https://www.brookings.edu/research/systems-thinking-to-transform-schools-identifying-levers-that-lift-educational-quality/

Ghaffarzadegan, N., Larson, R., & Hawley, J. (2017). Education as a complex system. *Systems Research and Behavioral Science, 34*(3), 211–215. doi:10.1002res.2405 PMID:28522920

Grinis, I. (2019). The STEM requirements of "Non-STEM" jobs: Evidence from UK online vacancy postings. *Economics of Education Review, 70,* 144–158. doi:10.1016/j.econedurev.2019.02.005

Hernandez-Gantes, V. M., Fletcher, E. C. Jr, & Keighobadi, S. (2019). A case study of the development of a career academy: Good intentions not enough? *Journal of Research in Technical Careers, 3*(2), 114–135. doi:10.9741/2578-2118.1059

Kelley, T. R., & Knowles, J. G. (2016). A conceptual framework for integrated STEM education. *International Journal of STEM Education*, *3*(1), 1–11. doi:10.118640594-016-0046-z

Kelley, T. R., Sung, E., Han, J., & Knowles, J. G. (2022). Impacting secondary students' STEM knowledge through collaborative STEM teacher partnerships. *International Journal of Technology and Design Education*. doi:10.100710798-022-09783-w

Ladson-Billings, G. (2013). Lack of achievement or loss of opportunity? In P. L. Carter & K. G. Welner (Eds.), *Closing the opportunity gap: What America must do to give all children an even chance* (pp. 11–24). Oxford University Press. doi:10.1093/acprof:oso/9780199982981.003.0002

Litow, S. S., & Kelley, T. (2021). *Breaking barriers: How P-TECH schools create a pathway from high school to college to career*. Teachers College Press.

Margolis, J. (2008). *Stuck in the shallow end: Education, race, and computing*. MIT Press.

Meadows, D. H., & Wright, D. (2008). *Thinking in systems: A primer*. Chelsea Green Pub.

Mitra, D. L. (2009). The role of intermediary organizations in sustaining student voice initiatives. *Teachers College Record*, *111*(7), 1834–1870. doi:10.1177/016146810911100708

Mitra, D. L. (2018). *Educational change and the political process*. Routledge, Taylor & Francis Group.

Montoya, J., Lundell, R., Peterson, F., Tarantino, S., Ramsey, M., Katz, G., Baldini, R., Fruchter, R., & Fischer, M. (2018). Building sustainable communities: A project-based learning approach to modify student perceptions of the building industry. *ACEEE Summer Study on Energy Efficiency in Buildings*, *11*, 1–13. https://www.aceee.org/files/proceedings/2018/index.html#/paper/event-data/p342

National Center for Science and Engineering Statistics. (2023). *Diversity and STEM: Women, Minorities, and Persons with Disabilities* (NSF 23-315). National Science Foundation. https://ncses.nsf.gov/pubs/nsf23315/

National Research Council. (2012). *A Framework for K-12 science education: Practices, crosscutting concepts, and Core ideas*. The National Academies Press.

National Science Board. (2021). *The STEM labor force of today: Scientists, engineers, and skilled technical workers* (NSB-2021-2; Science and Engineering Indicators 2022). National Science Foundation. https://ncses.nsf.gov/pubs/nsb20212

Oakes, J. (1990). *Multiplying inequalities: The effects of race, social class, and tracking on opportunities to learn mathematics and science*. Rand Corp.

Orr, M., Hughes, K., & Karp, M. (2003). *Shaping postsecondary transitions: Influences of the National Academy Foundation career academy* (No. 29). Columbia University, Institute on Education and the Economy. https://files.eric.ed.gov/fulltext/ED477819.pdf

Parking, C., Scott, S., & Geddes, A. (2020). Snowball sampling. In P. Atkinson, S. Delamont, A. Cernat, J. Sakshaug, & R. Williams (Eds.), *SAGE Research Methods Foundations*. SAGE Publications Ltd., doi:10.4135/9781526421036831710

Radinsky, J., Bouillion, L., Lento, E. M., & Gomez, L. M. (2001). Mutual benefit partnership: A curricular design for authenticity. *Journal of Curriculum Studies*, *33*(4), 405–430. doi:10.1080/00220270118862

Rios, R., Artigas, A., & Casulleras, R. P. (2014). Establishing and maintaining STEM-related school-industry partnerships: Opportunities and obstacles. In C. P. Constantinou, N. Papadouris, & A. Hadjigeorgiou (Eds.), *E-Book Proceedings of the ESERA 2013 Conference: Science Education Research for Evidence-based Teaching and Coherence in Learning*. ESERA. https://www.esera.org/publications/esera-conference-proceedings/esera-2013#strand-9

Schneider, B., Krajcik, J., Lavonen, J., Salmela-Aro, K., Klager, C., Bradford, L., Chen, I.-C., Baker, Q., Touitou, I., Peek-Brown, D., Dezendorf, R. M., Maestrales, S., & Bartz, K. (2022). Improving science achievement—Is it possible? Evaluating the efficacy of a high school chemistry and physics project-based learning intervention. *Educational Researcher*, *51*(2), 109–121. doi:10.3102/0013189X211067742

Shernoff, D. J., Csikszentmihalyi, M., Shneider, B., & Shernoff, E. S. (2003). Student engagement in high school classrooms from the perspective of flow theory. *School Psychology Quarterly*, *18*(2), 158–176. doi:10.1521cpq.18.2.158.21860

Sinatra, G. M., Heddy, B. C., & Lombardi, D. (2015). The challenges of defining and measuring student engagement in science. *Educational Psychologist*, *50*(1), 1–13. doi:10.1080/00461520.2014.1002924

Smit, R., Robin, N., De Toffol, C., & Atanasova, S. (2019). Industry-school projects as an aim to foster secondary school students' interest in technology and engineering careers. *International Journal of Technology and Design Education*. Advance online publication. doi:10.100710798-019-09538-0

Strobel, J., & Sun, Y. (2020). Current state and suggestions for the K-12 STEM school industry partnership in the United States. In M. J. de Vries, S. Fletcher, S. Kruse, P. Labudde, M. Lang, I. Mammes, C. Max, D. Munk, B. Nicholl, J. Strobel, & M. Winterbottom (Eds.), *The impact of technology education: International insights* (Vol. 3, pp. 143–162). Waxmann Verlag.

Tan, V., Nicholas, C., Scribner, J. A., & Francis, D. C. (2019). Enhancing STEM learning through an interdisciplinary, industry-generated project. *Technology and Engineering Teacher*, *79*(1), 26–31.

The Australian Industry Group. (2017). *Strengthening school-industry STEM skills partnerships: Final project report*. The Australian Industry Group. https://cdn.aigroup.com.au/Reports/2017/AiGroup_OCS_STEM_Report_2017.pdf

Watters, J. J., & Diezmann, C. M. (2013). Community partnerships for fostering student interest and engagement in STEM. *Journal of STEM Education: Innovations and Research*, *14*(2), 47–55.

Webber, K., Robert, J., Williams Moore, M., & Burg, T. (2020). Measuring the impact of business engagement on STEM education. In C. C. Johnson, M. J. Mohr-Schroeder, T. J. Moore, & L. D. English (Eds.), *Handbook of research on STEM education* (pp. 459–469). Routledge. doi:10.4324/9780429021381-42

Yin, R. K. (2014). *Case study research: Design and methods* (5th ed.). SAGE.

Zhang, H., Couch, S., Estabrooks, L., Perry, A., & Kalainoff, M. (2023). Role models' influence on student interest in and awareness of career opportunities in life sciences. *International Journal of Science Education, Part B*, 1–19.

ENDNOTE

1 All references to the school and community are pseudonyms to ensure personally identifiable information about the interview participants is not disclosed.

APPENDIX

Guiding questions for the semi-structured interview protocol

Describing the Actors and Their Role

1. Can you begin by sharing the organization you work for and your role in the organization?
2. What is your involvement with planning school-industry partnerships? How long have you been involved and how did you, personally, become involved? [be sure to probe and differentiate between planning involvement and day-to-day involvement]
3. How did student-industry partnerships begin at [school/organization]? Who were the key players in getting this work off the ground?
4. Who do you work with directly over the course of the year to support these student experiences? These can be individuals at your organization or partnering organizations [probe for names and roles].
5. Is there a particular student story that stands out to you?

Aims, Values and Goals of Partnership

6. In your view, what has been the most significant positive outcome of the mentorships? Is this perspective shared by students? Is this perspective shared by the cooperating [industry/education] partners?
7. What do you think your [industry/education] partners see as the most important reason for participating?
8. What are some of the challenges you have faced or are currently facing to implement meaningful student-industry mentoring experiences? How have you overcome them? [if not mentioned be sure to probe about getting started with between organization collaborations]
9. Your [organization] has maintained these partnerships for many years. *What* or *who* do you attribute this success and longevity to?
10. How does participating in the student-industry mentorship program connect with your personal values or your organization's values around what is important for STEM education?

Following-Up

11. Are there any people you suggest we should contact for an interview that has a unique or important perspective to share? Can you provide an introduction? [If no, can you their contact information?]
12. Is there anything else you would like to share about these experiences I have not mentioned?

Exemplar PBL: Designing Enrichment Toys for Zoo Animals

One of the first PBLs youth experience at Mid-America STEM High School derives from a long-running partnership with the local Zoo. Learners are tasked with designing enrichment toys for animals at the local zoo. This project begins with an introduction from zoo professionals to describe the problems faced by animals. During class time, learners choose an animal at the zoo, research what kind of enrichment that animal needs, and participate in a virtual "ask the expert" session as part of their research process.

In the next phase of the unit, learners create a low-fidelity prototype of their proposed enrichment toy. They post videos on Flip showing their work in progress and receive additional feedback from the experts. The feedback is used to revise prototypes and submit final designs for Zoo staff to judge. Zoo staff selects a small number of winning projects which are made into prototypes the animals can play with! Learners visit the Zoo in the spring becoming the first visitors to witness animals interacting with the winning enrichment toys. In years past, learners created hammocks for newborn snow leopards and enrichment "beehives" for bears. The entire unit spans about three weeks of class time.

In addition to meeting the state curriculum standards in science, this project is explicitly designed to introduce students to the skills and processes they will use and develop while at Mid-America STEM High School. These skills are outlined in the district's Portrait of a Graduate and include problem solving, critical thinking, and contributing to the local community.

The table outlines the process students follow in the capstone course. This design is the result of many iterations by the capstone learning coach. This version has been edited to retain participant anonymity.

Table 1. Map and roles of key actors in the student capstone course

Stage (months)	Learner Activity	Coach Activity	External Partner Involvement (e.g., community, technical mentors, supporters)	Deliverable(s)
Identify (summer)	Identify a problem to solve about which they are passionate.	Reviewing "problems" and providing feedback.	*None at this step*	Research proposal
Investigate (Aug/Sept)	Conduct academic research, interview users/experts, and observe the problem in action.	Support and adjustment, as needed.	Public presentation of Progress: learners discuss what they've learned and ask for feedback.	Annotated Analytical Bibliography
Interpret (Oct)	Use human-centered design practices to identify design principles to drive product development.	Support and adjustment, as needed.	Public presentation of Progress: learners present their deliverables; asking partners for any ideas for solutions to start off their next step.	Ecosystem Map Empathy Map Root Cause Analysis Interpret Statement
Ideate (Nov)	Brainstorm and identify what they need to learn to create it, and then actually learn the STEM concepts behind it.	Support and adjustment, as needed.	Public presentation of Progress: learners present their ideas and ask for feedback or other perspectives.	Brainstorming STEM concepts
Implement (Nov-Feb)	create a low-fidelity prototype for users to interact with; form a focus group of users and test their low-fidelity prototype for feedback, make changes; create a high-fidelity prototype/ product that they name and brand.	Support and adjustment, as needed.	Submit to External State competition for judging. Presentation of Progress: learners present their prototypes at a few different stages and ask for feedback.	Low-fidelity prototype Commercialization plan High-fidelity prototype Marketing plan (packaging, product name, logo)
Inform (March)	Reflect on the hard and soft skills learned during their high school experience, identifying when/where they learned it and how each came in handy with this culminating project. Create a documentary of their design process and skills to solve a problem.	Support and adjustment, as needed.	Judges at the presentation of learning: learners screen their film, demo their product, and answer questions.	Documentary Final presentation of Learning

Compilation of References

Abdi, A. (2014). The effect of inquiry-based learning method on students' academic achievement in science course. *Universal Journal of Educational Research, 2*(1), 37–41. doi:10.13189/ujer.2014.020104

Acaira, E., Vile, J., & Reisner, E. R. (2010). *Citizen Schools: Achieving high school graduation: Citizen Schools' youth outcomes in Boston.* Policy Studies Associates, Inc. https://static1.squarespace.com/static/57542d1b0442628dccd81967/t/5898ae5f9f7456131b0b420d/1486401121825/PSA-Citizen-Schools-Youth-Outcomes-in-Boston.pdf

Afroogh, S., Esmalian, A., Donaldson, J. P., & Mostafavi, A. (2021). Empathic design in engineering education and practice: An approach for achieving inclusive and effective community resilience. *Sustainability, 13*, 0-0. doi:10.3390/su13074060

Afterschool Alliance. (2013). *Defining youth outcomes for STEM learning in afterschool.* Afterschool Alliance. http://www.afterschoolalliance.org/STEM_Outcomes_2013.pdf

Afterschool Alliance. (2015). *Afterschool programs: Inspiring students with a connected learning approach.* Afterschool Alliance. http://afterschoolalliance.org/documents/Afterschool_and_Connected_Learning.pdf

Ahmed, M. W., Anderson, Y., Gerald-Goins, T., Hollowell, G. P., Salim, E. T., Sangutei, T., Simpson, B., Spence, P., Whittington, D., & White, S. L. (2021). Promoting STEM-literacy by designing decision-driven interdisciplinary courses for non-science majors. *Journal of STEM Education: Innovations and Research, 21*(3), 28–34.

Ainley, M. (2017). Interest: Knowns, unknowns, and basic processes. In P. A. O'Keefe & J. M. Harackiewicz (Eds.), *The Science of Interest* (pp. 3–24). Springer International Publishing., doi:10.1007/978-3-319-55509-6_1

Akiba, M., LeTendre, G. K., & Scribner, J. P. (2007). Teacher quality, opportunity gap, and national achievement in 46 countries. *Educational Researcher, 36*(7), 369–387. doi:10.3102/0013189X07308739

Alamance-Burlington School System. (2016). *7/11/16 Closing the achievement GAP.* Alamance-Burlington School System.

Ali, S. R., & Menke, K. A. (2014). Rural Latino youth career development: An application of social cognitive career theory. *The Career Development Quarterly, 62*(2), 175–186. doi:10.1002/j.2161-0045.2014.00078.x

Allen, J. M., Muragishi, G. A., Smith, J. L., Thoman, D. B., & Brown, E. R. (2015). To Grab and To Hold: Cultivating communal goals to overcome cultural and structural barriers in first generation college students' science interest. *Translational Issues in Psychological Science, 1*(4), 331–341. doi:10.1037/tps0000046 PMID:26807431

Alrubail, R. (2015, June 2). Teaching empathy through design thinking. *Edutopia.* https://www.edutopia.org/blog/teaching-empathy-through-design-thinking-rusul-alrubail

American Association for the Advancement of Science. (2011). *Vision and change in undergraduate biology education: A call to action.* AAAS.

Anderson, J. L., Ellis, J. P., & Jones, A. M. (2014). Understanding early elementary children's conceptual knowledge of plant structure and function through drawings. *CBE Life Sciences Education*, *13*(3), 375–386. doi:10.1187/cbe.13-12-0230 PMID:25185222

Andrea, L., & Cupita, L. (2016). Just in time teaching: A strategy to encourage students' engagement. *HOW*, *23*(2), 89–105. doi:10.19183/how.23.2.163

Angle, J., Colston, N., French, D., Gustafson, J., O'Hara, S., & Shaw, E. (2016). Addressing the call to increase high school students' STEM awareness through a collaborative event hosted by science and education faculty: A how-to approach. *Science Educator*, *25*(1), 43–50.

Archer, L., DeWitt, J., Osborne, J., Dillon, J., Willis, B., & Wong, B. (2010). 'Doing' science versus 'being' a scientist: Examining 10/11-year-old schoolchildren's constructions of science through the lens of identity. *Science Education*, *94*(4), 617–639. doi:10.1002ce.20399

Aschbacher, P. R., Li, E., & Roth, E. J. (2009). Is science me? High school students' identities, participation and aspirations in science, engineering, and medicine. *Journal of Research in Science Teaching*, *47*(5), 564–582. doi:10.1002/tea.20353

Assaraf, O. B.-Z., & Orion, N. (2005). Development of system thinking skills in the context of earth system education. *Journal of Research in Science Teaching*, *42*(5), 518–560. doi:10.1002/tea.20061

Assaraf, O. B.-Z., & Orion, N. (2010). System Thinking Skills at the Elementary School Level. *Journal of Research in Science Teaching*, *47*(5), 540–563.

Assessing Women & Men in Engineering . (2008). Pre-college out- reach STEM activity pre-college middle school science surveys. The Pennsylvania State University and University of Missouri. http://aweon- line.org/onlinesurveys.html

Assessing Women & Men in Engineering. (2010). *Pre-college out-reach STEM activity upper elementary school surveys.* The Pennsylvania State University and University of Missouri. http://aweonline.org/online- surveys.html

Baines, A. M., DeBarger, A., De Vivo, K., Warner, N., Santos, S., Brinkman, J., Udall, D., Zuckerbrod, N., Felsen, K., & Urban, R. (2021). *Key principles for project-based learning*. Lucas Education Research. https://www.lucasedresearch.org/wp-content/uploads/2021/02/Key-Principles-for-PBL-White-Paper.pdf

Balemen, N., & Keskin, M. Ö. (2018). The effectiveness of project-based learning on science education: A meta-analysis search. *International Online Journal of Education & Teaching*, *5*(4), 849–865.

Ball, C., Huang, K. T., Cotten, S. R., & Rikard, R. V. (2017). Pressurizing the STEM pipeline: An expectancy-value theory analysis of youths' STEM attitudes. *Journal of Science Education and Technology*, *26*(4), 372–382. doi:10.100710956-017-9685-1

Ballen, C. J., Wieman, C., Salehi, S., Searle, J. B., & Zamudio, K. R. (2017). Enhancing diversity in undergraduate science: Self-efficacy drives performance gains with active learning. *CBE—Life Sciences Education, 16*(4). doi:10.1187/cbe.16-12-0344

Bandura, A. (1998). Self-Efficacy. In V. S. Ramachaudran (Ed.), *Encyclopedia of Human Behavior* (Vol. 4, pp. 71–81). Academic Press.

Bandura, A. (1999). Social cognitive theory: An agentic perspective. *Asian Journal of Social Psychology*, *2*(1), 21–41. doi:10.1111/1467-839X.00024

Bandura, A. (2000). Self-efficacy: The foundation of agency. In W. J. Perrig & A. Groub (Eds.), *Control of human behavior, mental processes, and consciousness: Essays in honor of the 60th birthday of August Flammer* (pp. 16–31). Taylor & Francis.

Bandura, A. (2005). The evolution of social cognitive theory. In K. G. Smith & M. A. Hitt (Eds.), *Great Minds in Management* (pp. 9–35). Oxford University Press. doi:10.5465/amr.2007.23467624

Bandura, A. (2006). Guide for constructing self-efficacy scales. In F. Pajares & T. Urdan (Eds.), *Self-efficacy beliefs of adolescents* (Vol. 5, pp. 307–337). Information Age Publishing.

Bandura, A., & Locke, E. A. (2003). Negative self-efficacy and goal effects revisited. *The Journal of Applied Psychology, 88*(1), 87–99. doi:10.1037/0021-9010.88.1.87 PMID:12675397

Barraza, L. (1999). Children's drawings about the environment. *Environmental Education Research, 5*(1), 49–66. doi:10.1080/1350462990050103

Basu, S. J., & Barton, A. C. (2007). Developing a sustained interest in science among urban minority youth. *Journal of Research in Science Teaching, 44*(3), 466–489. doi:10.1002/tea.20143

Becker, J., & Smith, D. B. (2017). The need for cross-sector collaboration. *Stanford Social Innovation Review, 16*, C2–C3. doi:10.48558/8473-JD42

Belanger, A. L., Diekman, A. B., & Steinberg, M. (2017). Leveraging communal experiences in the curriculum: Increasing interest in pursuing engineering by changing stereotypic expectations. *Journal of Applied Social Psychology, 47*(6), 305–319. doi:10.1111/jasp.12438

Belanger, A. L., Joshi, M. P., Fuesting, M. A., Weisgram, E. S., Claypool, H. M., & Diekman, A. B. (2020). Putting Belonging in Context: Communal Affordances Signal Belonging in STEM. *Personality and Social Psychology Bulletin, 46*(8), 1186–1204. doi:10.1177/0146167219897181 PMID:31928327

Bempechat, J., Kenny, M., Blustein, D., & Seltzer, J. (2014). Fostering positive youth development through work-based learning: The Cristo Rey model. *Teachers College Record, 116*(13), 232–252. doi:10.1177/016146811411601310

Benish, S. (2018). Meeting STEM workforce demands by diversifying STEM. *Journal of Science Policy and Governance, 13*(1), 0-0. https://www.sciencepolicyjournal.org/uploads/5/4/3/4/5434385/benish.pdf

Berger, A., Turk-Bicakci, L., Garet, M., Song, M., Knudson, J., Haxton, C., Zeiser, K., Hoshen, G., Ford, J., Stephan, J., Keating, K., & Cassidy, L. (2013). *Early college, early success: Early College High School initiative impact study.* American Institutes for Research. https://eric.ed.gov/?id=ED577243

Bevan, B., Dillon, J., Hein, G. E., Macdonald, M., Michalchik, V., Miller, D., Root, D., Rudder, L., Xanthoudaki, M., & Yoon, S. (2010). *Making science matter: Collaborations between informal science education organizations and schools.* Center for Advancement of Informal Science Education.

Bhattarai, B., Davis, E. W., Lakin, J. M., Davis, V. A., & Marghitu, D. (2021, July), *Poster: Use of Augmented Reality (AR) and Virtual Reality(VR) to tackle 4 amongst the "14 Grand Challenges for Engineering in the 21st Century" identified by National Academy of Engineering.* Paper presented at 2021 ASEE Virtual Annual Conference Content Access, Virtual Conference. https://peer.asee.org/37589

Bielefeldt, A. R. (2017). Is engineering about caring for and helping others? In B. Hernandez (Ed.), *Focus on educational research: Practices, challenges and perspectives* (pp. 89–118). Nova Science Publishers.

Black, P., & Wiliam, D. (2010). Inside the black box: Raising standards through classroom assessment. *Phi Delta Kappan, 92*(1), 81–90. doi:10.1177/003172171009200119

Blikstein, P., & Krannich, D. (2013). The makers' movement and fablabs in education: Experiences, technologies, and research. *Proceedings of the 12th International Conference on Interaction Design and Children.* Association for Computing Machinery. 10.1145/2485760.2485884

Blotnicky, K. A., Franz-Odendaal, T., French, F., & Joy, P. (2018). A study of the correlation between STEM career knowledge, mathematics self-efficacy, career interests, and career activities on the likelihood of pursuing a STEM career among middle school students. *International Journal of STEM Education, 5*(1), 1–15. doi:10.118640594-018-0118-3 PMID:30631712

Bouillion, L. M., & Gomez, L. M. (2001). Connecting school and community with science learning: Real world problems and school–community partnerships as contextual scaffolds. *Journal of Research in Science Teaching, 38*(8), 878–898. doi:10.1002/tea.1037

Bowen, B., & Peterson, B. (2018). Exploring authenticity through an engineering-based context in a project-based learning mathematics activity. *Journal of Pre-College Engineering Education Research (J-PEER), 9*(1), 0-0. doi:10.7771/2157-9288.1073

Bowers, E. P., Li, Y., Kiely, M. K., Brittian, A., Lerner, J. V., & Lerner, R. M. (2010). The five Cs model of positive youth development: A longitudinal analysis of confirmatory factor structure and measurement invariance. *Journal of Youth and Adolescence, 39*(7), 720–735. doi:10.100710964-010-9530-9 PMID:20397040

Boyer, E. L. (1996). The scholarship of engagement. *Bulletin - American Academy of Arts and Sciences. American Academy of Arts and Sciences, 49*(7), 18–33. https://www.jstor.org/stable/3824459. doi:10.2307/3824459

Boys and Girls Club of Durham and Orange Counties. (2017). *2017 Impact report.* Boys and Girls Clubs of America.

Brown, R. H. (2013). *Hear Our Truths: The Creative Potential of Black Girlhood.* University of Illinois Press. doi:10.5406/illinois/9780252037979.001.0001

Bryk, A. S., Gomez, L. M., Grunow, A., & LeMahieu, P. G. (2015). *Learning to improve: How America's schools can get better at getting better.* Harvard Education Press.

Bureau of Labor and Statistics. (September 4, 2019). *Table 1.11: Employment in STEM occupations, 2018 and projected 2028.* United States Department of Labor. https://www.bls.gov/emp/tables/stem-employment.htm

Burke, A., Okrent, A., & Hale, K. (2022, January 18). *The State of U.S. Science and Engineering 2022.* National Science Board. https://ncses.nsf.gov/pubs/nsb20221

Burns, H., & Lesseig, K. (2017). Infusing empathy into engineering design: Supporting under-represented student interest and sense of belongingness. *Proceedings of the American Society for Engineering Education ASEE Annual Conference & Exposition.* Monolith. https://monolith.asee.org/public/conferences/78/papers/19857/view

Butler, T. T. (2018). Black girl cartography: Black girlhood and place-making in education research. *Review of Research in Education, 42*(1), 28–45. doi:10.3102/0091732X18762114

Bystydzienski, J. M., Eisenhart, M., & Bruning, M. (2015). High School Is Not Too Late: Developing Girls' Interest and Engagement in Engineering Careers. *The Career Development Quarterly, 63*(1), 88–95. doi:10.1002/j.2161-0045.2015.00097.x

Calabrese Barton, A., Tan, E., & Greenberg, D. (2017). The Makerspace Movement: Sites of Possibilities for Equitable Opportunities to Engage Underrepresented Youth in STEM. *Teachers College Record, 119*(6), 1–44. doi:10.1177/016146811711900608

Capobianco, B. M., Diefes-dux, H. A., Mena, I., & Weller, J. (2011). What is an engineer? Implications of elementary school student conceptions for engineering education. *Journal of Engineering Education, 100*(2), 304–328. doi:10.1002/j.2168-9830.2011.tb00015.x

Capobianco, B. M., & Yu, J. H. (2014). Using the construct of care to frame engineering as a caring profession toward promoting young girls' participation. *Journal of Women and Minorities in Science and Engineering*, *20*(1), 21–33. doi:10.1615/JWomenMinorScienEng.2014006834

Carlone, H. B., & Johnson, A. (2007). Understanding the science experiences of successful women of color: Science identity as an analytic lens. *Journal of Research in Science Teaching*, *44*(8), 1187–1218. doi:10.1002/tea.20237

Carlsen, M. H., Halvorsen, B. L., Holte, K., Bøhn, S. K., Dragland, S., Sampson, L., Willey, C., Senoo, H., Umezono, Y., Sanada, C., Barikmo, I., Berhe, N., Willett, W. C., Phillips, K. M., Jacobs, D. R. Jr, & Blomhoff, R. (2010). The total antioxidant content of more than 3100 foods, beverages, spices, herbs, and supplements used worldwide. *Nutrition Journal*, *9*(1), 3. doi:10.1186/1475-2891-9-3 PMID:20096093

Carter Andrews, D. J., Brown, T., Castro, E., & Id-Deen, E. (2019). The impossibility of being "perfect and white": Black girls' racialized and gendered schooling experiences. *American Educational Research Journal*, *56*(6), 2531–2572. doi:10.3102/0002831219849392

CAST. (2018). *Universal design for learning guidelines version 2.2*. UDL Guidelines. http://udlguidelines.cast.org

Cech, E. (2013). Ideological wage inequalities? The technical/social dualism and the gender wage gap in engineering. *Social Forces*, *91*(4), 1147–1182. doi:10.1093fot024

Cech, E. (2015). Engineers and engineeresses? Self-conceptions and the development of gendered professional identities. *Sociological Perspectives*, *58*(1), 56–77. doi:10.1177/0731121414556543

Chan, A. D., & Fishbein, J. (2009). A global engineer for the global community. *The Journal of Policy Engagement*, *1*(2), 4–9.

Chu, L., Ting, Y. L., & Tai, Y. (2020, July). Building STEM capability in a robotic arm educational competition. In *International Conference on Human-Computer Interaction* (pp. 408-421). Springer. 10.1007/978-3-030-50506-6_28

Clark, K., & Sheridan, K. (2010). Game Design Through Mentoring and Collaboration. *Journal of Educational Multimedia and Hypermedia*, *19*(2), 5–22.

Coburn, C. E., & Penuel, W. R. (2016). Research–practice partnerships in education: Outcomes, dynamics, and open questions. *Educational Researcher*, *45*(1), 48–54. doi:10.3102/0013189X16631750

Cohen, E. J., Bravi, R., & Minciacchi, D. (2021). Assessing the Development of Fine Motor Control in Elementary School Children Using Drawing and Tracing Tasks. *Perceptual and Motor Skills*, *128*(2), 605–624. doi:10.1177/0031512521990358 PMID:33496640

Collaborative for Academic, Social, and Emotional Learning. (2020). *CASEL's SEL Framework*. CASEL. https://casel.org/wp-content/uploads/2020/12/CASEL-SEL-Framework-11.2020.pdf

Collins, K. H. (2018). Confronting color-blind STEM talent development: Toward a contextual model for Black student STEM identity. *Journal of Advanced Academics*, *29*(2), 143–168. doi:10.1177/1932202X18757958

Colston, N., Thomas, J., Ley, M. T., Ivey, T., & Utley, J. (2017). Collaborating for early-age career awareness: A comparison of three instructional formats. *Journal of Engineering Education*, *106*(2), 326–344. doi:10.1002/jee.20166

Committee on STEM Education. (2018). *Charting a course for success: America's strategy for STEM education*. National Science and Technology Council. https://trumpwhitehouse.archives.gov/wp-content/uploads/2018/12/STEM-Education-Strategic-Plan-2018.pdf

Costa, V. B. (1995). When science is 'another world': Relationships between worlds of family, friends, school, and science. *Science Education*, *79*(3), 313–333. doi:10.1002ce.3730790306

Council of State Governments. (2017). *Federal funding for state employment and training programs covered by the workforce innovation and opportunity act.* Capitol Research.

Creamer, E. G. (2017). *An introduction to fully integrated mixed methods research.* Sage Publications.

Crenshaw, K. (1990). Mapping the margins: Intersectionality, identity politics, and violence against women of color. *Stanford Law Review*, *43*(6), 1241–1299. doi:10.2307/1229039

Creswell, J. W., Plano Clark, V. L., Gutmann, M. L., & Hanson, W. E. (2003). Advanced mixed methods research designs. In A. Tashakkori & C. Teddlie (Eds.), *Handbook of mixed methods in social and behavioral research* (pp. 209–240). Sage.

Creswell, J., & Poth, C. (2018). *Qualitative inquiry and research design: Choosing among five approaches* (4th ed.). Sage.

Crooks, N., Sosina, W., Debra, A., & Donenberg, G. (2022). The Impact of COVID-19 Among Black Girls: A Social-Ecological Perspective. *Journal of Pediatric Psychology*, *47*(3), 270–278. doi:10.1093/jpepsy/jsab139 PMID:35182154

Cuban, L. (2013). *Unlocking the Black Box of the Classroom.* Harvard Education Press.

Cuevas, P., Lee, O., Hart, J., & Deaktor, R. (2005). Improving science inquiry with elementary students of diverse backgrounds. *Journal of Research in Science Teaching*, *42*(3), 337–357. doi:10.1002/tea.20053

Cunningham, C., Lachapelle, C., & Lindgren-Streicher, A. (2006, June), *Elementary Teachers' Understandings Of Engineering And Technology* Paper presented at 2006 Annual Conference & Exposition, Chicago, Illinois. 10.18260/1-2--200

D'Costa, A. R. S., & Mark, A. (2013). Scaffolded Instruction Im- proves Student Understanding of the Scientific Method & Experimental Design. *BioOne*, *75*(1), 18–28.

Daily, G. C., Alexander, S., Ehrlich, P. R., Goulder, L., Lubchenco, J., Matson, P. A., Mooney, H. A., Postel, S., Schneider, S. H., Tilman, D., & Woodwell, G. M. (1997). *Ecosystem services: Benefits supplied to human societies by natural ecosystems.* Ecological Society of America.

Dalton, B., & Musetti, K. (2018). Tactile picture book making and multimodal composition: Students design for equity in English Language Arts. *Best Practices in Teaching Digital Literacies: Literacy Research. Practice and Evaluation*, *9*, 195–213.

Danley, A., McCoy, A., & Weed, R. (2006). Exit Tickets Open the Door to University Learning. *InSight: A Journal of Scholarly Teaching, 11*, 48-58.

Darche, S., Nayar, N., & Bracco, K. R. (2009). *Work-Based Learning in California.* James Irvine Foundation. https://www2.wested.org/www-static/online_pubs/workbasedlearning.pdf

Davis, R. (2020). New Study Shows Black Women Are Among The Most Educated Group In The United States. *Essence.* https://www.essence.com/news/new-study-black-women-most-educated/

de Graaf, H. J., Musters, C. J. M., Noordervliet, M. A. W., & ter Keurs, W. J. (2005). An instrument for interactive setting development goals of a socio-environmental system. *Ecological Economics*, *52*(1), 19–29. doi:10.1016/j.ecolecon.2004.05.006

DeBacker, T., & Nelson, R. (1999). Variations on an expectancy-value model of motivation inscience. *Contemporary Educational Psychology*, *24*(2), 71–94. doi:10.1006/ceps.1998.0984 PMID:10072310

Deemer, E. D., Thoman, D. B., Chase, J. P., & Smith, J. L. (2014). Feeling the threat: Stereotype threat as a contextual barrier to women's science career choice intentions. *Journal of Career Development*, *41*(2), 141–158. doi:10.1177/0894845313483003

Denton, A., & Gandhi, E. (2020). *Scaling out-of-school time STEM programming: A national scan.* Overdeck Family Foundation and Education Northwest. https://educationnorthwest.org/sites/default/files/overdeck-ost-stem-report.pdf

Deobald, J. (2020, May 1). Coding and empathy: Foundational skills for the future of work. *Forbes*. https://www.forbes.com/sites/forbesagencycouncil/2020/05/01/coding-and-empathy-foundational-skills-for-the-future-of-work/?sh=6727bffc62f5

Diekman, A. B., Brown, E. R., Johnston, A. M., & Clark, E. K. (2010). Seeking congruity between goals and roles: A new look at why women opt out of science, technology, engineering, and mathematics careers. *Psychological Science*, *21*(8), 1051–1057. doi:10.1177/0956797610377342 PMID:20631322

Diekman, A. B., Clark, E. K., Johnston, A. M., Brown, E. R., & Steinberg, M. (2011). Malleability in communal goals and beliefs influences attraction to stem careers: Evidence for a goal congruity perspective. *Journal of Personality and Social Psychology*, *101*(5), 902–918. doi:10.1037/a0025199 PMID:21859224

Dixson, D. D., & Worrell, F. C. (2016). Formative and summative assessment in the classroom. *Theory into Practice*, *55*(2), 153–159. doi:10.1080/00405841.2016.1148989

Dorph, R., Bathgate, M. E., Schunn, C. D., & Cannady, M. A. (2018). When I grow up: The relationship of science learning activation to STEM career preferences. *International Journal of Science Education*, *40*(9), 1034–1057. doi:10.1080/09500693.2017.1360532

Durlak, J. A., Weissberg, R. P., & Pachan, M. (2010). A meta-analysis of after-school program that seek to promote personal and social skills in children and adolescents. *American Journal of Community Psychology*, *45*(3-4), 294–309. doi:10.100710464-010-9300-6 PMID:20300825

Eacott, S. (2008). Strategy in educational leadership: In search of unity. *Journal of Educational Administration*, *46*(3), 353–375. doi:10.1108/09578230810869284

Easterday, M. W., Lewis, D. R., & Gerber, E. M. (2014). *Design-based research process: Problems, phases, and applications.* International Society of the Learning Sciences.

Eccles, J. S., & Wigfield, A. (2020). From expectancy-value theory to situated expectancy-value theory: A developmental, social cognitive, and sociocultural perspective on motivation. *Contemporary Educational Psychology, 61,* 0-0. doi:10.1016/j.cedpsych.2020.101859

Eccles, J. S., & Wigfield, A. (2002). Motivational beliefs, values, and goals. *Annual Review of Psychology*, *53*(1), 109–132. doi:10.1146/annurev.psych.53.100901.135153 PMID:11752481

Eguchi, A. (2014). Learning experience through RoboCupJunior: Promoting STEM education and 21st century skills with robotics competition. In *Society for Information Technology & Teacher Education International Conference*, (pp. 87-93). Association for the Advancement of Computing in Education (AACE).

Eguchi, A. (2016). RoboCupJunior for promoting STEM education, 21st century skills, and technological advancement through robotics competition. *Robotics and Autonomous Systems*, *75*, 692–699. doi:10.1016/j.robot.2015.05.013

Epstein, J. L. (2018). School, family, and community partnerships in teachers' professional work. *Journal of Education for Teaching*, *44*(3), 397–406. doi:10.1080/02607476.2018.1465669

Epstein, R., Blake, J., & González, T. (2017). *Girlhood interrupted: the erasure of black girls' childhood.* SSRN., doi:10.2139srn.3000695

Estrada, M., Burnett, M., Campbell, A. G., Campbell, P. B., Denetclaw, W. F., Gutiérrez, C. G., Hurtado, S., John, G. H., Matsui, J., McGee, R., Okpodu, C. M., Robinson, T. J., Summers, M. F., Werner-Washburne, M., & Zavala, M. (2016). Improving Underrepresented Minority Student Persistence in STEM. *CBE Life Sciences Education*, *15*(3), 1–10. doi:10.1187/cbe.16-01-0038 PMID:27543633

Estrada, M., Woodcock, A., Hernandez, P. R., & Schultz, P. W. (2011). Toward a model of social influence that explains minority student integration into the scientific community. *Journal of Educational Psychology*, *103*(1), 206–222. doi:10.1037/a0020743 PMID:21552374

Ewell, E., Haglid, H., Truszkowski, E., Walicki, C., De Meulder, P., De Meulder, M., Stephens, T., Zhang, X., Trama, C., Hamilton, A., Mo, D., Wohner, J., Furrey, R., Labowsky, M., Perry, A., Hu, S., & Labowsky, H. L. (2022). Invention as a complement to high school chemistry. *Journal of Chemical Education*, *99*(5), 2012–2019. doi:10.1021/acs.jchemed.2c00135

Exploratorium Teacher Institute. (n.d.). *Ocean Acidification in a Cup*. Exploratorium. https://www.exploratorium.edu/snacks/ocean-acidification-in-cup

Felten, P., & Clayton, P. H. (2011). Service-learning. *New Directions for Teaching and Learning*, *2011*(128), 75–84. doi:10.1002/tl.470

Ferry, T. R., Fouad, N., & Smith, P. L. (2000). The role of family context in a social cognitive model for career-related choice behavior: A math and science perspective. *Journal of Vocational Behavior*, *57*(3), 348–364. doi:10.1006/jvbe.1999.1743

Field, A. (2013). *Discovering statistics using IBM SPSS statistics* (4th ed.). Sage Publications.

Fila, N. D., Hess, J., Hira, A., Joslyn, C. H., Tolbert, D., & Hynes, M. M. (2014). The people part of engineering: Engineering for, with, and as people. *Proceedings of the IEEE Frontiers in Education Conference (FIE)*. IEEE. 10.1109/FIE.2014.7044106

Finley, A. P., & McNair, T. (2013). *Assessing underserved students' engagement in high-impact practices*. Association of American Colleges and Universities.

Fitzgerald, H. E., Bruns, K., Sonka, S. T., Furco, A., & Swanson, L. (2016). The centrality of engagement in higher education: Reflections and future directions. *Journal of Higher Education Outreach & Engagement*, *20*(1), 245–253. https://files.eric.ed.gov/fulltext/EJ1097202.pdf

Flores, L. Y., Navarro, R. L., & Ali, S. R. (2017). The state of SCCT research in relation to social class: Future directions. *Journal of Career Assessment*, *25*(1), 6–23. doi:10.1177/1069072716658649

Foresite Group. (2014). *How to teach your students about civil engineering in 5 short steps*. Foresitre Group. http://www.foresitegroup.net/civil-engineering-class/

Forsyth, S., Penzkover, K., Dalton, B., & Yang, G. (2020b). Build a Better Book: The value of adding meaning to the making. *Connected Science Learning, 2*(3), 0-0. https://www.nsta.org/connected-science-learning/connected-science-learning-july-september-2020/build-better-book

Forsyth, S., Penzkover, K., Davis, P., & Matty, I. (2020). Design thinking in public library makerspaces. *Public Libraries*, *59*(4), 58–64.

Fortus, D., & Vedder-Weiss, D. (2012). Measuring students' continuing motivation for science learning. *Journal of Research in Science Teaching*, *4*(51), 497–522. doi:10.1002/tea.21136

Fouad, N. A., & Santana, M. C. (2017). SCCT and underrepresented populations in STEM fields: Moving the needle. *Journal of Career Assessment*, *25*(1), 24–39. doi:10.1177/1069072716658324

Francis, J. J., Johnston, M., Robertson, C., Glidewell, L., Entwistle, V., Eccles, M. P., & Grimshaw, J. M. (2010). What is an adequate sample size? Operationalising data saturation for theory-based interview studies. *Psychology & Health, 25*(10), 1229–1245. doi:10.1080/08870440903194015 PMID:20204937

Frenzel, A. C., Goetz, T., Pekrun, R., & Watt, H. M. G. (2010). Development of mathematics interest in adolescence: Influences of gender, family, and school context. *Journal of Research on Adolescence, 20*(2), 507–537. doi:10.1111/j.1532-7795.2010.00645.x

Friday Institute for Educational Innovation. (2012a). *Student Attitudes toward STEM Survey-Middle and High School Students*. Author.

Friday Institute for Educational Innovation. (2012b). *Student Attitudes Toward STEM (S-STEM) Survey: Tips for Using Your Data*. Author.

Fuesting, M. A., Diekman, A. B., & Hudiburgh, L. (2017). From classroom to career: The unique role of communal processes in predicting interest in STEM careers. *Social Psychology of Education, 20*(4), 875–896. doi:10.100711218-017-9398-6

Fuller, B., & Kim, H. (2022). *Systems thinking to transform schools: Identifying levers that lift educational quality*. Brookings Institution. https://www.brookings.edu/research/systems-thinking-to-transform-schools-identifying-levers-that-lift-educational-quality/

Furco, A. (1996). Service learning: A balanced approach to experiential education. In B. Taylor & Corporation for National Service (Eds.), Expanding boundaries: Serving and learning (pp. 2-6). Corporation for National Service.

Furtak, E. M. (2006). The problem with answers: An exploration of guided scientific inquiry teaching. *Science Education, 90*(3), 453–467. doi:10.1002ce.20130

Garfin, D. R. (2020). Technology as a coping tool during the coronavirus disease 2019 (COVID-19) pandemic: Implications and recommendations. *Stress and Health, 36*(4), 555–559. doi:10.1002mi.2975 PMID:32762116

Garner, P. W., Gabitova, N., Gupta, A., & Wood, T. (2018). Innovations in science education: Infusing social emotional principles into early STEM learning. *Cultural Studies of Science Education, 13*(4), 889–903. doi:10.100711422-017-9826-0

Gasbarra, P., & Johnson, J. (2008). *Out Before the Game Begins Hispanic Leaders Talk about What's Needed to Bring More Hispanic Youngsters Into Science, Technology and Math Professions* (Public Agenda Report). IBM.

George Mason University College of Science. (n.d.). *Learning assistants*. GMU. https://science.gmu.edu/academics/special-study-options/learning-assistants

Gershenson, S., Hart, C. M., Hyman, J., Lindsay, C., & Papageorge, N. W. (2018). The long-run impacts of same-race teachers (No. w25254). National Bureau of Economic Research.

Ghaffarzadegan, N., Larson, R., & Hawley, J. (2017). Education as a complex system. *Systems Research and Behavioral Science, 34*(3), 211–215. doi:10.1002res.2405 PMID:28522920

Ghazzawi, D., Pattison, D., & Horn, C. L. (2021). Persistence of Underrepresented Minorities in STEM Fields: Are Summer Bridge Programs Sufficient? *Frontiers in Education, 6*, 630529. Advance online publication. doi:10.3389/feduc.2021.630529

Gholson, M. L.Maisie L. Gholson. (2016). Clean corners and algebra: A critical examination of the constructed invisibility of black girls and women in mathematics. *The Journal of Negro Education, 85*(3), 290–301. doi:10.7709/jnegroeducation.85.3.0290

Gholson, M. L., & Martin, D. B. (2019). Blackgirl face: Racialized and gendered performativity in mathematical contexts. *ZDM*, *51*(3), 391–404. doi:10.100711858-019-01051-x

Gilbert, A., & Yerrick, R. (2001). Same school, separate worlds: A sociocultural study of identity, resistance, and negotiation in a rural, lower track science classroom. *Journal of Research in Science Teaching*, *38*(5), 574–598. doi:10.1002/tea.1019

Gilmartin, S. K., Li, E., & Aschbacher, P. (2006). The relationship between interest in physical science/engineering, science class experiences, and family contexts: Variations by gender and race / ethnicity among secondary students. *Journal of Women and Minorities in Science and Engineering*, *12*(2-3), 179–207. doi:10.1615/JWomenMinorScienEng.v12.i2-3.50

Girls Inspired and Ready to Lead, Incorporated. (n.d.). *About GIRL*. GIRL. https://girlsinc.org/about-us/

Glatthorn, A. A. (1993). Perspectives and imperatives - Outcomes-based education: Reform and the curriculum process. *Journal of Curriculum and Supervision*, *8*(4), 354–363.

Glatthorn, A. A. (1999). Curriculum alignment revisited. *Journal of Curriculum and Supervision*, *15*(1), 26–34.

Goel, A. K., Rugaber, S., & Vattam, S. (2009). Structure, behavior, and function of complex systems: The structure, behavior, and function modeling language. *Artificial Intelligence for Engineering Design, Analysis and Manufacturing*, *23*(1), 23–35. doi:10.1017/S0890060409000080

Gomoll, A., Hmelo-Silver, C. E., Šabanović, S., & Francisco, M. (2016). Dragons, ladybugs, and softballs: Girls' STEM engagement with human-centered robotics. *Journal of Science Education and Technology*, *25*(6), 899–914. doi:10.100710956-016-9647-z

Goodwin, T. (2016). Educating for ecological literacy. *The American Biology Teacher*, *78*(4), 287–291. doi:10.1525/abt.2016.78.4.287

Goudie, C. A. (2009). Institutionalizing ecological literacy: A critical component of the new "green" economy. *Business Perspectives*, *19*(3), 16–21.

Greenberg Motamedi, J., & Singh, M. (2016). *Oregon MESA: Improving grades in science and mathematics*. Education Northwest. https://files.eric.ed.gov/fulltext/ED606319.pdf

Greenberg Motamedi, J., Serrano, D., & Hanson, H. (2020). *Oregon MESA: Increasing the odds of high school graduation*. Education Northwest. https://files.eric.ed.gov/fulltext/ED606318.pdf

Green, K., Roller, C., & Cubeta, M. (2019). A plethora of fungi: Teaching a middle school unit on fungi. *Science Activities*, *56*(2), 52–62. doi:10.1080/00368121.2019.1682961

Grigg, W. S., Lauko, M. A., & Brockway, D. M. (2006). *The Nation's Report Card, Science 2005: Assessment of Student Performance in Grades 4, 8, and 12*. U.S. Department of Education, National Center for Education Statistics.

Grinis, I. (2019). The STEM requirements of "Non-STEM" jobs: Evidence from UK online vacancy postings. *Economics of Education Review*, *70*, 144–158. doi:10.1016/j.econedurev.2019.02.005

Haak, D. C., HilleRisLambers, J., Pitre, E., & Freeman, S. (2011). Increased structure and active learning reduce the achievement gap in introductory biology. *Science*, *332*(6034), 1213–1216. doi:10.1126cience.1204820 PMID:21636776

Halliday, A. S. (2019). *The Black Girlhood Studies Collection*. Women's Press.

Halverson, E. R., & Sheridan, K. (2014). The maker movement in education. *Harvard Educational Review*, *84*(4), 495–504. doi:10.17763/haer.84.4.34j1g68140382063

Han, J., Kelley, T., & Knowles, J. G. (2021). Factors influencing student STEM learning: Self-efficacy and outcome expectancy, 21st century skills, and career awareness. *Journal for STEM Education Research*, *4*(2), 117–137. doi:10.100741979-021-00053-3

Harding, S. (1992). After the neutrality ideal: Science, politics, and" strong objectivity. *Social Research*, *59*(3), 567–587.

Harding, S. G. (2004). *The feminist standpoint theory reader: Intellectual and political controversies*. Psychology Press.

Hargrave, C. (2015). Counter Space: Analysis of Educational Structures of an After-School Program that Fosters Black Academic Success Narratives. *The Journal of Negro Education*, *84*(3), 348–361. doi:10.7709/jnegroeducation.84.3.0348

Harper, S. R., & Griffin, K. A. (2011). Opportunity beyond affirmative action: How low-income and working class Black male achievers access highly selective, high-cost colleges and universities. *Harvard Journal of African American Public Policy*, *17*(1), 43–60.

Hayden, K., Ouyang, Y., Scinski, L., Olszewski, B., & Bielefeldt, T. (2011). Increasing student interest and attitudes in STEM: Professional development and activities to engage and inspire learners. *Contemporary Issues in Technology & Teacher Education*, *11*(1), 47–69.

Heisey, N., & Kucan, L. (2010). Introducing science concepts to middle childhood students through read-alouds: Interactions and multiple texts make the difference. *The Reading Teacher*, *63*(8), 666–676. doi:10.1598/RT.63.8.5

Hernandez-Gantes, V. M., Fletcher, E. C. Jr, & Keighobadi, S. (2019). A case study of the development of a career academy: Good intentions not enough? *Journal of Research in Technical Careers*, *3*(2), 114–135. doi:10.9741/2578-2118.1059

Hernández-Hernández, A. M., Viga-de Alva, D., Huerta-Quintanilla, R., Canto-Lugo, E., Laviada-Molina, H., & Molina-Segui, F. (2016). Friendship concept and community network structure among elementary school and university students. *PLoS One*, *11*(10), 1–17. doi:10.1371/journal.pone.0164886 PMID:27760171

Hess, J. L., & Fila, N. D. (2016). The development and growth of empathy among engineering students. *Proceedings of the American Society for Engineering Education ASEE Annual Conference & Exposition.* ASEE. 10.18260/p.26120

Hidi, S., & Renninger, K. A. (2006). The four-phase model of interest development. *Educational Psychologist*, *41*(2), 111–127. doi:10.120715326985ep4102_4

Hinton, A. O., Termini, C. M., Spencer, E. C., Rutaganira, F. U. N., Chery, D., Roby, R., Vue, Z., Pack, A. D., Lillian, J. B., Garza-Lopez, E., Marshall, A. G., Lewis, S. C., Shuler, H. D., Taylor, B. L., & McReynolds, M. R. (2020). Patching the Leaks: Revitalizing and Reimagining the STEM Pipeline. *Cell*, *183*(3), 568–575. doi:10.1016/j.cell.2020.09.029 PMID:33125882

Hite, R. (2020). *Curriculum needs assessment for formal settings* [EDCI 5321- Curriculum Theory: Design and Development]. Texas Tech University.

Hite, R. L., & Spott, J. L. (2022). Improving Parents' and Teachers' Perceptions of Girls' STEM Activities and Interests before and after an Informal STEM Intervention. *Journal of STEM Outreach*, *5*(1), 1–13. doi:10.15695/jstem/v5i1.01

Hite, R., Spott, J., Johnson, L., & Sobehrad, L. (2020). STEM Challenge: Two Years of Community-Engaged Engineering. *Journal of Research In Innovative Teaching & Learning*, *13*(1), 57–82. doi:10.1108/JRIT-12-2019-0080

Hoffman, A. J., McGuire, L., Rutland, A., Hartstone-Rose, A., Irvin, M. J., Winterbottom, M., Balkwill, F., Fields, G. E., & Mulvey, K. L. (2020). The relations and role of social competencies and belonging with math and science interest and efficacy for adolescents in informal stem programs. *Journal of Youth and Adolescence*, *50*(2), 314–323. doi:10.100710964-020-01302-1 PMID:32804294

Honig, M. & McDonald, M. (2015). *Youth Makerspace Playbook.* Maker Ed. http://makered.org/wp-content/uploads/2015/09/Youth-Makerspace-Playbook_FINAL.pdf

Honig, M., & McDonald, M. (2005). From promise to participation: Afterschool programs through the lens of socio-cultural learning theory. *Afterschool Matters, Occasional Paper Series.* WCW. https://niost.wcwonline.org/pdf/afterschoolmatters/asm_2005_op5_fall/asm_2005_op5_fall-1.pdf

Hooks, B. (1994). *Teaching to Transgress.* Routledge.

Hora, M. T., & Ferrare, J. J. (2013). Instructional systems of practice: A multidimensional analysis of math and science undergraduate course planning and classroom teaching. *Journal of the Learning Sciences, 22*(2), 212–257. doi:10.1080/10508406.2012.729767

Howard, A., Swalwell, K., & Adler, K. (2018). Making class: Children's perceptions of social class through illustrations. *Teachers College Record, 120*(7), 1–44. doi:10.1177/016146811812000704

Hoyer, K. M., & Sparks, D. (2017). *Instructional time for third and eighth graders in public and private schools: School year 2011-12.* Department of Education.

Huang, G., Taddese, N., & Walter, E. (2000). Entry and persistence of women and minorities in college science and engineering education (Research and Development Report NCES 2000-601). National Center for Education Statistics.

Hynes, M., & Maxey, K. R. (2018, June 23). Investigating the fit between students' personal interests and their perceptions of engineering in a National Society of Black Engineers (NSBE) pre-college summer workshop. *ASEE Annual Conference & Exposition, Salt Lake City, Utah.* ASEE. https://peer.asee.org/30727

Hynes, M., Joslyn, C., Hira, A., Holly, J., & Jubelt, N. (2016). Exploring diverse pre-college students' interests and understandings of engineering to promote engineering education for all. *International Journal of Engineering Education, 32*, 2318–2327.

Inda, M., Rodríguez, C., & Peña, J. V. (2013). Gender differences in applying social cognitive career theory in engineering students. *Journal of Vocational Behavior, 83*(3), 346–355. doi:10.1016/j.jvb.2013.06.010

International Society for Technology in Education. (2016). *ISTE National Educational Technology Standards for Students.* Author.

Jackson, C., Mohr-Schroeder, M. J., Bush, S. B., Maiorca, C., Roberts, T., Yost, C., & Fowler, A. (2021). Equity-oriented conceptual framework for K-12 STEM literacy. *International Journal of STEM Education, 8*(1), 1–16. doi:10.118640594-021-00294-z

Jagers, R. J., Skoog-Hoffman, A., Barthelus, B., & Schlund, J. (2021). Transformative social emotional learning: In pursuit of educational equity and excellence. *American Educator, 45*(2), 12–17, 39.

Januszyk, R., Miller, E. C., & Lee, O. (2016). Addressing student diversity and equity: The next generation science standards are leading a new wave of reform. *Science Scope, 39*(8), 16–19. https://www.jstor.org/stable/43827310. doi:10.2505/4s16_039_08_16

Javidi, G., & Sheybani, E. (2017). Empowering teachers to raise career awareness in computing: Lessons learned. Systemics. *Cybernetics and Informatics, 15*(3), 10–15.

Jones, K. L., & Hite, R. (2021). A Global Comparison of the Circumscription and Compromise Theory of Career Development in Science Career Aspirations. *School Science and Mathematics, 121*(7), 381–394. doi:10.1111sm.12492

Jones, K. L., & Hite, R. (2022). Exploring When, Why and How Students Circumscribe and Compromise Science Careers. *The Career Development Quarterly*. doi:10.1002/cdq.12311

Jones, K. S., Newman, D. A., Su, R., & Rounds, J. (2021). Black-White differences in vocational interests: Meta-analysis and boundary conditions. *Journal of Business and Psychology*, *36*(4), 589–607. doi:10.100710869-020-09693-5

Jones, M. G., Chesnutt, K., Ennes, M., Mulvey, K. L., & Cayton, E. (2021). Understanding science career aspirations: Factors predicting future science task value. *Journal of Research in Science Teaching*, *58*(7), 937–955. doi:10.1002/tea.21687

Jung, Y. J., Zimmerman, H. T., & Land, S. M. (2019). Emerging and developing situational interest during children's tablet-mediated biology learning activities at a nature center. *Science Education*, *103*(4), 900–922. doi:10.1002ce.21514

Kafai, Y., Fields, D., & Searle, K. (2014). Electronic textiles as disruptive designs: Supporting and challenging maker activities in schools. *Harvard Educational Review*, *84*(4), 532–556. doi:10.17763/haer.84.4.46m7372370214783

Kampe, J., & Oppliger, D. E. (2011, June). *Effectiveness of team-based STEM project learning to recruit minority high school students to STEM. 2011 ASEE Annual Conference & Exposition*, Vancouver, BC. 10.18260/1-2--17820

Kandlhofer, M., & Steinbauer, G. (2016). Evaluating the impact of educational robotics on pupils' technical-and social-skills and science related attitudes. *Robotics and Autonomous Systems*, *75*, 679–685. doi:10.1016/j.robot.2015.09.007

Kang, H., Calabrese Barton, A., Tan, E., Simpkins, S., Rhee, H., & Turner, C. (2018). How do middle school girls of color develop STEM identities? Middle school girls' participation in science activities and identification with STEM careers. *Science Education*, *103*(2), 418–439. doi:10.1002ce.21492

Karimi, Z., Lakin, J. M., & Davis, V. A. (2022). Water Quality: Adaptable Modules for Engaging K-16 Students. *2022 IEEE Frontiers in Education Conference*. IEEE. 10.1109/FIE56618.2022.9962639

Kasza, P., & Slater, T. F. (2017). A survey of best practices and key learning objectives for successful secondary school STEM academy settings. *Contemporary Issues in Education Research*, *10*(1), 53–66. doi:10.19030/cier.v10i1.9880

Kelley, T. R., Knowles J. G., Holland, J. D., & Han, J. (2020). Increasing high school teachers self-efficacy for integrated STEM instruction through a collaborative community of practice. *International Journal of STEM Education, 7*(14). doi:10.1186/s40594-020-00211-w

Kelley, T. R., & Knowles, J. G. (2016). A conceptual framework for integrated STEM education. *International Journal of STEM Education*, *3*(1), 1–11. doi:10.118640594-016-0046-z

Kelley, T. R., Sung, E., Han, J., & Knowles, J. G. (2022). Impacting secondary students' STEM knowledge through collaborative STEM teacher partnerships. *International Journal of Technology and Design Education*. doi:10.100710798-022-09783-w

Kennedy, T. J., & Odell, M. R. (2014). Engaging students in STEM education. *Science Education International*, *25*(3), 246–258.

Khoo, H. E., Azlan, A., Tang, S. T., & Lim, S. M. (2017). Anthocyanidins and anthocyanins: Colored pigments as food, pharmaceutical ingredients, and the potential health benefits. *Food & Nutrition Research*, *61*(1), 1361779–1361779. doi:10.1080/16546628.2017.1361779 PMID:28970777

Kier, M. W. (2013). *Examining the effects of a STEM career video intervention on the interests and STEM professional identities of rural, minority middle school students* [Doctoral dissertation, North Carolina State University].

Kier, M. W., & Blanchard, M. R. (2020). Eliciting students' voices through STEM career explorations. *International Journal of Science and Mathematics Education*, *19*(1), 151–169. doi:10.100710763-019-10042-z

Kier, M. W., Blanchard, M. R., Osborne, J. W., & Albert, J. L. (2014). The development of the STEM career interest survey (STEM-CIS). *Research in Science Education*, *44*(3), 461–481. doi:10.100711165-013-9389-3

Kim, A. Y., & Sinatra, G. M. (2018). Science identity development: An interactionist approach. *International Journal of STEM Education*, *5*(51), 1–6. doi:10.118640594-018-0149-9 PMID:30631740

King, N. S., & Pringle, R. M. (2019). Black girls speak STEM: Counterstories of informal and formal learning experiences. *Journal of Research in Science Teaching*, *56*(5), 539–569. doi:10.1002/tea.21513

Kisselburgh, L. G., Berkelaar, B. L., & Buzzanell, P. M. (2009). Discourse, gender, and the meaning of work rearticulating science, technology, and engineering careers through communicative lenses. *Annals of the International Communication Association*, *33*(1), 259–299. doi:10.1080/23808985.2009.11679089

Knapp, T. R. (2016). Why Is the one-group pretest–posttest design still used? *Clinical Nursing Research*, *25*(5), 467–472. doi:10.1177/1054773816666280 PMID:27558917

Knight, D. B., Mappen, E. F., & Knight, S. L. (2011). A review of the literature on increasing the representation of women undergraduates in STEM disciplines through civic engagement pedagogies. *Science Education & Civic Engagement: An International Journal*, *3*(1), 36–47.

Krishnamurthi, A., Bevan, B., Rinehart, J., & Coulon, V. R. (2013). What Afterschool STEM Does Best: How Stakeholders Describe Youth Learning Outcomes. *Afterschool Matters*, *18*, 42–49.

Kurz, M. E., Yoder, S. E., & Zu, L. (2015). Effects of exposure on attitudes towards STEM interests. *Education*, *136*(2), 229–241. doi:10.1643/0045-8511(2002)002

Ladson-Billings, G. (1995). Toward a Theory of Culturally Relevant Pedagogy. *American Educational Research Journal*, *32*(3), 465–491. doi:10.3102/00028312032003465

Ladson-Billings, G. (2013). Lack of achievement or loss of opportunity? In P. L. Carter & K. G. Welner (Eds.), *Closing the opportunity gap: What America must do to give all children an even chance* (pp. 11–24). Oxford University Press. doi:10.1093/acprof:oso/9780199982981.003.0002

Lakin, J. M., Wittig, A. H., Davis, E. W., & Davis, V. A. (2020). Am I an engineer yet? Perceptions of engineering and identity among first year students. *European Journal of Engineering Education*, *45*(2), 214–231. doi:10.1080/03043797.2020.1714549

Lakin, J., Marghitu, D., Davis, V., & Davis, E. (2021). Introducing engineering as an altruistic STEM Career. *Science Teacher (Normal, Ill.)*, *88*(4), 29–35.

Lakin, J., Mastrogiovanni, M., Davis, E., Marghitu, D., & Davis, V. (2022). It makes me wanna go out and help people": Enhancing African American Youth's Perceptions of Engineering. *Journal of Women and Minorities in Science and Engineering*, *28*(1), 41–74. doi:10.1615/JWomenMinorScienEng.2021036696

Lane, T. B., & Id-Deen, L. (2023). Nurturing the capital within: A qualitative investigation of Black women and girls in STEM summer programs. *Urban Education*, *58*(6), 1298–1326. doi:10.1177/0042085920926225

Larnell, G. V., Bullock, E. C., & Jett, C. C. (2016). Rethinking teaching and learning mathematics for social justice from a critical race perspective. *Journal of Education*, *196*(1), 19–29. doi:10.1177/002205741619600104

Leake, E. (2019). Empathy as research methodology. In P. Liamputtong (Ed.), *Handbook of research methods in health social sciences* (pp. 237–252). Springer., doi:10.1007/978-981-10-5251-4_65

Lederman, N. G., & Abell, S. K. (Eds.). (2014). *Handbook of Research on Science Education*. Routledge.

Lemke, J. L. (2001). Articulating communities: Sociocultural perspectives on science education. *Journal of Research in Science Teaching, 38*(3), 296–316. doi:10.1002/1098-2736(200103)38:3<296::AID-TEA1007>3.0.CO;2-R

Lemus, J. D., Bishop, K., & Walters, H. (2010). QuikSCience: Effective linkage of competitive, cooperative, and service learning in science education. *American Secondary Education, 38*(3), 40–61.

Lent, R. W. (2013). Social cognitive career theory. In S. Brown & R. Lent (Eds.), *Career development and counseling: Putting theory and research to work.* John Wiley & Sons.

Lent, R. W., & Brown, S. D. (2017). Social cognitive career theory in a diverse world: Guest editors' introduction. *Journal of Career Assessment, 25*(1), 3–5. doi:10.1177/1069072716657811

Lent, R. W., Brown, S. D., & Hackett, G. (1994). Toward a unifying social cognitive theory of career and academic interest, choice, and performance. *Journal of Vocational Behavior, 45*(1), 79–122. doi:10.1006/jvbe.1994.1027

Lent, R. W., Brown, S. D., Sheu, H.-B., Schmidt, J., Brenner, B. R., Gloster, C. S., Wilkins, G., Schmidt, L. C., Lyons, H., & Treistman, D. (2005). Social cognitive predictors of academic interests and goals in engineering: Utility for women and students at historically black universities. *Journal of Counseling Psychology, 52*(1), 84–92. doi:10.1037/0022-0167.52.1.84

Leonard, J., Chamberlin, S., Johnson, J., & Verma, G. (2016). Social Justice, Place, and Equitable Science Education: Broadening Urban Students' Opportunities to Learn. *The Urban Review, 48*(3), 355–379. doi:10.100711256-016-0358-9

Lesseig, K., Nelson, T. H., Slavit, D., & Seidel, R. A. (2016). Supporting middle school teachers' implementation of STEM design challenges. *School Science and Mathematics, 116*(4), 177–188. doi:10.1111sm.12172

Letourneau, S. M., & Bennett, D. (2020). Using narratives to evoke empathy and support girls' engagement in engineering. *Connected Science Learning, 2*(3), 0-0. https://www.nsta.org/connected-science-learning/connected-science-learning-july-september-2020/using-narratives-evoke

Litow, S. S., & Kelley, T. (2021). *Breaking barriers: How P-TECH schools create a pathway from high school to college to career.* Teachers College Press.

Luo, W., Wei, H. R., Ritzhaupt, A. D., Huggins-Manley, A. C., & Gardner-McCune, C. (2019). Using the S-STEM survey to evaluate a middle school robotics learning environment: Validity evidence in a different context. *Journal of Science Education and Technology, 28*(4), 429–443. doi:10.100710956-019-09773-z

Lydian Hill Foundation & Geena Davis Institute on Gender in Media. (2018). *Portray her: Representations of women STEM characters in media.* See Jane. https://seejane.org/wp-content/uploads/portray-her-full-report.pdf

Lyon, G. H., Jafri, J., & Louis, K. S. (2013). Beyond the pipeline: STEM Pathways for youth development. *Afterschool Matters, 16*, 48–57.

Maietta, R. C. (2006). State of the art: Integrating software with qualitative analysis. In L. Curry, R. Shield, & T. Wetle (Eds.), *Improving aging and public health research: Qualitative and mixed methods* (pp. 117–139). American Public Health Association and the Gerontological Society of America.

Maltese, A., & Tai, R. (2011). Pipeline persistence: Examining the association of educational experiences with earned degrees in STEM among US students. *Science Education, 95*(5), 877–907. doi:10.1002ce.20441

Margolis, J. (2008). *Stuck in the shallow end: Education, race, and computing.* MIT Press.

Markle, R. S., Williams, T. M., Williams, K. J., deGravelles, K. H., Bagayoko, D., & Warner, I. M. (2022). Supporting Historically Underrepresented Groups in STEM Higher Education: The Promise of Structured Mentoring Networks. *Frontiers in Education, 7*, 674669. doi:10.3389/feduc.2022.674669

Martin, L., & Dixon, C. (2013). *Youth conceptions of making and the Maker Movement.* Exploratorium. https://www.exploratorium.edu/sites/default/files/pdfs/IDC_2013_Martin_Dixon.pdf

Master, A. H., & Meltzoff, A. N. (2020). Cultural stereotypes and sense of belonging contribute to gender gaps in STEM. *International Journal of Gender, Science and Technology*, *12*, 152–198.

McClure, E., Guernsey, L., Clements, D., Bales, S., Nichols, J., Kendall-Taylor, N., & Levine, M. (2017). How to integrate stem into early childhood education. *Science and Children*, *55*(2), 8–10. doi:10.2505/4c17_055_02_8

McCreedy, D., & Dierking, L. D. (2014). *Cascading Influences: Long term impacts of informal STEM experiences for girls.* Paper presented at the 27th Annual Visitor Studies Association Conference, Albuquerque, New Mexico.

Meadows, D. H., & Wright, D. (2008). *Thinking in systems: A primer.* Chelsea Green Pub.

Melchior, A., Burack, C., Gutbezahl, J., Hoover, M., & Marcus, J. (2015). *FIRST longitudinal study: Summary of preliminary findings–year 2.* The Center for Youth and Communities at Brandeis University.

Meyer, A., Rose, D. H., & Gordon, D. (2014). *Universal design for learning: Theory and practice.* CAST Professional Publishing.

Michigan Department of Natural Resources. (2023). *Salmon in the classroom.* MDNR. https://www.michigan.gov/dnr/education/educators/sic#:~:text=The%20Michigan%20DNR%20Salmon%20in,great%20natural%20resources%20each%20year

Miller, K., Sonnert, G., & Sadler, P. (2018). The influence of students' participation in STEM competitions on their interest in STEM careers. *International Journal of Science Education. Part B*, *8*(2), 95–114.

Milner, H. R. (2021). *Start where you are, but don't stay there: Understanding diversity, opportunity gaps, and teaching in today's classrooms.* Harvard Education Press.

Mitra, D. L. (2018). *Educational change and the political process.* Routledge, Taylor & Francis Group.

Mitra, D. L. (2009). The role of intermediary organizations in sustaining student voice initiatives. *Teachers College Record*, *111*(7), 1834–1870. doi:10.1177/016146810911100708

Mohtar, R. H., & Daher, B. (2012). Water, energy, and food: The ultimate nexus. In Encyclopedia of Agricultural, Food, and Biological Engineering (2nd ed., pp. 1-5). CRC Press, Taylor, and Francis Group. https://doi.org/ doi:0.1081/E-EAFE2-120048376

Montoya, J., Lundell, R., Peterson, F., Tarantino, S., Ramsey, M., Katz, G., Baldini, R., Fruchter, R., & Fischer, M. (2018). Building sustainable communities: A project-based learning approach to modify student perceptions of the building industry. *ACEEE Summer Study on Energy Efficiency in Buildings, 11*, 1–13. https://www.aceee.org/files/proceedings/2018/index.html#/paper/event-data/p342

Morel, R. P., Coburn, C., Catterson, A. K., & Higgs, J. (2019). The multiple meanings of scale: Implications for researchers and practitioners. *Educational Researcher*, *48*(6), 369–377. doi:10.3102/0013189X19860531

Moreno, N., Tharp, B., Vogt, G., Newell, A., & Burnett, C. (2016). Preparing Students for Middle School Through After-School STEM Activities. *Journal of Science Education and Technology*, *25*(6), 889–898. doi:10.100710956-016-9643-3

Morris, M. W. (2015). The relevance of sacred inquiry in the education of delinquent Black girls. *International Journal of Human Resources Development and Management*, *15*(2-4), 185–193. doi:10.1504/IJHRDM.2015.071156

Morris, M. W. (2016). *Pushout: The criminalization of Black girls in schools.* The New Press.

Morris, M. W. (2019). *Sing a rhythm, dance a blues: Education for the liberation of Black and Brown girls*. The New Press.

Musters, C. J. M., de Graaf, H. J., & ter Keurs, W. J. (1998). Defining socio-environmental systems for sustainable development. *Ecological Economics*, *26*(3), 243–258. doi:10.1016/S0921-8009(97)00104-3

Myers, R. E. III, & Fouts, J. T. (1992). A cluster analysis of high school science classroom environments and attitude toward science. *Journal of Research in Science Teaching*, *29*(9), 929–937. doi:10.1002/tea.3660290904

Nadelson, L., Callahan, J., Pyke, P., Hay, A., & Schrader, C. (2009). A SySTEMic solution: Elementary teacher preparation in STEM expertise and engineering awareness. *American Society for Engineering Education*, *106*(2), 157–168. doi:10.18260/1-2--5080

National Academies of Sciences, Engineering, and Medicine. (2019). *Minority serving institutions: America's underutilized resource for strengthening the STEM workforce*. The National Academies Press. doi:10.17226/25257

National Academies of Sciences, Engineering, and Medicine. (2021). *Science and Engineering in Preschool Through Elementary Grades*. National Academies Press. doi:10.17226/26215

National Academy of Engineering. (2008). *Changing the conversation: Messages for improving public understanding of engineering*. The National Academies Press.

National Academy of Sciences. National Academy of Engineering, & Institute of Medicine. (2011). *Expanding underrepresented minority participation: America's science and technology talent at the crossroads.* The National Academies Press. https://doi.org/doi:10.17226/12984

National Center for Education Statistics. (2006). *National Education Longitudinal Study of 1988.* NCES. doi:10.3886/ICPSR09389.v1

National Center for Education Statistics. (2008). *Average number of hours in the school day and average number of days in the school year for public schools, by state: 2007-08.* NCES. https://nces.ed.gov/surveys/sass/tables/sass0708_035_s1s.asp

National Center for Science and Engineering Statistics. (2019). *Women, minorities, and persons with disabilities in science and engineering: 2019.* N. S. Foundation. https:// www.nsf.gov/statistics/wmpd

National Center for Science and Engineering Statistics. (2023). *Diversity and STEM: Women, Minorities, and Persons with Disabilities* (NSF 23-315). National Science Foundation. https://ncses.nsf.gov/pubs/nsf23315/

National Governors Association Center for Best Practices, Council of Chief State School Officers. (2010). Common Core State Standards (English Language Arts Standards and Mathematics Standards for Middle School. National Governors Association Center for Best Practices, Council of Chief State School Officers.

National Research Council. (2012). *A Framework for K-12 science education: Practices, crosscutting concepts, and Core ideas*. The National Academies Press.

National Research Council. (2013). *Next Generation Science Standards: For States, By States*. The National Academies Press. doi:10.17226/18290

National Research Council. (2015). *Identifying and supporting productive STEM programs in out-of-school settings.* National Academies Press.

National Science and Technology Council. (2018). *Charting a course for success: America's strategy for STEM education.* Whitehouse. https://www.whitehouse.gov/ostp

National Science Board. (2018). *Science & engineering indicators.* National Science Foundation. https://nsf.gov/statistics/2018/nsb20181/report/sections/science-and-engineering-labor-force/women-and-minorities-in-the-s-e-workforce#minorities-in-the-s-e-workforce

National Science Board. (2021). *The STEM labor force of today: Scientists, engineers, and skilled technical workers* (NSB-2021-2; Science and Engineering Indicators 2022). National Science Foundation. https://ncses.nsf.gov/pubs/nsb20212

National Science Board. (2022). *Science and engineering indicators 2022: The state of U.S. science and engineering.* (NSB-2022-1). National Science Foundation. https://ncses.nsf.gov/pubs/nsb20221

National Science Foundation (2015). *America's Future: Environmental Research and Education for A Thriving Century. A 10 Year Outlook: A Report by the NSF Advisory Committee for Environmental Research and Education.* NSF.

National Science Foundation. (n.d.). Inclusion across the Nation of Communities of Learners of Underrepresented Discoverers in Engineering and Science. *NSF INCLUDES.* https://www.nsf.gov/news/special_reports/big_ideas/includes.jsp

Neild, R. C., Wilson, S. J., & McClanahan, W. (2019a). *Afterschool programs: A review of evidence under the Every Student Succeeds Act.* Research for Action. https://www.researchforaction.org/publications/afterschool-programs-a-review-of-evidence-under-the-every-student-succeeds-act/

Neild, R. C., Wilson, S. J., & McClanahan, W. (2019b). *Afterschool evidence guide: A companion to Afterschool programs. A review of evidence under the Every Student Succeeds Act.* Research for Action. https://www.researchforaction.org/publications/afterschool-evidence-guide-a-companion-to-a-review-of-evidence-under-the-every-student-succeeds-act/

Next Gen Science. (2013). *Next Generation Science Standards: For States, By States.* The National Academies Press. https://www.nextgenscience.org/

Nolte, K. (2008). *Red Cabbage.* University of Arizona. https:// cals.arizona.edu/fps/sites/cals.arizona.edu.fps/files/cotw/Red_Cabbage.pdf

Noonan, R. (2017). *STEM jobs: 2017 update.* Office of the Chief Economist, Economics and Statistics Administration (ESA Issue Brief #02-17).

Novak, G. M. (2011). Just-in-time teaching. *New Directions for Teaching and Learning, 2011*(128), 63–73. doi:10.1002/tl.469

Nugent, G., Barker, B., White, A., & Grandgenett, N. (2011, June). The impact of robotics competitions on youth STEM learning, attitudes and 21st century workplace skills. In EdMedia+ Innovate Learning (pp. 3614-3619). Association for the Advancement of Computing in Education (AACE).

Nugent, G., Barker, B., Grandgenett, N., & Welch, G. (2016). Robotics camps, clubs, and competitions: Results from a US robotics project. *Robotics and Autonomous Systems, 75*, 686–691. doi:10.1016/j.robot.2015.07.011

Nugent, G., Barker, B., Welch, G., Grandgenett, N., Wu, C., & Nelson, C. (2015). A model of factors contributing to STEM learning and career orientation. *International Journal of Science Education, 37*(7), 1067–1088. doi:10.1080/09500693.2015.1017863

O'Bryan, L., Bohn, D., & Shehab, S. (2022). A Human-Centered Design Tool Kit for STEM-Based Capstone Courses. *Journal of College Science Teaching, 51*(6), 23–28.

Oakes, J. (1990). *Multiplying inequalities: The effects of race, social class, and tracking on opportunities to learn mathematics and science.* Rand Corp.

Office for Civil Rights. (2017). *Civil rights data collection [Database].* U.S. Department of Education. https://ocrdata.ed.gov/

Onyeka-Crawford, A., Patrick, K., & Chaudhry, N. (2017). *Let her learn: Stopping school pushout for girls of color.* National Women's Law Center.

Orr, M., Hughes, K., & Karp, M. (2003). *Shaping postsecondary transitions: Influences of the National Academy Foundation career academy* (No. 29). Columbia University, Institute on Education and the Economy. https://files.eric.ed.gov/fulltext/ED477819.pdf

Ortiz, N. A., Morton, T. R., Miles, M. L., & Roby, R. S. (2019). What about us? Exploring the challenges and sources of support influencing Black students' STEM identity development in postsecondary education. *The Journal of Negro Education, 88*(3), 311–326. doi:10.7709/jnegroeducation.88.3.0311

Ottinger, R. (2022). *Why education investors shouldn't ignore afterschool STEM.* Forbes Business Council. https://www.forbes.com/sites/forbesbusinesscouncil/2022/03/01/why-education-investors-shouldnt-ignore-afterschool-stem/?sh=2cd2377aafa6

Pajares, F. (2006). Self-efficacy during childhood and adolescence. In F. Pajares & T. Urdan (Eds.), *Self-efficacy beliefs of adolescents* (Vol. 5, pp. 339–367). Information Age Publishing.

Parking, C., Scott, S., & Geddes, A. (2020). Snowball sampling. In P. Atkinson, S. Delamont, A. Cernat, J. Sakshaug, & R. Williams (Eds.), *SAGE Research Methods Foundations.* SAGE Publications Ltd., doi:10.4135/9781526421036831710

Penuel, W. R., Fishman, B. J., Haugan Cheng, B., & Sabelli, N. (2011). Organizing research and development at the intersection of learning, implementation, and design. *Educational Researcher, 40*(7), 331–337. doi:10.3102/0013189X11421826

Petrich, M., Wilkinson, K., & Bevan, B. (2013). It looks like fun, but are they learning? In M. Honey & D. Kanter (Eds.), *Design, Make, Play: Growing the Next Generation of Innovators* (pp. 50–70). Routledge.

Piburn, M. D., & Baker, D. R. (1993). If I were the teacher... qualitative study of attitude toward science. *Science Education, 77*(4), 393–406. doi:10.1002ce.3730770404

Pinkard, N., Erete, S., Martin, C., & McKinney de Royston, M. (2017). Digital Youth Divas: Exploring Narrative-Driven Curriculum to Spark Middle School Girls' Interest in Computational Activities. *Journal of the Learning Sciences, 26*(3), 477–516. doi:10.1080/10508406.2017.1307199

Playton, S. C. (2021). *Primary STEM Career Interest and Awareness: Adaptation and Validation of a Survey for Measuring STEM Career Interventions with Primary Students.* [Embargoed doctoral dissertation, Texas Tech University].

Playton, S. C., Hite, R., & Childers, G. (2023). *Measuring STEM Career Awareness and Interest in Middle Childhood STEM Learners: Validation of the STEM Futures Survey.*

Playton, S. C., Hite, R., & Leach, P. K. (2021). A case study of a researcher-practitioner partnership in teaching STEM+C to rural elementary students. *Electronic Journal For Research In Science & Mathematics Education, 25*(2), 30–46.

Playton, S., Hite, R., & Klonowski, P. (2021). A case study of a researcher-practitioner partnership in teaching STEM+C to rural elementary students. *Electronic Journal for Research in Science & Mathematics Education, 25*(2), 30–46.

Porfeli, E. J., & Lee, B. (2012). Career development during childhood and adolescence. *New Directions for Youth Development, 2012*(134), 11–22. doi:10.1002/yd.20011 PMID:22826162

President's Council of Advisors on Science and Technology (U.S.). (2012). *PCAST engage to excel: Producing one million additional college graduates with degrees in science, technology, engineering, and mathematics.* President's Council of Advisors on Science and Technology (PCAST).

Purzer, Ş. (2011). The relationship between team discourse, self-efficacy, and individual achievement: A sequential mixed-methods study. *Journal of Engineering Education, 100*(4), 655–679. doi:10.1002/j.2168-9830.2011.tb00031.x

Radinsky, J., Bouillion, L., Lento, E. M., & Gomez, L. M. (2001). Mutual benefit partnership: A curricular design for authenticity. *Journal of Curriculum Studies, 33*(4), 405–430. doi:10.1080/00220270118862

Rahm, J., Martel-Reny, M.-P., & Moore, J. C. (2005). The role of afterschool and community science programs in the lives of urban youth. *School Science and Mathematics, 105*(6), 283–291. doi:10.1111/j.1949-8594.2005.tb18129.x

Rahm, J., & Moore, J. (2015). A Case Study of Long-Term Engagement and Identity-in-Practice: Insights into STEM Pathways of Four Underrepresented Youths. *Journal of Research in Science Teaching, 53*(5), 768–801. doi:10.1002/tea.21268

Ray, R. (2022). School as a Hostile Institution: How Black and Immigrant Girls of Color Experience the Classroom. *Gender & Society, 36*(1), 88–111. doi:10.1177/08912432211057916

Regan, E., & DeWitt, J. (2014). Attitudes, interest and factors influencing STEM enrolment behaviour: An overview of relevant literature. In E. K. Henriksen, J. Dillon, & J. Ryder (Eds.), *Understanding Student Participation and Choice in Science and Technology Education* (pp. 63–88). Springer.

Rios, R., Artigas, A., & Casulleras, R. P. (2014). Establishing and maintaining STEM-related school-industry partnerships: Opportunities and obstacles. In C. P. Constantinou, N. Papadouris, & A. Hadjigeorgiou (Eds.), *E-Book Proceedings of the ESERA 2013 Conference: Science Education Research for Evidence-based Teaching and Coherence in Learning.* ESERA. https://www.esera.org/publications/esera-conference-proceedings/esera-2013#strand-9

Robbins, T., Stagman, S., & Smith, S. (2012). *Young children at risk: National and State Prevalence of Risk Factors Fact Sheet.* National Center for Children in Poverty: Mailman School of Public Health Columbia University. https:// www.nccp.org/publication/young-children-at-risk/

Rose, D. H., & Meyer, A. (2002). *Teaching every student in the digital age: Universal design for learning.* Association for Supervision and Curriculum Development.

Roth, W.-M., & Lee, S. (2004). Science education as/for participation in the community. *Science Education, 88*(2), 263–291. doi:10.1002ce.10113

Royce, C. A., Morgan, E., & Ansberry, K. (2012). *Teaching science through trade books.* NSTA Press.

Sadler, P. M., Sonnert, G., Hazari, Z., & Tai, R. (2012). Stability and volatility of STEM career interest in high school: A gender study. *Science Education, 96*(3), 411–427. doi:10.1002ce.21007

Sahin, A. (2013). STEM clubs and science fair competitions: Effects on post-secondary matriculation. *Journal of STEM Education: Innovations and Research, 14*(1), 5–11.

Sahin, A., Ekmekci, A., & Waxman, H. C. (2017). The relationships among high school STEM learning experiences, expectations, and mathematics and science efficacy and the likelihood of majoring in STEM in college. *International Journal of Science Education, 39*(11), 1549–1572. doi:10.1080/09500693.2017.1341067

Salam, M., Awang Iskandar, D. N., Ibrahim, D. H. A., & Farooq, M. S. (2019). Service learning in higher education: A systematic literature review. *Asia Pacific Education Review, 20*(4), 573–593. doi:10.100712564-019-09580-6

Salonen, A., Kärkkäinen, S., & Keinonen, T. (2018). Career-related instruction promoting students' career awareness and interest towards science learning. *Chemistry Education Research and Practice*, *19*(2), 474–483. doi:10.1039/C7RP00221A

Sanders, W., & Rivers, J. (1996). *Cumulative and residual effects of teachers of future student academic achievement*. University of Tennessee Value-Added Research and Assessment Center.

Schad, M., & Jones, W. M. (2020). The maker movement and education: A systematic review of the literature. *Journal of Research on Technology in Education*, *52*(1), 65–78. doi:10.1080/15391523.2019.1688739

Schlegel, R. J., Chu, S. L., Chen, K., Deuermeyer, E., Christy, A. G., & Quek, F. (2019). Making in the classroom: Longitudinal evidence of increases in self efficacy and STEM possible selves over time. *Computers & Education*, *142*, 1–15. doi:10.1016/j.compedu.2019.103637

Schneider, B., Krajcik, J., Lavonen, J., Salmela-Aro, K., Klager, C., Bradford, L., Chen, I.-C., Baker, Q., Touitou, I., Peek-Brown, D., Dezendorf, R. M., Maestrales, S., & Bartz, K. (2022). Improving science achievement—Is it possible? Evaluating the efficacy of a high school chemistry and physics project-based learning intervention. *Educational Researcher*, *51*(2), 109–121. doi:10.3102/0013189X211067742

Schwartz, M. S., Sadler, P. M., Sonnert, G., & Tai, R. H. (2009). Depth versus breadth: How content coverage in high school science courses relates to later success in college science coursework. *Science Education*, *93*(5), 798–826. doi:10.1002ce.20328

Science Buddies. (2012). *Cabbage Chemistry--Finding Acids and Bases*. Scientific American. https://www.scientifi-cameri- can.com/article/bring-science-home-cabbage-chemistry/

Scott, K. A., & White, M. A. (2013). CompuGirls' Standpoint: Culturally Responsive Computing and Its Effect on Girls of Color. *Urban Education*, *48*(5), 657–681. doi:10.1177/0042085913491219

Sefton-Green, J. (2013). *Learning at not-school: a review of study, theory, and advocacy for education in non-formal settings*. MIT Press.

Shernoff, D. J., Csikszentmihalyi, M., Shneider, B., & Shernoff, E. S. (2003). Student engagement in high school classrooms from the perspective of flow theory. *School Psychology Quarterly*, *18*(2), 158–176. doi:10.1521cpq.18.2.158.21860

Simpson, A., Burris, A., & Maltese, A. (2020). Youth's engagement as scientists and engineers in an afterschool making and tinkering program. *Research in Science Education*, *50*(1), 1–22. doi:10.100711165-017-9678-3

Şimşek, P., & Kabapınar, F. (2010). The effects of inquiry-based learning on elementary students' conceptual understanding of matter, scientific process skills and science attitudes. *Procedia - Social and Behavioral Sciences, 2*(2), 1190-1194. https://doi.org/https://doi.org/10.1016/j.sb- spro.2010.03.170

Sinatra, G. M., Heddy, B. C., & Lombardi, D. (2015). The challenges of defining and measuring student engagement in science. *Educational Psychologist*, *50*(1), 1–13. doi:10.1080/00461520.2014.1002924

Singer, A., Montgomery, G., & Schmoll, S. (2020). How to foster the formation of STEM identity: Studying diversity in an authentic learning environment. *International Journal of STEM Education*, *7*(57), 1–12. doi:10.118640594-020-00254-z

Smith C. E. & Cooper-Hewitt Museum. (2007). *Design for the other 99%*. Smithsonian Cooper-Hewitt National Design Museum and Assouline Publishing.

Smith-Doerr, L., Alegria, S. N., & Sacco, T. (2017). How diversity matters in the US science and engineering workforce: A critical review considering integration in teams, fields, and organizational contexts. *Engaging Science, Technology, and Society*, *3*, 139–153. doi:10.17351/ests2017.142

Smit, R., Robin, N., De Toffol, C., & Atanasova, S. (2019). Industry-school projects as an aim to foster secondary school students' interest in technology and engineering careers. *International Journal of Technology and Design Education*. Advance online publication. doi:10.100710798-019-09538-0

Sobehrad, L., & Sobehrad, S. (2022). *Medieval History in the Modern Classroom: Using Project-Based Learning to Engage Today's Learners*. Arc Humanities Press.

Spence, P. L., Gerald-Goins, T., & Goins, G. (2021). *Food, Energy and Water Learning Modules (FEWLM) Workbooks*. The DISCUSSION Network, QUBES Educational Resources. https://qubeshub.org/publications/2287/?v=1

Spence, P. L., Phillips, R. S., McAllister, A. R., White, S. L., & Hollowell, G. P. (2020). Student-scientist curriculum: integrating inquiry-based research experiences and professional development activities into an introductory biology laboratory course. *Journal of Microbiology & Biology Education, 21*(3). https://doi.org/doi:10.1128/jmbe. v21i3.2225

Stake, J. E., & Mares, K. R. (2005). Evaluating the impact of science-enrichment programs on adolescents' science motivation and confidence: The splashdown effect. *Journal of Research in Science Teaching*, *42*(4), 359–375. doi:10.1002/tea.20052

Stanley, J. C., & Campbell, D. T. (1980). *Experimental and Quasi-experimental Designs for Research*. Rand McNally.

STEMNoire. (n.d.). *Home*. STEMNoire. https://www.stemnoire.org/

Stevens, S., Andrade, R., & Page, M. (2016). Motivating Young Native American Students to Pursue STEM Learning Through a Culturally Relevant Science Program. *Journal of Science Education and Technology*, *25*(6), 947–960. doi:10.100710956-016-9629-1

Stewart, C. O. (2022). STEM Identities: A Communication Theory of Identity Approach. *Journal of Language and Social Psychology*, *41*(2), 148–170. doi:10.1177/0261927X211030674

Stone, B. (2016). Playing around in science: How self-directed inquiry benefits the whole child. *International Journal of the Whole Child*, *1*(1), 1–10.

Strand, K., Marullo, S., Cutforth, N., Stoecker, R., & Donohue, P. (2003). Principles of best practice for community-based research. *Michigan Journal of Community Service Learning*, *9*(3), 5–15. http://hdl.handle.net/2027/spo.3239521.0009.301

Strobel, J., & Sun, Y. (2020). Current state and suggestions for the K-12 STEM school industry partnership in the United States. In M. J. de Vries, S. Fletcher, S. Kruse, P. Labudde, M. Lang, I. Mammes, C. Max, D. Munk, B. Nicholl, J. Strobel, & M. Winterbottom (Eds.), *The impact of technology education: International insights* (Vol. 3, pp. 143–162). Waxmann Verlag.

Sullivan, J. F. (2006). *Broadening Engineering's Participation — A Call for K-16 Engineering Education*. National Academies Press.

Swamidass, P., & Schnittka, C. (2017). Finding and preparing teachers to meet the needs of US student innovators in the making. *Technology and Innovation*, *18*(4), 331–342. doi:10.21300/18.4.2017.331

Syed, M., Azmitia, M., & Cooper, C. R. (2011). Identity and academic success among underrepresented ethnic minorities: An interdisciplinary review and integration. *The Journal of Social Issues*, *67*(3), 442–468. doi:10.1111/j.1540-4560.2011.01709.x

Tai, R. H., & Liu, Q. (2006). Planning early for careers in science. *Science*, *312*(5777), 1143–1144. doi:10.1126cience.1128690 PMID:16728620

Tan, V., Nicholas, C., Scribner, J. A., & Francis, D. C. (2019). Enhancing STEM learning through an interdisciplinary, industry-generated project. *Technology and Engineering Teacher*, *79*(1), 26–31.

TERC. (2023). *TERC selected to lead new informal STEM learning equity resource center* (ISL-ERC). TERC. https://www.terc.edu/news/terc-selected-to-lead-new-informal-stem-learning-equity-resource-center-isl-erc/

Texas Education Agency. (n.d.). *Texas Essential Knowledge and Skills*. TEA. https://tea.texas.gov/academics/curriculum-standards/teks/texas-essential-knowledge-and-skills

Than, W. W., & Kyaw, W. H. (2021). School environment, student engagement and educational aspiration of high school students. *Academia Letters*, *1712*, 1–12. doi:10.20935/AL1712

The Afterschool Alliance. (2008). *Evaluations backgrounder: A summary of formal evaluations of the academic impact of afterschool programs*. ASA.

The Australian Industry Group. (2017). *Strengthening school-industry STEM skills partnerships: Final project report*. The Australian Industry Group. https://cdn.aigroup.com.au/Reports/2017/AiGroup_OCS_STEM_Report_2017.pdf

Thiry, H. (2017). *External evaluation of the Build a Better Book program, University of Colorado Boulder, 2016-17* [Unpublished manuscript]. University of Colorado Boulder.

Thiry, H. (2018). *External evaluation of the Build a Better Book program, University of Colorado Boulder, 2017-18* [Unpublished manuscript]. University of Colorado Boulder.

Thiry, H. (2020). *External evaluation of the 2020 Build a Better Book librarian professional development workshop* [Unpublished manuscript]. University of Colorado Boulder.

Thomas, A. (2021). Out of the box: Countering the silence of Black girls with memory-work. In D. Apugo, L. Mawhinney, & A. Mbilishaka (Eds.), *Strong Black Girls: Reclaiming Schools in Their Own Image* (pp. 9–19). Teachers College Press.

Thomas, J. A., & Strunk, K. K. (2017). Expectancy-value and children's science achievement: Parents matter. *Journal of Research in Science Teaching*, *54*(6), 693–712. doi:10.1002/tea.21382

Trafton, A. (2007, October 24). *Cardboard Boaters sink or skim*. Massachusetts Institute of Technology. https://news.mit.edu/2007/regatta-tt1024

Trauth-Nare, A. (2015). Influence of an intensive, field-based life science course on preservice teachers' self-efficacy for environmental science teaching. *Journal of Science Teacher Education*, *26*(5), 497–519. doi:10.100710972-015-9434-3

Trytten, D. A., Foor, C. E., Murphy, T. J., Shehab, R. L., Walden, S. E., & Pan, R. (2014). UNSEEN DIFFERENCES. *ASEE Prism, 24*(3), 0-0. https://www.proquest.com/scholarly-journals/unseen-differences/docview/1628648512/se-2

Tsui, L. (2007). Effective Strategies to Increase Diversity in STEM Fields: A Review of the Research Literature. *The Journal of Negro Education*, *76*(4), 555–581.

U.S. Commission on Civil Rights. (2010, October). *Encouraging minority students to pursue science, technology, engineering and math careers*. USCCR.

U.S. Department of Education. (2019). *Innovation spotlights: Nine dimensions for supporting powerful STEM learning with technology*. USDE. https://eric.ed.gov/?q=interdisciplinary+stem&ff1=dtySince_2017&ff2=pubInformation+Analyses&id=ED603418

Unfried, A., Faber, M., Stanhope, D. S., & Wiebe, E. (2015). The development and validation of a measure of student attitudes toward science, technology, engineering, and math (S-STEM). *Journal of Psychoeducational Assessment*, *33*(7), 622–639. doi:10.1177/0734282915571160

United States Department of Education. (n.d.). *Education Innovation and Research.* USDE. https://oese.ed.gov/

United States Department of Health and Human Services. (2021). *Office of Minority Health Home.* United States Department of Health and Human Services. https://minorityhealth.hhs. gov/

Usart, M., Schina, D., Esteve-Gonzalez, V., & Gisbert, M. (2019, May). Are 21st Century Skills Evaluated in Robotics Competitions? The Case of First LEGO League Competition. In *the Proceedings of the 11th International Conference on Computer Supported Education* (pp. 445-452).

Utica College. (n.d.) *Bloom's Taxonomy of Measurable Verbs.* Utica College. https://www.utica.edu/academic/Assessment/new/Blooms%20Taxonomy%20-%20Best.pdf

van den Hurk, A., Meelissen, M., & van Langen, A. (2019). Interventions in education to prevent STEM pipeline leakage. *International Journal of Science Education, 41*(2), 150–164. doi:10.1080/09500693.2018.1540897

Vanderbilt University. (2018). *Report on the Faculty: Composition and Demographics of Provost Reporting Vanderbilt University Employed Faculty Members.* Vanderbilt University.. https://cdn.vanderbilt.edu/vu-wp0/wp-content/uploads/sites/241/2018/05/08164600/ReportOnTheFaculty2018.pdf

Vanderbilt University. (2023a). *Intersectional Study of Black Women and Girls in Society.* Vanderbilt University. https://www.vanderbilt.edu/ibwgs/

Vanderbilt University. (2023b). *Trans-Institutional Programs (TIPs).* Vanderbilt University. https://www.vanderbilt.edu/strategicplan/trans-institutional-programs/

Vanderbilt University. (2023c). *Vanderbilt at a Glance: Diversity and Inclusion.* Vanderbilt University. https://admissions.vanderbilt.edu/profile/

Vespa, J., Armstrong, D., & Medina, L. (2020). *Demographic Turning Points for the United States: Population Projections for 2020 to 2060.* USCB. https://www.census.gov/library/publications/2020/demo/p25-1144.html

Vest, C. M. (2008). Context and challenge for twenty-first century engineering education. *Journal of Engineering Education, 97*(3), 235–236. doi:10.1002/j.2168-9830.2008.tb00973.x

Virginia Department of Education. (n.d.a). *Profile of a Virginia graduate.* DoE. https://www.doe.virginia.gov/instruction/graduation/profile-grad/index.shtml

Virginia Department of Education. (n.d.b). *Standards of learning (SOL) and testing.* DoE. https://www.doe.virginia.gov/testing/index.shtml

Vossoughi, S., & Bevan, B. (2014). *Making and Tinkering: A review of the literature.* National Research Council. https://sites.nationalacademies.org/cs/groups/dbassesite/documents/webpage/dbasse_089888.pdf

Vossoughi, S., Hooper, P. K., & Escudé, M. (2016). Making through the lens of culture and power: Toward transformative visions for educational equity. *Harvard Educational Review, 86*(2), 206–232. doi:10.17763/0017-8055.86.2.206

Vygotsky, L. S. (1978). *Mind in society: The development of higher mental processes.* Harvard University Press.

Wallace, E., Perry, J., Ferguson, R., & Jackson, D. (2015). The Careers in Health and Medical Professions Program (CHAMPS): An Impact Study of a University-Based STEM+H Outreach Program. *Journal of Science Education and Technology, 24*(4), 484–495. doi:10.100710956-014-9536-2

Walsh, B. (2017). Texture, buttons, sound and code: Modal preference and the formation of expert identities. *7th Annual Conference on Creativity and Fabrication in Education (FabLearn17)*, Palo Alto, CA. 10.1145/3141798.3141819

Wang, M., Eccles, J. S., & Kenny, S. (2013). Not lack of ability but more choice: Individual and gender differences in choice of careers in science, technology, engineering, and mathematics. *Psychological Science, 24*(5), 770–775. doi:10.1177/0956797612458937 PMID:23508740

Watters, J. J., & Diezmann, C. M. (2013). Community partnerships for fostering student interest and engagement in STEM. *Journal of STEM Education: Innovations and Research, 14*(2), 47–55.

Webber, K., Robert, J., Williams Moore, M., & Burg, T. (2020). Measuring the impact of business engagement on STEM education. In C. C. Johnson, M. J. Mohr-Schroeder, T. J. Moore, & L. D. English (Eds.), *Handbook of research on STEM education* (pp. 459–469). Routledge. doi:10.4324/9780429021381-42

Weiner, S., Lande, M., & Jordan, S. S. (2018). What have we 'learned' from maker education research? A learning sciences-base review of ASEE literature on the maker movement. *American Society for Engineering Education Annual Conference & Exposition.* Salt Lake City, UT.

Weld, J., Heiden, E., & Kemis, M. (2015). Characteristics and effects of a statewide STEM program. *K-12 STEM Education, 1*(4), 167–180.

Wilkins-Yel, K. G., Simpson, A., & Sparks, P. D. (2019). Persisting despite the odds: Resilience and coping among women in engineering. *Journal of Women and Minorities in Science and Engineering, 25*(4), 353–368. doi:10.1615/JWomenMinorScienEng.2019026945

Wilson, E., & Mukhopadhyaya, P. (2022). Role of empathy in engineering education and practice in North America. *Education Sciences, 12*(6), 1–12. doi:10.3390/educsci12060420

Wilson, J., Krakowsky, A. M., & Herget, C. J. (2010). Starting early: Increasing elementary (K-8) student science achievement with retired scientists and engineers. *IEEE Transactions on Education, 53*(1), 26–31. doi:10.1109/TE.2009.2022546

Wolter, B. H., Lundeberg, M. A., & Bergland, M. (2013). What makes science relevant?: Student perceptions of multimedia case learning in ecology and health. *Journal of STEM Education: Innovations and Research, 14*(1), 26–35.

Yerdelen, S., Kahraman, N., & Taş, Y. (2016). Low socioeconomic status students' STEM career interest in relation to gender, grade level, and stem attitude. *Journal of Turkish Science Education, 13*, 59–74. doi:10.12973/tused.10171a

Yerrick, R., & Ridgeway, M. (2017). Culturally responsive pedagogy, science literacy, and urban underrepresented science students. *International Perspectives on Inclusive Education, 11*, 87–103. doi:10.1108/S1479-363620170000011007

Yin, R. K. (2014). *Case study research: Design and methods* (5th ed.). SAGE.

Yin, R. K. (2018). *Case study research and applications: Design and methods* (6th ed.). Sage.

Zhang, H., Couch, S., Estabrooks, L., Perry, A., & Kalainoff, M. (2023). Role models' influence on student interest in and awareness of career opportunities in life sciences. *International Journal of Science Education, Part B*, 1–19.

Zirkel, S. (2002). Is there a place for me? Role models and academic identity among white students and students of color. *Teachers College Record, 104*(2), 357–376. doi:10.1177/016146810210400206

Zohrabi, M. (2013). Mixed method research: Instruments, validity, reliability and reporting findings. *Theory and Practice in Language Studies, 3*(2), 254–262. doi:10.4304/tpls.3.2.254-262

Zoltowski, C. B., Oakes, W. C., & Cardella, M. E. (2012). Students' Ways of Experiencing Human-Centered Design. *Journal of Engineering Education, 101*(1), 28–59. doi:10.1002/j.2168-9830.2012.tb00040.x

Zuniga, K., Olson, J. K., & Winter, M. (2005). Science education for rural Latino/a students: Course placement and success in science. *Journal of Research in Science Teaching, 42*(4), 376–402. doi:10.1002/tea.20064

About the Contributors

Jessica L. Spott is the Director of the STEM Center for Outreach, Research & Education at Texas Tech University, USA. Her research studies focus on underrepresented students in STEM, STEM outreach in formal and informal contexts and university, community college and K-12 community interactions. In addition, she loves to study MotherScholars and mothers in STEM in an effort to increase the success and diversity of the STEM atmosphere.

Lane J. Sobehrad is the Coordinator of Research and Innovation for Lubbock Independent School District. He is responsible for all major projects involving alternative money including discretionary grants, research projects, community partnerships, and teacher incentive programs. His research studies focus on pedagogy and instructional practices in undergraduate settings, specifically project-based learning. In addition, he commonly works with district partners on engaged scholarship that brings additional opportunities, awareness, and access to district stakeholders.

Rebecca Hite is Associate Professor of Science Education in the Department of Curriculum and Instruction at Texas Tech University, USA. Her research studies focus on curricular and instructional interventions, augmented by emerging technologies, that occur within in/formal science and engineering learning spaces as well as professional pathways for K–12 STEM teacher leadership in policy-advocacy.

***

Gina Childers is an assistant professor of STEM Education at Texas Tech University. Her research interests encompass informal, non-formal, and community-based STEM learning contexts as well as technology utilization in STEM and cybersecurity education. She currently serves as the STEM Education Track Program Coordinator at Texas Tech University.

Fanni Coward started as a faculty member at Texas Tech University (TTU) in 2009. In 2013, Dr. Coward added additional duties as a Site Coordinator, serving as a liaison between the school districts and TTU's teacher preparation program and coaching pre-service teacher candidates in the field. She received tenure and promotion to Associate Professor in 2015. Additionally, she assumed a leadership role in the College, first serving as the Associate Chair in 2016, and then later as the Department Chair from 2017 to 2019. Dr. Coward is dedicated to her teaching and has earned numerous teaching awards during her time as a faculty member. Her current research interests center around Teacher Identity, Teacher Recruitment, and Diversity Issues in Schooling and Teacher Education.

Danielle Craddock holds a Bachelor of Arts in Communication from Old Dominion University, a Master of Arts in Higher Education-Student Services from George Mason University, and a certificate in Financial Success for Nonprofits from Cornell University. She is the founder and owner of Danielle B. Craddock LLC where she serves as a consultant for start-up nonprofits and as a speaker on the topics of mentoring or career exploration. She founded and directed the nonprofit, Girls Inspired & Ready to Lead, Inc. (GIRL).

Claudette P. Davis received her Ph.D. in Cell, and Molecular Biology from the Graduate School and University Center, CUNY, in 2007 and is currently an Associate Professor at LaGuardia Community College, CUNY. She teaches Anatomy and Physiology, and biology for non-majors and serves as a research mentor to undergraduate STEM majors and high school students. In addition, she is the coordinator for the City University of New York's STEM Research Academy program for high school students at LaGuardia. She is a co-coordinator of the Q-STEM Research and Summer Academy program, whose goal is to provide LaGuardia's STEM majors with transferable skills as they prepare to complete their Bachelor's degree at a 4-year college. She is also the LaGuardia Humanitarian Initiative's professional development leader interested in providing faculty with valuable tips and tools to implement experiential learning in their classroom.

Edward W. Davis received his Ph.D. from the University of Akron in 1996. He worked in the commercial plastics industry for 11 years, including at Shell Chemicals in Louvain-la-Nueve, Belgium and EVALCA in Houston, TX. He joined the faculty at Auburn University in 2007, where he regularly taught courses in three different engineering departments: Chemical Engineering, Civil Engineering, and Polymer and Fiber Engineering. In 2014 he was promoted to Senior Lecturer, and in 2015 he moved to the Materials Engineering program as an Assistant professor. Currently an Associate professor, his research focus is environmental and biological applications of polymeric nanomaterials. He is also very interested in improving STEM education. He is particularly interested in how societal challenges can be used to motivate the study of engineering and promote consideration of engineering as a career by high school students.

Virginia A. Davis is a Professor in Auburn University's Department of Chemical Engineering. She conducts interdisciplinary research on producing functional materials with unique properties from nanomaterial dispersions and polymers. Research topics include 1D and 2D nanomaterials, liquid crystals, rheology, additive manufacturing, and polymer recycling. In addition, Davis conducts collaborative research focused on understanding STEM identity formation, particularly among students from historically underrepresented groups. She has served on several university diversity, equity, and inclusion committees and is an inaugural member of the Alabama STEM Council.

Ashlie Denton is a principal researcher at Education Northwest. Dr. Denton has nearly ten years of experience working on multidisciplinary STEM research, evaluation, and strategic planning projects. In her work, she walks beside educators and other practitioners as they work to improve access and opportunity for students traditionally underserved in STEM. Dr. Denton supports out-of-school time programs in understanding and continuously improving their implementation and student outcomes through mixed-methods research and evaluation design and analysis. Prior to joining Education Northwest, Ashlie was a National Science Foundation Integrative Graduate Education and Research Traineeship fellow at Portland State University.

Kimberlyn A. Ellis is a doctoral student in the Human Genetics Training Program at the Vanderbilt Genetics Institute. She studies the interplay of genetic, behavioral, and social determinants of health and how it contributes to disease onset and outcomes in minority populations.

Christy L. Erving is an Associate Professor in the Department of Sociology and Population Research Center at The University of Texas at Austin. Using theories, concepts, and perspectives from various disciplines, her program of research focuses on clarifying and explaining status distinctions in health. Her primary research areas explore how race, ethnicity, gender, and immigrant status intersect to produce health differentials; psychosocial determinants of Black women's health; and the Black–White mental health paradox. She received a B.A. in Sociology and Hispanic Studies at Rice University, and M.A. and Ph.D. in Sociology at Indiana University, Bloomington, with a minor in Social Science Approaches to Health and Healing Systems. Prior to joining the faculty at UT-Austin, Dr. Erving was a faculty member at Vanderbilt University.

Stacey Forsyth is the director of CU Science Discovery, a K-12 STEM education outreach organization at the University of Colorado Boulder, and Co-PI of the NSF ITEST-funded Build a Better Book project. In her role, she oversees informal STEM education programs for K-12 students and teachers, including summer and after-school STEM classes, in-school programs, teacher professional development workshops and community outreach programs. She is passionate about inspiring and supporting curious kids to ask questions and find creative solutions for real-world problems and diversifying the future STEM workforce by expanding opportunities for youth to explore a variety of STEM fields and careers. She holds a B.S. in Ecology, Behavior & Evolution from the University of California, San Diego, and a Ph.D. in Ecology & Evolutionary Biology from the University of Arizona.

Tonya Gerald-Goins is an Associate Professor in the Department of Chemistry and Biochemistry at her undergraduate alma mater, North Carolina Central University (NCCU) where she received her B.S. in Chemistry with an ACS concentration. She received her Ph.D. in Pharmacology and Experimental Therapeutics from the University of Maryland at Baltimore School of Medicine (UM, B SOM). Her post-doctoral training was at the NCCU Julius L. Chambers – Biomedical - Biotechnology Research Institute in the Neuroscience/Drug Abuse program Her passion is to increase the number of underrepresented minority groups in STEM careers while her research focus is the effects of emerging contaminants on the gene expression of the organisms exposed to these chemicals. To follow her passion and research focus, Dr. Gerald-Goins assist in the management of grants funded by NOAA and NSF, such as a past Co-PI for the NSF INCLUDES DISCUSSION Network grant. Dr. Gerald-Goins is married to Dr. Gregory D. Goins, Professor at NCAT, and mother to two sons, Daniel, and Kendall.

Gregory D. Goins is a Professor and Chair of Natural Resources and Environmental Design at North Carolina A&T State University. Dr. Goins' research interests include rapid detection and remediation of xenobiotic chemicals in local water systems. His lab utilizes flowing solution culture hydroponics research to better understand food crop improvement. He is the co-Director of the Small Farms Resource and Innovation Center (SFRIC) located with the tri-partite mission design of the College of Agriculture and Environmental Sciences (CAES). The SFRIC is an institutionalized space at NCA&T dedicated to small farms that maximizes CAES' synergy between experiential, technical, and learning expertise. The SFRIC initiative represents a network of integrated technical assistance resources pertaining to USDA priorities in climate smart agricultural production, digital agriculture, heirs' property retention, and agricultural entrepreneurship.

Jason Greenberg Motamedi, is a managing researcher at Education Northwest. His research supports educators and policymakers as they use data and evidence to improve outcomes among language learners and underrepresented students, and to increase the skills and diversity of teachers. He has led multiple researcher-practitioner partnerships. As an educational anthropologist, Greenberg Motamedi has extensive experience conducting and leading ethnographic studies, and has developed expertise in mixed-methods research design, and analysis.

Levi Johnson is a mathematics teacher at Lubbock High School in the Lubbock Independent School District located in Lubbock, TX. She is passionate about working with students, teaching them about approaching problems in new and innovative ways.

Kaylee Johnston is a mathematics teacher at Lubbock High School in the Lubbock Independent School District located in Lubbock, TX. She is passionate about working with students, teaching them about approaching problems in new and innovative ways.

Nicole Joseph is an Associate Professor of Mathematics Education at the Peabody College of Education and Human Development at Vanderbilt University. She studies Black girls and their identity development in mathematics learning spaces.

Kelly Knight, a 2020 University Teaching Excellence Award winner, is an associate professor with the George Mason University Forensic Science Program and a STEM Accelerator. As a STEM Accelerator, she mentors STEM students and leads K-12 STEM outreach programs. She is the co-founder and director of Females of Color and those Underrepresented in STEM (FOCUS), a program which makes STEM exciting and accessible for middle and high school students. Through her various advisory roles and participation in various DEI committees across campus such as the President's Anti-Racism and Inclusive Excellence taskforce, Professor Knight's work in higher education centers the development and implementation of equitable practices for both STEM students and faculty.

Joni Lakin is Professor of Educational Studies at the University of Alabama. She studies educational measurement issues, including test validity and fairness with a particular interest in the influence of fairness and validity of assessments on the identification of students with academic talents. Her informal STEM research and educational evaluation work has mainly focused on "altruistic engineering" frameworks and addressing the educational needs of students in high poverty, predominantly African American, and often rural school districts in Alabama.

Claire Lannoye-Hall is a Ph.D. candidate in Curriculum and Instruction and Texas Tech University. She has more than two decades of experience working in the informal science field, focusing on creating equitable educational experiences for students of all ages and backgrounds.

Daniela Marghitu is a faculty member in the Computer Science and Software Engineering Department at Auburn University, where she has worked since 1996. Her research areas include STEM K12 Inclusive Computing Research and Outreach; Web Applications Design and Development; Education and Assistive Technology; Software Engineering; Web and Software Engineering Usability and Accessibility. Dr. Marghitu has mentored over fifteen hundred k12, computing undergraduate, graduate students including representatives of underserved/underrepresented communities, women, and people with disabilities. She has published extensively across peer-reviewed publications and has given numerous presentations at national and international conferences.

James Marshall is a Professor and Senior Director of the Doctorate In Education program in the Department of Educational Leadership at San Diego State University. His scholarship, teaching, and consulting revolve around learning initiatives and hastening their predictive impact through careful planning, design, and continuous improvement. With over 200 publications to his credit, Marshall's work encompasses a diverse range of works that include empirical research, program evaluation efforts, and policy development. His evaluation endeavors are particularly significant and include more than 250 individual studies of funded projects and program investments totaling over $120 million dollars. From assessing strengths and needs, to conceptualizing strategy and program initiatives, and then measuring return on investment, Marshall's unique approach relies on a proven mix of assessment and evaluation, appreciative inquiry, and empathic understanding that predictably yields quantifiable results. His book, Right from the Start: The Essential Guide to Implementing School Initiatives, summarizes lessons learned through evaluation of hundreds of programs in both the public and private sectors.

Anthony Perry is a postdoctoral associate at MIT's Institute for Data, Systems, and Society. His research focuses on creating engaging and equitable STEM learning experiences. This research is informed by more than a decade of experience as an education practitioner. He holds a Ph.D. in Curriculum & Instruction from Texas Tech University, an M.S.Ed. from Northwestern University, and B.S. in Astronomy-Physics from the University of Wisconsin.

Stephanie C. Playton holds a Ph.D. in Curriculum & Instruction and is currently a STEM Learning Specialist serving underserved K-12 youth and in-service teachers in rural Virginia through providing inquiry and hands-on learning experiences with a focus on computer science and engineering concepts. Her research interests include STEM learning, career awareness and interest, and computational thinking.

Edward Price is a Professor in the Department of Physics and is the Director of the Center for Research and Engagement in STEM Education at California State University San Marcos. Dr. Price co-leads the CSUSM Mobile Making program, which operates after school making programs at local schools and community centers, with undergraduate students as facilitators and near-peer mentors. He has a Ph.D. in physics from the University of California San Diego and a Bachelor of Science in physics from Brown University.

Padmanabhan Seshaiyer is a Professor of Mathematical Sciences at George Mason University where he has served in multiple leadership positions including the Associate Dean for Academic Affairs, Director of the STEM Accelerator Program and Director of the Center for Outreach in Mathematics Professional Learning and Educational Technology. During the last decade, he has initiated and directed a variety of educational programs including faculty development, post-graduate, graduate and undergraduate research, K-12 outreach, teacher professional development, and enrichment programs to foster the interest of students and teachers in STEM at all levels. He is a.co-founder of Females of Color and those Underrepresented in STEM (FOCUS), a program which makes STEM exciting and accessible for middle and high school students. He is the director of TEAM: Teacher Engagement Advocacy and Mentorship, a program to empower teacher participants with the latest scientific findings on the learning characteristics of underrepresented populations, awareness of cultural differences, understanding of students with multiple learning styles, and the use of equitable and non-biased assessments.

Jessica Sickler is Principal of J. Sickler Consulting in Pittsburgh, PA. She is an evaluator and researcher who specializes in helping organizations understand what works and why in their informal education efforts. In her work, she explores the variety of ways learning happens within museums, gardens, zoos, and public engagement programs, centered on giving educators the data they need to spark insights for improvement and innovation. She holds an M.S.Ed. from Bank Street College of Education in museum education and K-6 teaching.

Sinem Siyahhan is an Associate Professor in the School of Education, and the Associate Director for the Center for Research and Engagement in STEM Education at California State University San Marcos.

Porché L. Spence, is an Assistant Professor in the Department of Natural Resources and Environmental Design at North Carolina Agricultural and Technical State University. Dr. Spence currently serves as the co-program director for North Carolina State University College of Natural Resources Forestry High School Summer Research Program and the external evaluator for the North Carolina State University National Needs Fellowship Forestry Program. Dr. Spence holds a Ph.D. in Soil Science from North Carolina State University. She received a Master of Science in Earth Science and Bachelor of Science in Environmental Science with a concentration in Biology from North Carolina Central University. She completed her post-doctoral training as an Oak Ridge Institute for Science and Education (ORISE) fellow at the US Environmental Protection Agency in the Gulf Ecology Division and subsequently as a Howard Hughes Medical Institute (HHMI) teaching fellow at North Carolina Central University in the Center for Science, Math, and Technology Education. She designs and implements inquiry-based curricula to teach the scientific process and promotes community engagement emphasizing STEM learning. Dr. Spence has co-designed, co-implemented, and co-evaluated low-cost, inquiry-based Food Energy and Water Nexus Learning Modules for elementary and middle school youth. Her research focuses on water quality, environmental pollution, and developing and evaluating STEM curriculum and programs; however, her passion is to inspire students from underrepresented minority communities to become informed citizens and to pursue careers in STEM, conservation, and natural resources.

Gigi Yang joined the Build a Better Book team in 2018 as the project's Network Coordinator. In this role, she supports the work of BBB network partners and teen interns across the country. She supports the project's research goals, manages the BBB website, and curates and disseminates information about the project through online communications. Prior to joining the Build a Better Book team, Ms. Yang worked for many years as a librarian, managing Young Adult Services at a public library. She holds a B.S. in Textiles and Apparel Design from Cornell University and a M.S. in Library and Information Science from Emporia State University.

Tong Zhang is a first-generation American who immigrated from China as a small child. She attended Title I schools throughout her elementary years and credits the dedicated teachers and well-rounded education she received for shaping her career interests and fostering her academic success. Tong is passionate about expanding equitable STEM opportunities across the educational continuum in her home state of Oregon and beyond. She has 10+ years of experience in nonprofit and education leadership and working with organizations to create meaningful and sustained social impact. She was previously the executive director of Oregon MESA, an invention-based equity program serving underrepresented K-12 students. She was also a founding staff member and COO of Thread, an organization providing wrap-around support to high school students in Baltimore city. Prior to her work in nonprofit management, she worked on developing immunotherapy approaches for gynecologic cancers. She has a Ph.D. in Immunology from The Johns Hopkins University-School of Medicine and a B.S. in Biochemistry from the University of Washington.

Index

www.ingramcontent.com/pod-product-compliance
Lightning Source LLC
LaVergne TN
LVHW080645011026
849160LV00014B/1328